Spanish Mysticism

Spanish Mysticism

Editor

Cristóbal Serrán-Pagán y Fuentes

MDPI • Basel • Beijing • Wuhan • Barcelona • Belgrade • Manchester • Tokyo • Cluj • Tianjin

Editor
Cristóbal Serrán-Pagán y
Fuentes
Valdosta State University
USA

Editorial Office
MDPI
St. Alban-Anlage 66
4052 Basel, Switzerland

This is a reprint of articles from the Special Issue published online in the open access journal *Religions* (ISSN 2077-1444) (available at: https://www.mdpi.com/journal/religions/special_issues/Spanish_Mysticism).

For citation purposes, cite each article independently as indicated on the article page online and as indicated below:

LastName, A.A.; LastName, B.B.; LastName, C.C. Article Title. *Journal Name* **Year**, *Volume Number*, Page Range.

ISBN 978-3-0365-2127-5 (Hbk)
ISBN 978-3-0365-2128-2 (PDF)

Contents

About the Editor . . . vii

Cristóbal Serrán-Pagán y Fuentes
Introduction to "Spanish Mysticism"
Reprinted from: *Religions* **2021**, *12*, 731, doi:10.3390/rel12090731 . . . **1**

Gracia López-Anguita
Ibn 'Arabī's Metaphysics in the Context of Andalusian Mysticism: Some Akbarian Concepts in the Light of Ibn Masarra and Ibn Barrajān
Reprinted from: *Religions* **2021**, *12*, 40, doi:10.3390/rel12010040 . . . **3**

Pablo Beneito
Qurrat al-'Ayn, the Maiden of the Ka'ba: On the *Themenophany* Inspiring Ibn 'Arabī's *Tarjumān*
Reprinted from: *Religions* **2021**, *12*, 158, doi:10.3390/rel12030158 . . . **23**

Avishai Bar-Asher
The Ontology, Arrangement, and Appearance of Paradise in Castilian Kabbalah in Light of Contemporary Islamic Traditions from al-Andalus
Reprinted from: *Religions* **2020**, *11*, 553, doi:10.3390/rel11110553 . . . **41**

Julia Alonso
The Divine Feminine Presence in Ibn 'Arabi and Moses de Leon
Reprinted from: *Religions* **2021**, *12*, 156, doi:10.3390/rel12030156 . . . **53**

Federico Dal Bo
Giqatilla's Philosophical Poems on the Hebrew Vowels: Poetry, Philosophy, and Theology in Giqatilla's *Ginnat Egoz* and *Sefer ha-Niqqud*
Reprinted from: *Religions* **2021**, *12*, 554, doi:10.3390/rel12070554 . . . **77**

María M. Carrión
"One Kind of Water Brings Another." Teresa de Jesús and Ibn 'Arabi
Reprinted from: *Religions* **2020**, *11*, 542, doi:10.3390/rel11100542 . . . **97**

Daniel Dombrowski
St. John of the Cross and the Monopolar Concept of God in the Abrahamic Religions in Spain
Reprinted from: *Religions* **2020**, *11*, 372, doi:10.3390/rel11070372 . . . **111**

Cristóbal Serrán-Pagán y Fuentes
The Active Life and the Contemplative Life in St. John of the Cross: The Mixed Life in the Teresian Carmelite Tradition
Reprinted from: *Religions* **2020**, *11*, 509, doi:10.3390/rel11100509 . . . **127**

Luce López-Baralt
Don Quixote and Saint John of the Cross's Spiritual Chivalry
Reprinted from: *Religions* **2021**, *12*, 616, doi:10.3390/rel12080616 . . . **151**

About the Editor

Cristobal Serran-Pagan Y Fuentes is Full Professor in the Department of Philosophy and Religious Studies and has taught undergraduate students at Valdosta State University in Valdosta, Georgia, since his arrival in 2007. He previously taught at Goucher College (2006-2007), Illinois Wesleyan University (2004-2006), and Coker College (2000-2003). He received his Ph.D. in Religious Studies from Boston University (2003). The areas of specialization include philosophy of religion, world religions, and comparative mysticism with a research interest in Spanish mysticism, interfaith dialog, ecology, and the interface between religion and science. The title of his dissertation was, "Mystical Vision and Prophetic Voice in Saint John of the Cross: Towards a Mystical Theology of Final Integration". Dr. Serrán-Pagán has been a member of The International Thomas Merton Society since 1997. He has served as an advisor and co-chair of the Robert Daggy Scholarship Program for many years. He was a recipient of the Daggy Youth Scholarship program (1997) and the Shannon Fellowship (1998). He is the editor of the Thomas Merton series, *Merton and the Tao: Dialogues with John Wu and the Ancient Sag*es (Fons Vitae, 2013) and the author of *Saint John of the Cross: His Prophetic Mysticism in the Historical Context of Sixteenth-Century Spain* (Pacem in Terris Press, 2018). He is a regular contributor to Merton conferences in Europe (England, Spain, the Czech Republic, and Poland), Argentina, Canada, and the United States. His publications include articles on Merton and the Spanish mystics in "Thomas Merton: A Mind Awake in the Dark" (Three Peaks Press, 2002) and "Seeds of Hope: Thomas Merton's Contemplative Message" (Cistercium-Ciem, 2008), "Merton on Dr. King, D. T. Suzuki, and Thich Nhat Hanh", all published in The Merton Seasonal. He has two articles published in the peer-reviewed journal The Merton Annual for the International Thomas Merton Society (ITMS): one article is "Panentheism in Thomas Merton and Teilhard de Chardin" (2021) and the second article is "Divine Mercy in Merton and in St. John of the Cross" (2017). His most recent publications include articles on St. John of the Cross in *Religions* (2020), Merton and Zhuangzi in *Paidagogos—Society for Philosophy, Theory & Praxis of Education* (2017), seven entries in the Thomas Merton Encyclopedia (2015), translated into Spanish (now found in the Diccionario de Thomas Merton), one article honoring Merton's Centenary in We are Already One (2015), and the last article published on Merton and Thich Nhat Hanh by Studia Anselmiana 182/Analecta Monastica 21 in Rome (2020). He was for many years involved in recruiting students from all over Georgia and teaching summer courses such as Spanish mysticism and world religions as part of the Madrid, Spain study abroad program through the European Council. He is the guest editor for the special issue on Spanish Mysticism in the peer-reviewed international journal *Religions.*

religions

MDPI

Editorial

Introduction to "Spanish Mysticism"

Cristóbal Serrán-Pagán y Fuentes

Department of Philosophy and Religious Studies, Valdosta State University, Valdosta, GA 31698, USA; cserranpagan@valdosta.edu

As the Guest Editor for the special volume on "Spanish Mysticism", my experience working with experts in this field has been excellent and very rewarding, especially in these current times, where we are dealing with COVID-19. I am very grateful to have served as the Guest Editor and to have contributed with an article on St. John of the Cross. It has been a great privilege to exchange ideas with scholars from all over the world.

Until just recently, scholars in theology or religious studies often only associated Spanish mysticism with the great Christian mystics from St. Ignatius of Loyola to St. Teresa of Avila to St. John of the Cross. In the pluralistic and global world in which we live today, we must try our best to expand our knowledge and make the connections that existed in medieval and modern times among the three major mystical traditions in the West. The long history of conflicts, tensions, wars and, yes, religious coexistence is an integral part in the study of Spanish mysticism that has to be reckoned with.

Spanish mysticism has become a field of study in itself due to the rich history of kabbalistic, Christian, and Sufi mystics born in the Iberian Peninsula—from Moses de Leon to Abraham Abulafia, Ignatius of Loyola to Teresa of Avila and John of the Cross, and Ibn 'Abbad of Ronda to Ibn al-'Arabi. The three monotheistic religions in the West left us a major cultural, spiritual, and religious legacy in the so-called period of *convivencia* or coexistence in medieval and modern Spain.

In total, nine articles were published electronically in this special volume of *Religions*. The authors were scholars from the United States (Dombrowski, Serrán-Pagán, Carrión), Spain (López-Anguita, Alonso, Beneito), Puerto Rico (López-Baralt), Israel (Bar-Asher), and Germany (Dal Bo). I am very proud of the quality of their research and their major contributions to this volume. I believe this field of Spanish mysticism will open up even more avenues and opportunities after people read these articles covering Jewish, Christian, and Islamic mysticism in the context of the Iberian Peninsula.

The primary scope of the articles collected here in this special volume serves the purpose of contextualizing Spanish mystical writings in their historical times and to examine how their legacy in the Iberian soil continues to evolve over time. The purpose of this volume is to bring together the different fields of knowledge from religious studies, theology, philosophy, history, comparative literature, philology, psychology, sociology, and the arts to address the main question: Do Spanish mystics borrow symbols and narratives from the mystical traditions of Judaism, Christianity, and Islam?

This special topic on Spanish mysticism has attracted scholars from different disciplines to study the great Spanish mystics. The overall focus of this issue is to trace the mutual influences found in Jewish, Christian, and Muslim mystics and to examine their spiritual legacies in greater depth.

The aim of this volume is to expand on the existing, currently available literature and to bring together the disjoint pieces of the puzzle so we can better and more holistically understand the rich legacy of the Spanish mystics and the extent to which their mystical thoughts are intertwined in the long history of Spanish mystical literature.

To those of you interested in Jewish mysticism in the context of past interactions with either Muslims or Christians, I cordially invite you to read the articles found in this collection by Dombrowski, Bar-Asher, Alonso, and Dal Bo. In their articles, they cover a

Citation: Serrán-Pagán y Fuentes, Cristóbal. 2021. Introduction to "Spanish Mysticism". *Religions* 12: 731. https://doi.org/10.3390/rel12090731

Received: 20 August 2021
Accepted: 27 August 2021
Published: 7 September 2021

range of topics, from the monopolar concept of God in the Abrahamic religions in Spain to the question of paradise in Kabbalah in the context of al-Andalus to the divine feminine in the Zohar, and in the Sufi of Murcia to the philosophical poems of Giqatilla. To those readers interested in Christian mysticism, I recommend that you check out the articles by Dombrowski, Serrán-Pagán, and López-Baralt on St. John of the Cross, and Carrión on St. Teresa of Avila. Finally, to those of you more interested in Sufi or Islamic mysticism in al-Andalus, I refer you to the articles written by Carrión, López-Anguita, Alonso, and Beneito on Ibn al-'Arabi.

Nonetheless, each article makes comparisons to other mystical traditions found in the Spanish soil. The cultural, religious, and mystical ferment found among Jewish, Christian, and Muslim texts is undeniable. In this collection of articles, the reader will find a clear trend of comparative and interdisciplinary studies proving, once and for all, the innumerable interconnections and mutual influences exhibited by the great mystics in their classic writings. Moreover, this is just the next stage in the comparative study of Spanish mysticism that started with scholars like Asín Palacios and have flourished over the last few decades. I owe a great deal to the expertise of each contributor to this volume, and to many other scholars such as McGuinn, Perrin, Matt, Girón-Negrón, Jaoudi, Sancho Fermín, Vidal Castro, Gonzalez Costa, Gómez Aranda, Fenton, Wolfson, Idel, Garb, Dauber, Benarroech, Satz, Cócera, Feldmeir, Poveda, Beltrán Llavador, Velasco, de Pascual, and del Valle Rodríguez, among many others who were unable to submit an article to this volume but are contributing to this field of Spanish mysticism with their research and comparative studies.

I am very thankful for this opportunity to have collaborated with scholars from different continents and with staff from Asia and Europe. I again want to express my deepest and sincere gratitude to each one of the authors who contributed with an article to this special volume. Their articles are invaluable in contextualizing the important roles played by each mystic in their cultural and religious milieu and in raising critical questions and opening new possibilities to go further in their research while, at the same time, urging those who are interested in this comparative field of mysticism to bring new perspectives from their respective fields of knowledge, be it in political science, economics, architecture, or any other field of expertise that can enrich us through our fuller understanding of the Spanish mystics.

In closing, it is my personal hope that this collection of scholarly articles will continue the trend of conversation that currently exists among scholars coming from different religious traditions, cultures, and countries in order for the public reader to better understand the deep spiritual level that was attained by the great mystics in the historical context of the Iberian Peninsula. This alone could serve as a model for more in-depth comparative studies and for greater interfaith dialogues to be held among scholars across the different academic disciplines.

Funding: This research received no external funding.

Conflicts of Interest: The author declares no conflict of interest.

religions

MDPI

Article

Ibn Arabī's Metaphysics in the Context of Andalusian Mysticism: Some Akbarian Concepts in the Light of Ibn Masarra and Ibn Barrajān

Gracia López-Anguita

Facultad de Filología, Universidad de Sevilla, C/Palos de la Frontera s/n, 41004 Sevilla, Spain; glopezanguita@us.es

Abstract: The aim of this article is to trace the origins of some of the key concepts of Ibn Arabi's metaphysics and cosmology in earlier Andalusian Sufi masters. Within the context of the seminal works on Ibn Arabi's cosmology and metaphysics produced from the second half of the 20th century onwards and through a comparison of texts by the Sufi masters Ibn Masarra and Ibn Barrajān, we will see which elements are taken from previous sources and how they are transformed or re-interpreted by Ibn ʿArabī in a philosophical-mystical system that would become the point of reference for the later Eastern and Western Sufi tradition.

Keywords: Sufism; al-Andalus; Ibn Arabi; metaphysics; cosmology; islamic mysticism; Ibn Masarra; Ibn Barrajān

check for updates

Citation: López-Anguita, Gracia. 2021. Ibn ʿArabī's Metaphysics in the Context of Andalusian Mysticism: Some Akbarian Concepts in the Light of Ibn Masarra and Ibn Barrajān. *Religions* 12: 40. https://doi.org/10.3390/rel12010040

Received: 1 December 2020
Accepted: 30 December 2020
Published: 8 January 2021

1. Introduction

We believe that the optimal methodology for the study and interpretation of Ibn ʿArabī's work is self-exegesis, based on a comparison of his texts and on the reconstruction of his organic and dynamic thought, the keys to which are distributed throughout his extensive work. Commenting on Ibn ʿArabī through Ibn ʿArabī emerges as the only possible method to understand an author who is the creator of his own worldview and language. However, it is also evident that certain pre-existing ideas found in authors writing before the mystic may have been a point of reference for him, although establishing the origin and parentage of a certain idea or term is always a risky undertaking. A fundamental clue is provided logically, by the explicit mentions in his texts of works and authors that were reference points for him, which, however, does not apply where philosophers, not Sufis, are concerned. We can gain an idea of Ibn ʿArabī's learning background from the first part of his *Ījāza li-l-mālik al-Muẓaffar*[1] which reveals the fundamental role played by the science of the hadith in his education. Among the scholars included in this work are traditionists, exegetes, philologists, writers, faqihs, judges, etc.; philosophy is absent and scholastic theology (*kalām*) has a very minor presence. The most famous example of the

[1] Edited by (Badawi 1955). "Autobibliografía de Ibn ʿArabi". *Al-Andalus*, XX.2, pp. 107–28; Badawi. 1979. Paris; S. ʿAbd al-Fattāḥ. 1995. Cairo. A comprehensive list of Ibn ʿArabī's works is to be found in (Aladdin et al. 2009).

attitude that Ibn ʿArabī showed towards philosophy is probably the passage in *Futūḥāt* that indirectly alludes to Alfarabi, whilst avoiding explicitly saying his name[2].

Without entering here into the complex issue of the relationship between Akbarian thought and philosophy, in the sense of *falsafa* of Greek origin, it is sufficient to quote two passages from *Futūḥāt* that reflect quite concisely his opinion of this discipline, the usefulness and truth of which he underplays without completely rejecting it:

> Philosopher means lover of wisdom because the word Sophia (Sūfīyā), means wisdom in Greek; philosophy, therefore, means love of wisdom. Everyone who is endowed with intelligence loves wisdom. However, people who think reflectively are wrong more often than they are right with regard to divine sciences (*ilāhīyāt*), both if they are philosophers and if they are Muʿtazilites or Ashʿarites. (Ibn ʿArabī 1999a, vol. 4, pp. 227–28)

> Don't let yourself be turned off, when you come across a problem that was mentioned by a philosopher or a mutakallim or a speculative scholar in any discipline of knowledge, to such an extent that you would say about the person who mentions it and who is a truthful insightful (*muḥaqqiq*) Ṣūfī that he is a philosopher, just because the philosopher (*al-faylasūf*) mentioned that very problem and discussed and believed it. (And don't say) that the (Ṣūfī that discusses it) derived it from the philosophers, or that he has no religion, because the philosopher who had no religion (and was no Muslim) stated it earlier. Don't do that friend! It would be an inconsequential argument. For not all the philosopher's knowledge is untrue, and that particular problem may just involve some truth he possesses. (. . .) Your statement that the philosopher has no religion does not mean that everything he possesses (in the way of knowledge) is untrue. Every intelligent person would perceive that right away. (Ibn ʿArabī 1999a, vol. I, p. 56 and Rosenthal 1988, p. 12)

The influence of philosophy on the thought of the master from Murcia is a matter that has been discussed at great length. Philosophy was—initially at least—officially supported by the Almohad rulers (see Fierro 2016; Fierro and Fitz 2005) and Sufism was not persecuted unless the messianic intentions of its leaders posed a challenge or potential danger to the authorities (Ferhat 2005). At the time when Ibn ʿArabī began his journey on the mystical path and developed his doctrine, a cultured Sufism already existed in Al-Andalus, personified in the outstanding figures of Ibn al-ʿArīf (d. 536 H/1141 CE), Ibn Barrajān (d. 536 H/1141 CE), and Ibn Masarra (d. 319 H/931 CE). It appears that this feeling of rejection or reservation towards philosophy that we see in Ibn ʿArabī was widespread in Western Sufism at this time—Ibn al-ʿArīf would refer to this discipline in his *Miftāḥ al-saʿāda* as a "reprehensible doctrine" (*madhab madhmūn)* (see Shafik 2012 and Ibn al-ʿArīf 1993, pp. 90–91)—which does not mean that we cannot find philosophical elements in the work of these mystics: tools like syllogism[3] and notions such as the ten

2 "I have seen an infidel declare in a book called *The ideal city* (I had found this book which I had not seen previously, at the home of an acquaintance of mine in Marchena of the Olives had picked it up to see what it was about and came across the following chapter: 'In this chapter I wish to examine how to postulate [the existence of a divinity in this world'. He had not said God (Allah)! I was amazed and threw the book at its owner's face." (Addas 1993, p. 108 and Ibn ʿArabī 1911, III, p. 178). On the one hand, Addas (1993, p. 109) tells us that "to this lack of knowledge of Arab philosophy we must add a blatant ignorance of Greek philosophy."; and on the other, researchers like J. A. Pacheco Paniagua assert that "judging by his intercession in the debate on substance, accidents and their relationship with continuously renewed creation, Ibn Arabi raises the arguments adduced by the former [Ashʿarites and Muʿtazilites] and rejects them with a highly philosophical demonstration and with terminology with clear Aristotelian roots". (Pacheco Paniagua 2019, p. 227, transl. by the author; see also Pacheco Paniagua 2017). Ibn ʿArabī would frequently return to this "philosophy-illuminative experience" dialectic and works like *The alchemy of happiness* or even episodes in his life such as the supposed encounter with Averroes will act as allegories on the limitations of exclusively rational thought.

3 "The test always entails a triple composition and inevitably and because of this, two isolated elements and the meeting of them make up the third aspect, which must be found in each of the two premises (*muqaddimatayn*) to obtain the conclusion (*intāj*). For example, a = b and b = c, repeating b, the proof is made up of three elements a, b and c, and the unifying aspect is b as it is repeated in the two premises. The result is a = c." (Ibn ʿArabī 1999a, p. 104). On the use of syllogism and other philosophical and mathematical elements in Ibn ʿArabī, (Pacheco Paniagua 2007, 2012, transl. by the author).

Aristotelian categories[4] (*māqūlāt*), "generation and corruption"[5] (*al-takwīn wa-l-fasād/genesos kai fthoras*), first intellect[6] (*al-ʿaql al-awwal*), etc., can be found clearly, to a greater or lesser degree, in the work of Ibn ʿArabī. The works through which Aristotelian and, above all, Neoplatonic concepts and doctrines entered Al-Andalus may have been not strictly philosophical but rather theosophical in nature, such as the *Epistles of the Brethren of Purity*.

The master cited the most times in *Futūḥāt* is Abū Madyan[7] (d. 594 H/1194 CE), a shepherd from the town of Cantillana in Seville, who gained enormous recognition as a spiritual master in the Islamic West. His written work comprises poetry, prayers, an initiation manual (*Bidāyat al-murīd*), and a creed (*ʿaqīda*), i.e., devotional and wisdom works but not speculative writings. Ibn ʿArabī provides a brief biography of the Andalusian masters of whom he was a direct disciple in his *Rūḥ al-Quds*[8], a work written as an apologia for Western Sufism in the face of Eastern Sufism which may well be interpreted as a declaration of intent in favor of an illiterate but authentic Sufism as opposed to a more bookish and speculative yet superficial one. It could be said, therefore, that the Sufi masters who receive greater and more explicit recognition from Ibn ʿArabī are those who, a priori, lack education and philosophical intention.

Theology (*kalām*) was also considered by Sheikh al-akbar to be a limited science which, in trying to defend religion from attacks based on rationalist arguments, had resorted to using the same tool to respond (Ibn ʿArabī 1972, pp. 34–36). A theologian of the stature of Ghazālī (d. 1111 CE), however, is inevitably present in the work of Ibn Arabī. His presence is controversial at times because he reproaches him for his speculative arguments about God and His Names and Attributes, thus revealing significant differences between the two thinkers. He is, in fact, the theologian who probably had the greatest influence on his metaphysics[9].

Regarding the self-affiliation of Ibn ʿArabī to a specific discipline, W. Chittick (Chittick 2020) states that he never referred to himself as a Sufi—a term widely used by modern scholars—and that "he can be considered the greatest of all Muslim philosophers, provided we understand philosophy in the broad, modern sense and not simply as the discipline of *falsafa*". The term Sufi, however, is explicitly mentioned in the introduction to *Futūḥāt* (in the fragment quoted above) in which he asserts that the fact that a philosopher and a realized Sufi (*ṣūfī muḥaqqiq*) may concur on some matters does not mean that the latter derived from the former. It is logical to think that Ibn ʿArabī included himself in that category of *ṣūfī muḥaqqiq* and, indeed, his own disciples considered that to be the case, for example, ʿAbd al-Ganī al-Nābulusī (d. 1143 H/1731 CE) (Al-Nābulusī 1995, pp. 137–38). The *muḥaqqiq* is regarded in another passage in the same work as a higher level within the category of Sufi: "I mean by our companions the possessors of hearts, witnessings and unveilings, not the worshippers (*al-ʿubbād*), nor the pious renouncers (*al-zuhhād*), nor the Sufis without restriction -only those among them who are the people of realities and verification (*taḥqīq*)" (see Ibn ʿArabī 1972, p. 261 in Chittick 1989, p. 392)[10].

4 (Ibn ʿArabī 1911, v. III, p. 10 and Chittick 1998, p. 360).

5 (Ibn ʿArabī 1919, p. 71).

6 (Ibn ʿArabī 1919, pp. 72, 74).

7 See Cornell (1996). The Way of Abū Madyan. Doctrinal and Poetic Works of Abū Madyan Shuʿayb ibn al-Ḥusayn al-Anṣārī (c. 509/1115-16-594/1198). Cambridge: The Islamic Texts Society.

8 *Rūḥ al-quds fī muḥāsabat al-nafs*. 2004. Edited by Ali b. Ahmad Sasi. Tunis: Dār al-kutub li-l-kitāb. Translated by Asín Palacios in 1933 as *Vidas de santones andaluces: la "Epístola de la santidad" de Ibn Arabī de Murcia,* Madrid: Estanislao Mestre; English translation together with *Al-durrat al-Fākhira* by R. W. J. Austin, 1971. *Sufis of Andalusia: the Rūḥ al-Quds and the Dhurrat al-fākhira*, London: Allen and Unwin.

9 See (Zine 2004; Nakamura 1994; Addas 1993, pp. 102–3).

10 Unlike Ibn Masarra or Ibn Barrajān, Ibn ʿArabī did consider Ibn al-ʿArīf a *muḥaqqiq*. See (Addas 1992, p. 926).

2. Ibn Masarra

2.1. Cosmogony, Language and Prime Matter

Ibn Masarra al-Jabalī, from Córdoba, (d. 319 H/ 931 CE) is considered to be the first Sufi and, in general terms, the first speculative thinker in Al-Andalus. Although it was underpinned by Eastern foundations, Western Sufism, "in contradistinction to the psychologically-oriented and ethically-minded Sufi tradition that developed in the East, the discourse of Ibn Masarra and his Andalusī heirs can be defined as "theosophical", not in the modern-spiritual sense of the word, but rather as it is used in the academic study of Kabbalah" (Ebstein 2020, p. 41). In his indispensable *Mysticism and Philosophy in al-Andalus*, Ebstein (Ebstein 2014) argues that the flourishing of Sufism and Kabbalah in Spain in the 12th and 13th centuries is largely explained by the presence of the Ismaili tradition in the neighbouring lands of Africa where the Fatimid dynasty had established itself. Ismaili literature comprised the middle link between Eastern Neoplatonism, Hermeticism, and Pythagoreanism and the Andalusian mystics. One of the arguments advanced by Ebstein (Ebstein 2014, pp. 10, 11)[11] in favor of Shiite Ismaili influence on Ibn Masarra and Ibn ʿArabī is the fact that they were accused of being Shiites by some authors, most notably Ibn Khaldūn (d. 808 H/1406 CE). Caution should be exercised when evaluating accusations of this type because, as Alexander Knysh (Knysh 1999) showed, the controversy against Ibn ʿArabī occurred after his death and was instigated mainly by Ibn Taymiyya and, as regards Ibn Khaldūn, Knysh maintains that the Tunisian's stance towards Sufism became more radical, probably for political reasons, while he was holding important official posts in Egypt. In addition, some of the Akbarian doctrines criticised by Ibn Khaldūn such as Mahdism, were taken not from Ibn ʿArabī directly, but from supposed disciples of his such as Ibn Abī Wātil[12].

Leaving aside the extent of Fatimid political influence in the Islamic West, the truth is that Ismaili doctrines spread throughout al-Andalus due to the early arrival of both the *Epistles of the Brethren of Purity* (c. between C9 and C11 CE)—which were introduced, according to the idea most widely accepted by researchers, by al-Majrīṭī (d. c. 468 H/1007 CE)[13]—and from the corpus of alchemical texts attributed to Jābir b. Ḥayyān (d. 200 H/815 CE?)—a set of treatises that are also heavily steeped in Shiite, specifically Qarmation Ismaili, ideology[14].

Two works by Ibn Massara have survived to this day: *The Epistle on Interpretation*[15] and *The Book of the Properties of Letters, Their True Nature And Their Origin*[16]. It is in this domain, the Science of Letters, that Ibn ʿArabī acknowledges his debt to Ibn Masarra without referring to any specific titles (Stroumsa 2016), while also recognising his reservations regarding this discipline. There is at least one explicit mention of the *Book of The Properties of Letters* in *Futūḥāt* when he puts forward the idea of the Kaaba and the black stone as interpreters of the various levels of Revelation, an idea which, however, does not appear in the version of Ibn Masarra's text that has reached us (Stroumsa 2016, p. 87).

Ever since Asín Palacios published *Abenmasarra y su escuela* (Asín Palacios 1914) and attributed to the philosopher from Córdoba an intellectual affiliation with the work of the Pseudo-Empedocles, Ibn Masarra has been associated with many currents of thought: Muʿtazilisim, Bāṭinism, Neoplatonism, Sufism, Ismailism, etc., even after his manuscripts—which were considered lost—were discovered in the seventies. The rejection of Asín Palacios's hypothesis by most Arabists during the 20th century has given way to recog-

[11] See also (De Callataÿ 2014–15, Brown 2006; Al-Affifi 1964 and Tornero Poveda 1985).

[12] On Ibn Khaldūn's position on Sufism (Knysh 1999, pp. 184–97).

[13] Fierro (1996) maintains that they were introduced even earlier by Abū l-Qāsim Maslama b. Qāsim al-Qurṭubī, (353 H/964 CE). See Fierro, *Bāṭinism in al-Andalus*, pp. 106–8. On Jābir ibn Ḥayyān see (Lory 1989).

[14] See Lory (2004). *Alchimie et mystique en Terre d 'Islam.* Paris: Gallimard.

[15] Critical editions in (Kenny 2002 and Ibn Masarra 2007a).

[16] See English translation of the former in (Stroumsa and Sviri 2009), Spanish translation in (Garrido Clemente 2008c) and a summary of the contents of the latter in (Tornero Poveda 1993).

nition from a significant strand of studies over the last decade that, beyond the influence of Empedocles or pseudo-Empedocles, Asín was right in associating Ibn Masarra with Neoplatonic doctrines. The elements of the cosmos as it appears in Ibn Masarra's work are arranged in tiers, contemplation of which can allow humans to ascend through them and bear witness to the great chain that unites beings to the final link, the creator:

> The world, then, with all its creatures and signs is a ladder (*daraj*) by which those who contemplate ascend to the greatest signs of God on high. He who climbs, must climb from the lower to the higher. They climb by means of the intellects [...] Thereupon you will find your Lord and Creator; you will meet Him in yourself. (Stroumsa and Sviri 2009, p. 224)

For Ibn ʿArabī, in a similar way to Ibn Masarra, the world is "arranged in degrees (*marātib*)" and this orderly and hierarchical arrangement of the cosmos is specifically what makes it intelligible to human beings (Ibn ʿArabī 1919, p. 95), but with no "occasional or causal succession" between its elements (Ibn ʿArabī 1919, p. 49). On this subject, however, Ibn ʿArabī explains:

> We have been silent with regard to explaining the true nature of the causes so that anyone who speculates on them will not imagine that we are among those who attribute the action to someone other than God or those who attribute the action to God, associating the causes to Him. [...] He creates the thing by way of a cause if He wants to, or if He does not want to, He does not create a cause for it, because in His wisdom He has already planned to create it in this way, as we explained earlier. And it is impossible for it to be otherwise, because it is impossible for a thing to be different from how it is known [by God]. For this reason, we have not made special mention of anything relating to the causal relationship between the Pen and the Tablet because it has already been discussed by those who support the Revealed Law, the People of Truth, who consider the adherents of [the doctrine] of "the cause and the caused" to be ungodly. (Ibn ʿArabī 1919, p. 81)

Two of the fundamental elements of the Ibn Masarra's cosmogony are, on the one hand, Prime or Universal Matter (*habāʾ*) and, on the other, the Word or Logos (*kalima*). In Ibn Masarra's emanationist system, the first hypostasis was formed of "an intelligible, intangible prime Matter (ʿ*unṣur*), coeternal with God, from which God made all particular beings arise" (Lory 2006, p. 834), which is also identified with letters:

> Sahl al-Tustarī said: Letters are the Primordial Dust (*al-habāʾ*) and the origin of things and the beginning of their creation. From them was created order and the dominion became manifest." (Ibn Masarra 2007b, pp. 62–63)
>
> The letter *hāʾ* is the Primordial Dust, it is the totality of letters, from which things are created. It is located below *kun*. (Ibn Masarra 2007b, p. 68)

Without assigning it such a central value in the genesis of creation as Ibn Masarra, Ibn ʿArabī speaks cryptically and succinctly in his work of a Supreme Element (ʿ*unṣur aʿẓam*)[17], which may well relate to that of his predecessor from Córdoba:

> "[The Supreme Element] kept in the most hidden of the hidden (...) is the most perfect of created beings and were we not to have a pact of concealment that prevents us from explaining its essential reality, we would speak more extensively about it, showing how all creation (*mā siwà Allāh*) is united to it." (Ibn ʿArabī 1919, p. 50)

[17] Regarding this see López Anguita, Gracia (López-Anguita 2018). "Notas en torno al concepto de Elemento Supremo (ʿ*unṣur aʿẓam)* en Ibn ʿArabī y su escuela" in *Ibn Arabi y su época*. Edited by G. López-Anguita. University of Seville.

"We have made the centre [of the Universe] the receptacle of the Supreme Element as a warning that the higher rules over the lower."[18] (Ibn ʿArabī 1972, v. II, p. 317)

This Element, which is closely related to water[19] and the Divine Name al-Ḥayy al-Qayyūm (The Living One; the Self-Subsisting), is needed to produce life in creation and for the step from potential to action:

"Through the Supreme Element, the essences of the potential worlds [that remained] in a present with no before or after [i.e., outside the world of the contingent, unaffected by the passage of time] became manifest, until God decided to see them as concrete beings." (Ibn ʿArabī 1919, p. 44)

Although Ibn ʿArabī does not directly identify this element with the prime matter (*habāʾ/hayūlà/madda ūlà*, etc.), that is how some of his Eastern commentators and disciples understand it: "The Supreme Element is the Singular Prime Matter that balances the essences of the four elements. It is the matter of the heavens and the earth" (Qashānī 2005, p. 535)[20]; "God made manifest the form of the universe from the Dust and separated the heavens and the earth from the *ratq*[21] called the Supreme Element" (Qunawī 1983, p. 19). Considering that Ibn ʿArabī identifies the term *habāʾ* with the intermediate reality in which beings exist, potentially also called Muhammadan reality (*ḥaqīqa muḥammadiyya),* Breath of the Merciful (*nafas al-Raḥmān*), Cloud (*ʿamāʾ*) or even Immutable Entities (*aʿyān thābita*) or Divine Names, etc., it should be confined to a metaphysical reality more than the concept of Prime Matter as understood by alchemists, for example, in which case Supreme Element seems more appropriate[22].

The interrelation of the concept of the Word of God (*kalima*), the imperative *kun*[23] (be), the Divine command (*amr*), and the Divine will (*irāda*) as the beginning of creation in Ibn Masarra and Ibn ʿArabī and Ismaili writings, along with their Biblical and Greek forerunners, has been extensively studied by Ebstein (2014). The use of the term *dhikr*[24] is characteristic of Ibn Masarra and relates to both remembering and mentioning, evoking the Platonic theory of reminiscence. The *dhikr* is synonymous with First Intellect and ontologically is below the *kun* (see Ibn Masarra 2007b, p. 88). Creation through the Divine word is also very present in the work of Ibn ʿArabī, who uses the term *kalima*—among others like *kalām*—to refer to the Word of God, frequently placing it in a Christic context[25],

[18] Cf. this idea with Ibn Masarra, *Risālat al-iʿtibār*: "(...) the one who brings them together despite their differences and makes them perform contrary to their nature must be above them, encompassing them, higher and greater than them."; "the testimony of innate knowledge requires that he who governs them should be above them and encompass them." (Stroumsa and Sviri 2009, p. 220).

[19] See the cosmogonic importance of water in *Risālat al-iʿtibār* by Ibn Masarra: "The first thing to be created was the Throne and the water" (Stroumsa and Sviri 2009, p. 224).

[20] See also his glossary of sufi terms -*Iṣṭilāḥāt*- (Qashānī 1992)

[21] A Qur'anic concept (Qur'an 21:30) that literally means "stitched" and refers to the homogeneous whole that formed the earth and the heavens before God tore them apart (*fataqa*) and made them separate. The attribution of the two works referred to here (*Mirʾat al-ʿārifīn* and *Laṭāʾif al-iʿlām*) is disputed, but they undoubtedly belong to the Eastern Akabarian school.

[22] It seems difficult to draw a definitive conclusion concerning this concept; based on the Akbarian texts, the Prime Matter—expressed under many names—may conform to a greater or lesser extent to a philosophical, alchemical or another kind of definition. In his *Kitāb ʿAnqāʾ Mugrib* and *R. Ittiḥād al-kawnī* he uses the symbol of the griffin or phoenix (*ʿanqāʾ*) to refer to *habāʾ*, highlighting its spiritual and ineffable character: "I am the *ʿAnqāʾ Mugrib*, my home is in the West, in the middle station, on the shore of the Surrounding Ocean. Glory contains me from both sides and no finite essence reveals me." (Ibn ʿArabī 2006, p. 46). Ebstein (2014, p. 92 ff.) sees a clear influence of the work of the alchemist Jābir ibn Ḥayyān in the Akbarian concept of *habāʾ*. Pierre Lory also proposes points in common with Jābir although he considers that the doctrinal depth of Ibn ʿArabī goes far beyond the Jābirian identification between the name of a thing and its essence. (Lory 2004, p. 118).

[23] Reference to Qur'an 16:40: "The only words We say to a thing, when We desire it, is that We say to it: Be! (*kun*) and it is."

[24] It is worth noting that in the Ḥadīth we find the term *dhikr* with the meaning of Qur'an or Divine Register, close to that used by Ibn Masarra: "His throne was on the water, he wrote all things in the *dhikr* and created the heavens and the earth". (Al-kutub al-sitta 2000, Bukhārī, Bad'al-khalq, 3191).

[25] "Know that the existent beings are the worlds of Allāh which do not cease. [Allāh] exalted be He, said concerning the existence of Jesus, peace be upon him, that he is [the messenger of God] and His word (*kalima*) which He has cast unto Mary [Q 4:171]; this is Jesus, peace be upon him. So this is why we say that the existent beings are the word of Allāh". Ibn ʿArabī, II, p. 385, translation by (Ebstein 2014, p. 53).

identifying it with the imperative *kun*[26] or using it as a synonym for universal human being as microcosm (*kalima jāmi'a*, comprehensive Word)[27]. In other passages of *Futūḥāt* and *Fuṣūṣ al-ḥikam* the divine words (*kalimāt*) correspond to Immutable Entities, a specifically Akbarian concept that refers to beings in a potential state of creation (*mumkināt*)[28]: "'There is no changing the words of God' (Q 10:64), and the Words of God are no more than the Immutable Essences of things that have appeared in existence"[29]. The Christological dimension that we highlighted above is not merely confined to the identification of Jesus with the Word of God as expressed in the Holy Qur'an 4:171 but is also closely related to the articulation of letters through breathing. This subject is addressed in chapter 2 of *Futūḥāt*, in chapter 198 "The Breath of the All-Merciful" and in chapter 20 "On the science particular to Jesus". When he talks about vowels and how their existence facilitates articulation of the consonant *ductus*, he explains that they are what puts words into motion and, ultimately, gives them life. Giving life or causing resuscitation by exhaling air is, in Ibn 'Arabī's words, knowledge that corresponds to Jesus:

> Know—and may God help you in your search for knowledge—that the science particular to Jesus is the science of letters (*ḥurūf*). For this reason, Jesus received the power of breathing in life (*nafakh*)[30] which consists of the air that comes from the depths of the heart and is the spirit of life. Since breath makes stops on the path of exhalation to the mouth, we call these places [where the air] stops, letters, and that is where the entities inherent in the letters manifest. When these form[31], tangible life manifests in intelligible meanings (*ma'ānī*) and this is the first thing the Divine Presence manifests to the world. (Ibn 'Arabī 1999a, vol. 1, 256 cf. with Valsan 2016, p. 136).

Earlier we pointed out the transcendent qualities and macrocosmic implications of the letters of the Arabic alphabet; through the expression of His will God creates a universe that, in addition to being intelligible, is intellective and "speaking": "The cosmos, in its entirety, is rational, alive and rationally expressed (*nāṭiq*)" (Ibn 'Arabī 1999a, vol. I, p. 185); "The heavens were made endowed with reason (*'āqila*), hearing and obedient, and for each star and planet a path was drawn for it to follow." (Ibn 'Arabī 1999a, vol. 6, p. 179). This idea connects in a certain way with the text of Qur'an ayah 17:44: "The seven heavens and the earth, and whosoever in them is, extol Him. Nothing is, that does not proclaim His praise, but you do not understand their extolling. Surely He is All-clement, All-forgiving."

The science of letters is covered in general terms in chapters 2 and 198 of *Futūḥāt* and in *K. al-Mabādi wa-l-gayāt fī ma'ānī l-ḥurūf wa-l-ayāt* and, regarding specific letters in particular, in booklets such as *K. al-mīm wā wa-l-nūn* and *K. al-yā'/al-huwa*.[32] As we indicated above, the mysticism of the letters of the alphabet ranks among the most characteristic

---

26 "All existent things are the inexhaustible words of God (Q 18:120) that come from the divine Command *kun*, and *kun* is the word of God." *Fuṣūṣ*, 142 in (Hakīm 1981, p. 976).

27 Ibn 'Arabī, *Fuṣūṣ*, 50; *Futūḥāt* 146 in (Hakīm 1981, p. 977). This expression echoes the hadith "[Prophet Muḥammad] was given the synthesis of all-comprehensive words (*jawāmi' al-kilam*)" (Bukhārī 20/7099), in the sense that—according to Ibn 'Arabī- he received the whole of the Revelation including the former prophets, knowledge of the first and last people, everything that the human being has gained in this world; not only all the names of the patterns of the cosmos -adamic knowledge- but also their essences. Furthermore, he has the ability to synthesize words, whereby each law is manifested and all knowledge is inherent to Muḥammad, at all times, for every messenger and prophet, from Adam until the Day of Judgement (Chittick 1998, pp. 216, 222, 246). This idea has to do with Muḥammad´s role of Seal of Prophecy (Chittick 1989, p. 241).

28 Ibn 'Arabī, *Futūḥāt*, 65 in (Hakīm 1981, p. 976).

29 (Ibn 'Arabī 2009, p. 257).

30 Allusion to the Qur'anic episode in which Jesus gives life to some clay birds by breathing into them: Q 3:49. That is none other than the breath from the All-Merciful.

31 The verb that is used means to compose harmoniously *ta'allafa*. Cf. this use with Ibn Masarra (2007b, p. 63).

32 *K. al-mīm* and *K. al-yā'* published in the *Rasā'il* Ibn 'Arabī, Hayderabad, (Ibn 'Arabī 1948), the former re-published in an edition by Charles André Gilis, Beirut, 2002 and the latter by M. Fawzī al-Jabr, Beirut, 2004. The *K. al-mabādi* was edited by 'Abd al-Fattāḥ, Beirut, 2006. The *Kitāb al-alif* or *Kitāb al-aḥadiyya*, centring on divine unity symbolised in the first letter of the alphabet, could be included in this group. Compare with Ibn Masarra's identification of the letter *alif* with unity: "*Alif* is the first proof of divine unity (*tawḥīd*) as it is isolated at the beginning [of the word] and does not join with any of the other letters [that follow it]" (Ibn Masarra 2007b, p. 64).

contributions made by Ibn Masarra's thought and it is the one that gained him the most explicit recognition in later authors such as Ibn 'Arabī. In his *K. al-mīm wa-l-wāw wa-l-nūn (The Book of the letters mīm, waw and nūn)*, he quotes Ibn Masarra, cautiously distancing himself from the realm of magic or theurgy[33]: "Our discourse is about the secrets [of the letters] in the manner of Ibn Masarra al-Jabalī and others, and not about their operative virtues [*khawāṣṣ*], since a discourse on the operative virtues of things leads in most cases to the author being suspected of imposture." (Ibn 'Arabī 2001b, K. al-mīm, p. 87). Indeed, the relationship between lunar mansions and letters, the esoteric interpretation of isolated letters at the head of some surahs of the Qur'an and the symbolic interpretation of the pen strokes of letters in Ibn 'Arabī show a clear influence of *K. Khawāṣṣ al-ḥurūf* by Ibn Masarra. This basis, added to other influences and a brilliant reworking, take this idea of the universe as a "linguistic structure" even further. It is striking, however, that other texts that address this issue omit the name of Ibn Masarra and cite the famous Iraqi Sufi al-Ḥallāj[34] (d. 309 H/922 CE), a follower of Sahl al-Tustarī—the first reference in the Science of Letters—about whose teachings he provides little new information, other than regarding the doctrine of *kun*. This is the mention that Ibn 'Arabī makes of this author:

> When you hear someone on our path speak of letters and say that a certain letter is so many fathoms or so many spans in height or length, as al- Ḥallāj and others do, know that by "height" they mean operative virtues (*fi' l*) in the world of spirits and "length" refers to their operative force in the world of bodies (. . .) this technical terminology (*iṣṭilāḥāt*) was introduced by al- Ḥallāj. Those among the realised Sufis (*muḥaqqiqūn*) who understand the deep reality of *kun*, posses the science of Jesus (*'ilm 'isawī*)[35]. (Ibn 'Arabī 1972, vol. 3, p. 95 cf. transl. Valsan 2016, pp. 141–42)

Chapter 2 of *Futūḥāt* (see partial translation, commentary and study of the background to the mysticism of letters in Gril 2004) explains these doctrines in an exhaustive and intricate way, taking into account the mathematical implications of the combination of letters, cycles of time, and movements of the celestial spheres on the basis of which, according to his theory, letters are created. In *K. al-mīm* he cites Pythagoras (Fīthāgurus) as the precursor of this discipline that operates with letters and numbers, and the numerologists (*'adadiyyūn*). In the same way that human beings are made of spirit, soul and body, words also have a triple dimension: they can be thought, articulated, or written (Lory 2004, p. 119). This triplicity is recurrent and we also find it in the commentary on the isolated letters *alif, lām, mīm* at the head of the Surah of the Cow: "There are three archetypes of books: the "drawn" (*mastūr*) book, the "inscribed" (*marqūm*) book and the "unknown" (*majhūl*) book." (Gril 2004, p. 173)

In addition, Ibn 'Arabī's use of Arabic grammar rules to support explanations of certain metaphysical questions does not appear to be found in Ibn Masarra. In his prophetological work *Fuṣūṣ al-ḥikam* we find an example of this when he explains the hadith "I have been made to love three things in this world: women, perfume and prayer"[36] saying that Prophet Muḥammad gave priority to the feminine by mentioning the word three in the feminine form *(thalāth)*, instead of the masculine form *(thalātha)* which would have been correct, from the grammatical point of view, in an enumeration including a masculine element and he goes on to explain that whichever theological or philosophical school you follow, the terms that express metaphysical principles are feminine in gender; (divine attribute, *ṣifa*; capacity, *qudra*; cause, *'illa*; essence, *dhāt*, etc.) because they allude to

33 He would not show this critical attitude when speaking of his teacher Nuna Fāṭima bint al-Muthanna who mastered the science of letters to such a point that she had the power to press the Surah *Fātiḥa* into her service (Ibn 'Arabī 1971, p. 143).

34 For al-Ḥallāj the highest knowledge was that granted by the letters as they appear in the Qur'an: "The knowledge of all things resides in the Qur'ān, and the knowledge of the Qur'ān resides in the letters that are placed at the beginnings of the suras." (Al-Ḥallāj 1936, p. 95 in Gril 2004, p. 139).

35 The 1999a edition reads *'ilm 'ulwī* (science of the higher spheres) vol. 1, p. 257.

36 Hadith quoted with variants in *Al-kutub al-sitta* 2000 Nisā'ī, *Kitāb 'ashara al-nisā'*, 36: 1, hadith No. 3391.

the ontological priority of the feminine in creation (see Ibn ʿArabī 1946, pp. 214–15)[37]. Conversely, he derives grammatical rules from his metaphysics. In this regard, see the arguments regarding Arabic desinential inflection (*iʿrāb*) in *Futūḥāt*: the words that decline are *mutalawwin*, either complete—with three case vowels—or incomplete—diptotes—while the invariable words are *mutamakkin* (Gril 2004, p. 185), *talwīn* being the continuous passage from one spiritual state to another and *tamkīn* that of consolidation in a specific state.

With regard to precedents outside al-Andalus for this type of speculation about letters, Ebstein indicates that they cannot be found in the *Epistles of the Brethren of Purity*, but he attributes to them, nevertheless, an Ismaili origin: the North African works *Kitāb al-ʿālim wa-l-gulām* and *Kitāb al-Kashf*, of C10 CE or other writings by Rāzī and Sijistānī (Ebstein 2014, p. 237). In addition to the aforementioned authors, Ibn Arabi mentions other Eastern figures such as Ḥākim Tirmīdhī or the imam Jaʿfar al-Ṣādiq as references in the science of letters (see Ibn ʿArabī *K. al-mīm*, pp. 83, 86)[38].

2.2. Throne and Angelology

The First Intellect/Pen and the Universal Soul/Protected Tablet are two fundamental principles in Ibn ʿArabī's cosmology. From the Intellect, which represents the first being created by God, arises[39] the Soul, and from the interaction that takes place between the two by way of writing, the Pen being the active and luminous element that projects onto the Tablet, the passive and dark element, arises everything that occurs in creation. This same idea, referring to the intellect and the human soul, is expressed by Ibn Masarra through the metaphor of the sun and the moon: the rational human soul (*nafs nāṭiqa*) receives the light of the Intellect, as the moon receives it from the sun (Ibn Masarra 2007b, p. 69). Knowledge is found synthetically in the Intellect and analytically in the Tablet; the Tablet is "the receptacle (*maḥall*) of the dictate of the Intellect"[40]. Ibn Masarra had previously put forward this schema, and he added to this pair of opposite terms a third consisting of the Throne and the Footstool (*ʿarsh-kursī*), two classic elements of Islamic cosmology that symbolized the sphere that encompassed the entire universe beyond which lay the realm of the divine. Sometimes he equates both the Pen and the Tablet with the Intellect (see *K. Khawāṣ al-ḥurūf* ed. Jaʿfar pp. 154, 164 in Ebstein 2014, pp. 52, 53). "The *Dhikr* (word/remembrance) is the Universal Intellect designated by God—may He be exalted—for the Universal Soul, which is the Pedestal." (Ibn Masarra 2007b, p. 88).

The Throne (*ʿarsh*), finally, is also understood as a Tablet by Ibn Masarra: "[In the ayah] 'And it is He who created the heavens and the earth in six days, and His Throne was upon the waters.' (Q 11:7), 'He' is an allusion to His essence and 'Throne' is an allusion to the letter *lām* and the Tablet." (Ibn Masarra 2007b, p. 77). The Intellect, in the words of Ibn Masarra, is submerged (*mustagriq*) in the Soul and the latter, in turn, in the universal Body. For Ibn ʿArabī the Soul arises from the Intellect, part of which was breathed into it at the time of its creation (Ibn ʿArabī 1919, p. 55). For both of them, the Soul depends on the Intellect and there is a total correspondence between them and a continuous provision of knowledge from one to the other (Stroumsa and Sviri 2009, p. 237).

Despite the polyhedral cosmology that Ibn ʿArabī displays in his work, the Throne and the Footstool usually appear with a clear distinction and, unlike Ibn Masarra, the

37 See this type of analysis also in the *K. al-yāʾ* when he explains the relationship between the divine essence and the notion of self symbolised in the pronouns *huwa* (he) and *hiya* (she).

38 See also (Ebstein and Sviri 2011).

39 As we saw when causality was discussed, although it is true that in some passages of his work, Ibn ʿArabī implies that some elements of creation arise through or from others, at the same time he rejects the idea of intermediate cause and, we believe, avoids frequent use of the term "emanation" (*fayḍ*) when speaking of creation. When he does use it, he tries to distance himself from its Neoplatonic meaning by mentioning Divine Will immediately afterwards. "The Supreme Pen bestowed spirits [upon created beings, *inshaʾāt*] and God entrusted their custody to it. [These spirits] are a marvellous emanation (*fayḍ*), essential with regard to the Pen and voluntary as regards God—may he be exalted. The Tablet was expressly created by the will of God." (Ibn ʿArabī 1919, p. 56). "[The Pen] radiates emanations (*fayḍ*) from both sides: an emanation [relating to the] essence, and an emanation [relating to the] divine Will." (Ibn ʿArabī 1919, p. 51).

40 (Ibn ʿArabī 1919). *ʿUqlat al-mustawfiz*. Edited by Nyberg. Leiden.

Throne is identified with the Pen or Intellect but not with the Tablet (even though Intellect and Soul are usually considered realities above the Throne and Footstool):

> Therefore, from this point of view, [it is called] Intellect; from the point of view of the world of Inscription and Archetypal Writing, it is called Pen; from the point of view of the government of the world (*taṣarruf*), Spirit; from the point of view of the Divine Seat (*istiwāʾ*), Throne; and from the point of view of Divine Calculation (*iḥṣāʾ*), Evident Guide (*imām mubīn*)[41]. (Ibn ʿArabī 1919, p. 52).

Ibn ʿArabī uses the symbol of the Throne profusely in a number of treatises, relating it to various realities, various degrees of the same reality, or various modes of the manifestation of divine omnipotence. To the quintessential throne—the *ʿarsh al-Raḥmān* or Throne of the Merciful mentioned in the Qur'an—with its already varied symbology, the following uses of the term should be added: The Delimited Throne (*maḥdūd*), which is nothing but the human being itself, the Throne of the Spirit (*ʿarsh al-rūḥ*), referring to the rational human soul, the Throne of the Cloud (*al-ʿamāʾ*), the Throne of the Ruling and the Decree (*al-faṣl wa-l-qaḍāʾ*) which alludes to the throne on which God will reveal himself on Judgement Day (*yawm al-ḥashr*), the Throne of the Qur'an referring to the heart of the believer or the Throne of Manifestation (*takwīn*) that corresponds to the sphere that encompasses all created beings[42].

Speaking of the Throne-bearing angels in chapter 13 of *Futūḥāt* "On the Divine Throne" and in *ʿUqlat al-mustawfiz*, Ibn ʿArabī explicitly acknowledges that he is following Ibn Masarra:

> We have related, following Ibn Masarra al-Jabalī, who was one of the greatest men of the Sufi Way both in knowledge and in spiritual state and enlightenment (*kashf*), that the transported Throne refers to the dominion (*mulk*). And this is reduced to spirit, body, sustenance and grade. Adam and Israfīl bear the forms, Jibrīl and Muḥammad, the spirits, Mikāʾīl and Ibrāhīm, the provisions, and Mālik and Riḍwān, the eschatological rewards and punishments. (Ibn ʿArabī 1972, vol. 2, p. 348)

> He made—praise be upon Him—eight bearers to carry the Throne on the day of Resurrection. Today four angels carry it: One under the image of Isrāfīl, the second under the image of Gabriel, the third under the image of Michael, the fourth under the image of Riḍwān, the fifth under the image of Mālik, the sixth under the image of Adam, the seventh under the image of Abraham and the eighth under the image of Muḥammad—God's peace and blessings be upon him. These are the images of their spiritual station, not the images of their constitution. When Ibn Masarra al-Jabalī—God have mercy on him—mentioned them in the same way that we mention them, he said [the following] regarding this: Isrāfīl and Adam [are in charge] of the images, Muḥammad—God's peace and blessings be upon him—and Gabriel of the spirits, Mikāʾīl and Abraham of the favors (*arzāq*) [of Providence]. Riḍwān and Mālik [are in charge] of the reward and the threat [of punishment]. The Throne, in Ibn Masarra, is an expression that alludes to the dominion (*mulk*). (Ibn ʿArabī 1919, p. 58)

If we turn to the work of Ibn Masarra, we find that this idea appears in quite succinct form in the *K. Khawāṣṣ al-ḥurūf* and the distribution of angelological functions attributed to him by Ibn ʿArabī is not present:

> Four angels carry the Throne of God, and four carry the Footstool, this makes eight. Seven carry the seven heavens and the worlds. Each heaven has an angel

---

41 The identification (to be found in Ibn Barrajān as well as Ibn ʿArabī) of *imām mubīn* with divine calculation (*iḥṣāʾ*) is supported by Q. 36:12, "Lo! We it is Who bring the dead to life. We record that which they send before (them), and their footprints. And all things We have kept (*aḥṣaynāhu*) in a clear Register". Trans. Pickthal.

42 Ṣuʿād Ḥakīm lists fourteen types of thrones. (See Ḥakīm 1981, p. 791).

> that transports it, thanks to whom the sphere moves and who is in charge of establishing order in it. (Ibn Masarra 2007b, p. 84)

The fact that there are four now and eight in the Other Life suggests, as Asín Palacios deduced (Asín Palacios 1914, pp. 72–75), that each of these four functions has a dual reality: an exoteric one that manifests in this life and an esoteric one that will be shown in the Other Life. In the case of the first three pairs, at least, which combine a prophet (external reality) and an angel (internal reality), this theory seems to hold[43]. Regarding the identification of the dominion (*mulk*) with the Throne that Ibn ʿArabī also attributes to Ibn Masarra, we can ascertain that this identification occurs not with the Throne—to which the letter *lām* and the Tablet (*lawḥ*) correspond—but with the Footstool. Talking about the letter *mīm,* a letter associated with Will (*mashīʾa*), the dominion of the physical (*mulk*) and with place (*makān*), Ibn Masarra tells us that in every heaven there is a footstool (*kursī*), given that what makes the heavens and the earth stand firm are the attributes Al-Ḥaqq and al-Mulk (Ibn Masarra 2007b, p. 81)

Returning to chapter 13 of *Futūḥāt*, Ibn ʿArabī alludes once more to the master from Córdoba, stating that:

> What is told about the form of these bearers is approximately the same as that which Ibn Masarra stated. One is said to have a human form, the second that of a lion, the third an eagle, and the fourth a bull. This [bull] was the one the Samaritan saw, imagining that it was the god of Moses. That is why he built a calf for his people and said: "This is your God and the God of Moses," according to the story [in the Qur'an]. (Ibn ʿArabī 1972, p. 355)

Once more, the works by Ibn Masarra that have survived do not include this description—which is, however, found in Ibn Barrajān[44]—although this does not prevent us from thinking that they are in lost works or were transmitted orally by his disciples. This is the opinion held by Stroumsa (Stroumsa and Sviri 2009, pp. 96–97) who, moreover, has studied the biblical echo of this description in the vision of the chariot in Ezekiel and Jewish and Christian elaborations (Stroumsa 2006, pp. 103–5).

Ibn ʿArabī defines the Footstool as the place of the two feet [of God] using Ibn Masarra's term *mawḍiʿ*. The Word (*kalima/kalām*), according to Akbarian doctrine, is one on the Throne and divides into two on the Footstool, thereby introducing the principle of multiplicity in creation, and it continues to expand downwards, doubling itself: "His Word acquired a fourfold character with the creation of the fourth sphere [i.e., the sphere of the fixed stars]" (Ibn ʿArabī 1919, p. 69). The Footstool constitutes the differentiation necessary for the unity represented by the Throne of God to manifest. Each of the two feet is related to two groups of opposing divine attributes (Ibn ʿArabī 1919, p. 59). In the *Risāla al-iʿtibār* (*Epistle of contemplation*), Ibn Masarra speaks of the Throne as the first being created and, in this case, he does identify it with the Intellect.

43 This distribution of functions can be explained as follows: Adam, the first phenomenal manifestation of the body, and Isrāfīl, are in charge of bodies in the future life, to be understood in close connection with the trumpet—*ṣūr*- that sounds on the Day of Resurrection, since it shares a lexical root—ṢWR—with *ṣuwar* (images), as explained in *ʿUqla* (Ibn ʿArabī 1919, p. 86). The realm of creative imagination—understood not as a faculty but as a realm of creation—is identified with the trumpet of light that will be sounded by Isrāfīl on the Day of Resurrection. The Arabic word for trumpet coincides with the plural of the word image, thus, according to Ibn ʿArabī, when Isrāfīl blows the trumpet he will also breathe life into those images. The upper, broader part of the trumpet reaches to the Cloud (*al-ʿamāʾ*) and the lower part to the ground. Abraham's function as a provider seems more evident; described in the Qur'an as the intimate of God, he is also known for the biblical and Qur'anic episode in which he offers food to his guests without knowing that they are angels. He will be joined by Mikāʾīl, probably because he is responsible for the subsistence of the Self. Muḥammad—whose pre-existence in the form of Muḥammadan reality prior to Adam himself grants him pre-eminence in the spiritual world—joins Gabriel, the highest of the angels. The last pair consists of two angels, Mālik and Riḍwān, the guardians of hell and paradise, respectively. See also *The alchemy of happiness* (Ibn ʿArabī 2019) and chapter 167 of *Futūḥāt,* where the functions of the Throne-bearing angels are explained again. (López-Anguita 2014). Regarding the throne in Ibn Masarra see also (Garrido Clemente 2008a, 2008b).

44 "It is frequently said in the ancient books (*kutub mutaqaddima*) and primal knowledge (*ʿilm awwal*) that the carriers of the throne are four angels. One of them resembles a human, the others an ox, lion, and eagle." (Ibn Barrajān, Tanbīh in Casewit 2014). Could Ibn ʿArabī have derived this idea from Ibn Barrajān and not from Ibn Masarra?

We can say that in Ibn Masarra the denotation of divine sovereignty is shared by the Throne (ʻarsh) and by the Footstool (kursī), an identification fostered by the Qur'an itself, in which both terms are merged in the same meaning (Q 11:7; Q 2:55).

3. Ibn Barrajān

Thanks to the research carried out over the last two decades[45] that led, among other things, to the publication of previously unpublished works by Abū l-Ḥakam Ibn Barrajān of Seville (d. 536 H/1136 CE), we have an idea of not only the thought of one of Al-Andalus's most important and least known mystical authors[46] but also the intellectual environment from which thinkers such as Ibn ʻArabī developed. The available works by this master, dubbed "the Ghazālī of Al-Andalus"[47] in some sources, are a *Commentary on the Names of God (Sharḥ al-asmāʾ al-ḥusnà / Tarjumān lisān al-Ḥaqq)*, widely circulated in the East, which lists one hundred and forty Divine Names, and the commentaries on the Qur'an *Tafsīr al-Qurʾān* (or *Tanbīh*) and *Īḍāḥ al-ḥikma fī aḥkām al-ʻibra* (translated by Böwering and Casewit as *Elucidation of Wisdom)*. Ibn Barrajān is considered to have introduced into Al-Andalus a new way of doing hermeneutics, which some classical historians identify with the Sufi method[48].

We know that Ibn ʻArabī studied Ibn Barrajān's Qur'an commentary, *Īḍāḥ al-ḥikma* together with his master al-Mahdawī (d. 621H/1224 CE) in Tunis in 590 H/1194 CE although, according to Addas (Addas 1993, he may have had prior knowledge of this work through ʻAbd al-Ḥaqq al-Ishbilī (d. 581 H/1185 CE). Ibn ʻArabī refers to this *tafsīr* in a letter to Mahdawī as *The Book of Wisdom* (*K. al-Ḥikma*) (see Ṭāhir Ḥasanayn 1985, p. 12 and Elmore 2001, p. 611). Ibn Barrajān is cited explicitly in *Futūḥāt* when he speaks of the mystic making the Names of God his own[49] and as a precursor of the notion "The Truth through which creation exists" (Ibn ʻArabī 1911, vol. 2, p. 649; vol. 3, p. 77), and in *Mashāhid al-asrār, Mawāqiʻ al-nujūm* (González Costa 2009, p. 58) and *Tadbīrāt al-ilāhiyya* (Ibn ʻArabī 1919, p. 125) Ibn Barrajān is cited as a reference for other doctrines, in addition to these two. See for example: "That is what the Gnostic master Abū l-Ḥakīm Ibn Barrajān was referring to with the expression *imām mubīn* (evident guide) which is the Protected Tablet alluded to in the expression "everything" in verse 7:145 "we wrote for him on the Tablets of everything an admonition, and a distinguishing of everything." (Ibn ʻArabī 1919, p. 125).

---

[45] Of which we would single out the following: (Ibn Barrajān 2000). *Sharḥ asmāʾ Allāh al-Ḥusnà. Comentario sobre los nombres más bellos de Dios*. Edition and introductory study by Purificación de la Torre. Madrid: CSIC-AECI; González Costa, Amina. (González Costa 2009). "Un ejemplo de la hermenéutica sufí del Corán en Al-Andalus: *Īḍāḥ al-ḥikma* de Ibn Barraŷān de Sevilla (m. 536/1141)" in *Historia del Sufismo en Al-Andalus y el Magreb* edited by (González Costa and López-Anguita 2009). Córdoba: Almuzara; Bellver (2013). "Al-Ghazālī of al-Andalus: Ibn Barrajān, Mahdism and the Emergence of Learned Sufism on the Iberian Peninsula". *Journal of the American Oriental Society*, 133 (4): 659–81; Gril (2007). "La lécture supérieure du Coran selon Ibn Barraǧān et Ibn ʻArabî" in *Symbolisme et herméneutique dans la pensée de Ibn ʻArabī*. Edited by Bakri Aladdin. Damascus: IFPO. pp. 147–61; Küçük (2013a, 2013b). "Light Upon Light in Andalusian Sufism: Abū al-Ḥakam Ibn Barrajān (d. 536/1141) and Muḥyī l-Dīn Ibn al-ʻArabī (d. 638/1240) as Developer of His Hermeneutics. Part I: Ibn Barrajān's Life and Works," *ZDMG* 163 (2013a), pp. 87–116 and 2013b "Part II: Ibn Barrajān's Views and Legacy," ZDMG 162, pp. 383–409; Böwering, Gerhard and Yousef Casewit. 2015. A Qurʾān Commentary by Ibn Barrajān of Seville (d. 536/1141). *Īḍāḥ al-Ḥikma bi-aḥkām al-ʻibra (Wisdom Deciphered, the unseen Discovered)*; Yousef Casewit (2017). *The Mystics of al-Andalus. Ibn Barrajān and Islamic Thought in the Twelfth Century*. Cambridge University Press. For a full bibliographic review (Casewit 2012–13). On Ibn Barrajān´s use of the Bible in his *tafsīr* (Casewit 2016).

[46] Despite being educated as a jurist, he would be remembered as a mystic and an exegete. On sources from which to extract biographical information on Ibn Barrajān, (González Costa 2009, pp. 49–52; Küçük 2013a; Casewit 2017).

[47] With regard to his ascription to a specific school of thought, Küçük notes that "while Ibn Barrajān was presented by earlier researchers as a follower of Ibn Masarra, biographers or hagiographers say nothing about his "Bāṭinī" beliefs or his Muʻtazilī tendencies, as they do for Ibn Masarra." (Küçük 2013b, p. 405). For a study of the life and work of Ibn Barrajān, especially his exegesis, we would point to the exhaustive works of Casewit and Böwering (Casewit 2017; Ibn Barrajān 2015).

[48] Ibn Zubayr said about his *Tafsīr al-Qurʾān*: "It follows a method that has never had precedents, lingering over strange ayahs and invisible beings [questions]. He obscured expression in such a way that none can attain his meaning but those who know his words, his thousands of allusions and his inspiration" (Ibn Zubayr, Ṣila, nº 45, p. 32 in González Costa 2009, p. 57).

[49] In particular, the controversial doctrine of the adoption of the Names of God by the believer (*takhalluq bi-l-asmāʾ* called *taʻabbud* by Ibn Barrajān) (Ibn ʻArabī 1911, *Futūḥāt*, II, p. 649). On Ibn ʻArabī and Divine Names see (Ibn ʻArabī 1997).

The notion of contemplative interpretation, superior reading (*tilāwa ʿuliyà*, see Gril 2007) or symbolic transposition, ascending and in degrees, introduced into al-Andalus by Ibn Masarra, also constitutes the essence of Ibn Barrajān's hermeneutics. At first glance, this hermeneutics is included in the tradition, which had already begun in the East, of *taʾwīl* (esoteric or allegorical interpretation. Literally, the term means to remit an element to its beginning) or *tafsīr bi-l-ishāra* (exegesis by allusions) but for some researchers such as Casewit (2017, p. 207), it belongs to another type: the *ʿibra* or *iʿtibār*. The meaning that underlies this lexical root is that of crossing or passing from one shore to another, i.e., from the external form of a word, an ayah or any sign of Nature, to its interior meaning. Indeed, reading the book of nature and contemplating God's signs (*ayah*) with the intellect (*ʿaql*) enables those who contemplate to gradually ascend[50]. Hell and paradise, according to his dual vision of reality, underlie or coexist in this plane of existence: "Ibn Barrajān rejects the common understanding of the unseen world (*ghayb*) as a transcendent abode that is "out there" spatially, and "yet-to-come" temporally. (. . .) the visible world both conceals and reveals the invisible. This world signals the next world because it is an integral part of it."[51] (see Böwering and Casewit´s introduction to Ibn Barrajān 2015, p. 40) Compare with the Akbarian idea: "everything in this world is a model of the things in the other world" (Ibn ʿArabī 1911, vol. 4, p. 206). Ibn Barrajān suggests, moreover, that the meanings of the elements of nature are not univocal. Fire, for example, is a reminder of hell and, at the same time, of divine Mercy:

> Among the traces of this Book in existence is its allowing fire to exist, despite its burning power (. . .). But at times fire, by the wisdom that remains in the content, erupts and acts as a warning to servants who thus heed the warning and remember the house of the hereafter and, as such, gain knowledge and certainty about the existence of this house. And by contemplating how Mercy stops this exhalation of hell and how [Mercy and fire] then alternate, they understand wisdom and the mandate that is within it (...) The reason why punishment exists is that it is necessary for the Mercy (*raḥma*) that descend s to appease that anger to manifest itself; in the alternation of both is that which enables life. (Ibn Barrajān, *Īḍāḥ*, fol. 9a in González Costa 2013, pp. 259–60, transl. by author)

Ibn Masarra seems to prefigure and inspire Ibn Barrajān's concept of *iʿtibār* but, according to Casewit, the former places more emphasis on the ability of the human intellect to operate without revelatory guidance in its quest for the truth (Casewit 2017, p. 37)[52]. Far from being arbitrary, the exercise of *iʿtibār*, when we refer to the revealed text, does not disrespect literality or internal coherence. Neither is it univocal, since for each *muʿtabir* there will be a personal reading of the Koran. The idea of interpretation as crossing from one shore to another appears in one of the most markedly visionary Akbarian works, *Mashāhid*, where we find the expression "boat of interpretation" (Ibn ʿArabī 2001b, p. 81) in this sense. This sequence of observation (*naẓar, baṣar*) of Nature, meditation (*iʿtibār*), and ascent is present in a similar fashion in Ibn ʿArabī, Ibn Masarra and Ibn Barrajān:

> The delights are in the nourishments, the nourishments in the fruit, the fruit in the boughs [. . .] and the order issues from the Lordly Presence. Ascend from here, look (*unẓur*), enjoy yourself but do not speak. Then he said to me 'Preserve the intermediaries'. (Ibn ʿArabī 2001b, pp. 87, 92)

---

50 On *iʿtibār* as a ladder of cognitive ascension see (Altmann 1967; De Callataÿ 2014).

51 Nature as a book is an idea developed extensively by Ibn Barrajān: "If it is arranged thus and it enables life and our existence, this is because it contains a warning and a call to believers to remember through those signs what is within eternal life and to gain knowledge and certainty of the existence of the other world" (González Costa 2013, p. 259, transl. by author).

52 "Whereas the term *i´tibār* was used by Abū Naṣr al-Fārābī (d. 338/950), the Brethren of Purity, and Avicenna (d. 428/1037) to mean the inductive method that equips the philosopher with tools to demonstrate God's existence, Ibn Masarra's is a method of meditative ascension which differs from the purely cerebral process of discursive reasoning. Indeed, his conceptionof *iʿtibār* foreruns Ibn Ṭufayl's (d. 581/1185) autodidact, Ḥayy Ibn Yaqẓān, and is also Sufi-inspired since only the spiritually purified saints are endowed with this gift." (Casewit 2017, pp. 37–38).

> The observer may examine (*yanẓuru*) one of the three [genera]: animals, plants and inanimate beings. He observes the plant and sees an inanimate, [. . .] As he observes this nutrition, he sees that it ascends upwards and spreads sideways [and finally] he finds the place of the footstool and the place of the spirit to be permanent and encompassing. (Ibn Masarra 2007a, 93 transl. Stroumsa and Sviri 2009, pp. 219–21)

> [Water] descends from heaven, even though it is not manifest (*ẓāhir*) heaven itself today, it is therein in a non-manifest (*bāṭin*) manner. Just as creatures that are engendered from water are from heaven. (Ibn Barrajān, *Tanbīh*, vol. 5, p. 241 transl. Casewit 2014)

> The higher corresponds (*yantaẓim*) to the lower. (Ibn Barrajān, *Īḍāḥ* in Casewit 2014, p. 297)

Ḥaqq, khalq, raqīqa

One of the most seminal concepts in Ibn Barrajān's work is *al-ḥaqq al-makhlūq bihi al-ʿālam/al-samawāt wa-l-arḍ* (the truth through which is created the world / creation / the heavens and earth or The Real according to Creation is created) which appears in the three works mentioned above. This idea, based on various Qur'anic ayahs (Q 10:15; 15:85; 44:38–39) which imply that the *ḥaqq* is an instrument of God in creation, would find a significant echo in Ibn ʿArabī. This idea had already appeared in Ibn Masarra when, speaking of the letter *ḥā*, he explained that it represents *al-Ḥaqq*, the Truth or Reality with which God created the heavens and the earth (Ibn Masarra 2007b, pp. 68, 81). Truth, Ibn Masarra tells us, is the Name through which God creates the seven heavens and the earth, and is based on the ayah "It is He who created the heavens and the earth in truth; and the day He says 'Be', and it is; His saying is true. And His is the Kingdom the day the Trumpet is blown (. . .)" (Q 6:73) to explain that, as the two attributes that support and hold the heavens and the earth are *al-ḥaqq* and *al-mulk*, each heaven has a footstool (Ibn Masarra 2007b, p. 87). "According to Ibn Barrajān, the Quran, the universe and mankind are three aspects of the same reality (*al-Ḥaqq*), since *al-Ḥaqq* is everything." (Küçük 2013b, p. 385).

The attribute or Divine Name *al-Ḥaqq* (The Truth / The True / The Real) which in Ibn ʿArabī—probably, as we said earlier, under the influence of Ibn Barrajān—frequently appears linked to creation, designated in the first exegesis the action of God in the world. It was used as a synonym for *islām, dīn, hudà*, Qur´ān and was in opposition to *bāṭil* (vain, useless). It is also understood as something that ensues (*kā ´in*) and indeed is considered synonymous with the end of the world because, according to the Qur'an, it is something that necessarily must happen. Constructed with the preposition *bi* in front and in combination with verbs like create or reveal, it has the meaning that God did not create the world for nothing (Nwiya 1970, pp. 41–42). According to Akbarian cosmology, we can find multiple correspondences of *al-ḥaqq* with other elements: Truth equates to First Intellect (*al-ʿaql al-awwal*), Pen (al-qalam), Spirit (*Rūḥ*), Glorious Throne, Evident Guide (*al-imām al-mubīn*), Interpreting Spirit (*mutarjim*) of God, Universal Spirit (*Rūḥ kullī*), Justice (*ʿadl*), "The Truth through which Creation exists", Muḥammadan Reality, Spirit of Spirits, and even "every thing"[53]. It is also alluded to symbolically as Eagle (*ʿuqāb*) and White Pearl. The expression *al-haqq al-makhluq bi-hi* which, as we saw above, had been coined by Ibn Barrajān from the Qur'an ayah "We created not the heavens and the earth, and all that is between them, save in truth (*al-ḥaqq*)" (Q 15:85), is adopted by Ibn ʿArabī, acknowledging the explicit attribution to his master, in *Futūḥāt* (Ibn ʿArabī 1911, vol. 2, pp. 60, 104; vol. 3, p. 77).

This "Truth through which creation takes place" would become, in Akbarian ontology, a mode of expression of the *barzakh*, an indispensable doctrine to understand any aspect of the thought of Ibn ʿArabī. This isthmus or intermediate realm, which shares in the two

[53] (See Ḥakīm 1981, pp. 111–13; Ibn ʿArabī 2002, p. 109). In his *Īḍāḥ*, Ibn Barrajān mentions *ḥaqq mubīn* or *kitāb mubīn* rather than *imām mubīn* (González Costa 2013, p. 234).

realities that it separates, which can have a temporal, spatial, ontological, and epistemological meaning, is fundamental when it comes to relating the eternal and the contingent or God with creation. The following excerpt explains the ambiguous nature of this *ḥaqq* and the extent to which it is related to the Breath of the Merciful or the Cloud, denominations of the realm in which beings are brought from non-existence to existence: "The Cloud is the Real through whom takes place the creation of everything. It is called the Real (*Ḥaqq*) since it is identical with the Breath. And the breath is hidden within the Breather—for this is what one understands from "breath". Hence the Breath has the property of non manifest, but when it becomes manifest it has the property of the Manifest." (Ibn ʿArabī 1911, vol. 2, p. 310, translated by Chittick 1989, p. 134)

These two extremes *khalq* and *ḥaqq* are connected through *raqāʾiq*, subtle bonds that, according to some of Ibn al-ʿArabī's works, are in charge of uniting the Divine Names with each other and with created things, forming a kind of dialogue[54] and providing knowledge and existence continuously to created things (Ḥakīm 1981) This "cosmogonic drama" is set forth in *K. Inshāʾ al-dawāʾir* where it is explained that thirty bonds that extend from the Names of God, crossing each other and grouping together until they descend to the inhabitants of Hell and Paradise, keep creation alive. (Ibn ʿArabī 1919, pp. 40–41). The *raqāʾiq* are no more than relations, and they pre-exist created things: "God created rank (*makāna*) before He created place (*makān*). Then He stretched tenuities (*raqāʾiq*) from rank to specific places (…) then He brought into existence the things in their spaces" (Fut. II 528.26, trans. Chittick). These rays—as Ibn Arabi sometimes calls them—can introduce a principle of differentiation into beings: "Masculine and feminine, which are united in the *ḥaqīqa* but ordered in the circle of creation, have differentiated their ranks by means of the different *raqīqa*-s" (Ḥakīm 1981). It is in the perfect man and in the Supreme Element (*ʿunṣur aʿẓam*) where these threads will "tie together":

> God has made the human being as the sum of the subtle connections of the whole world. And from him to everything in the world a subtle relation is extended (*mumtamadd*). (Ḥakīm 1981, p. 536)
>
> That *raqīqa* between the servant and every part (*juzʾ*) of the world (…) exists according to an affinity/ what is appropriated (*yunāsib*) with the world and what has an affinity with him (*munāsib*). (Ḥakīm 1981, p. 536)
>
> [From the First Intellect] its subtle relations (*raqā'iq*) are extended to the Soul, the Prime Matter, the Body, the fixed stars, the center [of the Universe] and the elements, and through an ascending movement, [they reach] (…) engendered beings, the human being, and the Greatest Element -where [the subtle relations] tie together (*inʿiqād*), that is the origin of 46.656.000 subtle relations. (Ibn ʿArabī 1919, p. 52)
>
> In relation to the sphere of the world, the Greatest Element would be the [central] dot, and the circumference would be the Pen while the Table is what is between them. And in the same way the dot comprehends the circumference in its essence, this Element comprehends in its essence all the facets of the Pen which constitute those subtle relations we talked about before, they are one in the Element and they multiply themselves in the Intellect according to the diverse modes of reception of the [Element] in the [Intellect]. That's why the Greatest Element is stronger when recognising the unity of its Creator. (Ibn ʿArabī 1919, p. 82)

These subtle threads appear in the most diverse contexts. Discussing letters, he explains that the three letters of God (*alif, zayn, lam*) and the three letters of man (*nun, sad, dad*) multiplied by the three worlds of *mulk, malakūt,* and *jabarūt* result in nine spheres in which the knowledge of God is projected towards man (*aflāk al-ilqāʾ*) and nine spheres of reception that correspond to man (*aflāk al-talaqqī*) and "from each of the essential realities

[54] (Ibn ʿArabī 1999b, pp. 353–71).

of the nine spheres subtle bonds extend out toward the nine human spheres, and from this, reciprocally go back to the divine spheres." (Ibn ʿArabī 1911, vol. 1, p. 51 trans. Gril 2004, p. 156). This element of Akbarian cosmology also has angelological implications:

> There are subtle threads which extend from the Universal Soul to the Throne (. . .) these are like ladders (*maʿārij*) for the angels, while the meanings that descend in these tenuities are like angels. (Ibn ʿArabī 1911, vol. 3, p. 582, Trans. Chittick 1989, p. 406)

> (. . .) Wherever these bonds meet (*ijtamaʿ*), the angel itself is the meeting point and it is there where the angel comes to existence (*ḥadatha*). This newly arrived fact is thus the angel itself. If it bows with its whole being toward one of the sets of nine spheres, the other side attracts it. It thus comes and goes from one to another (*yataraddadu*). (Ibn ʿArabī 1911, vol. 1, p. 51, Trans. Gril 2004, p. 156)

This cosmological elaboration of the *raqīqa* does not appear to be present in Ibn Barrajān although the similar concept of *athar*—also used by Ibn ʿArabī—can be found, i.e., the effect or "trace" of a Divine Name in Creation as a way of connecting *ḥaqq* with *khalq*. These traces that "are in the whole existence" (Ibn Barrajān, *Iḍāḥ*, fol. 59b), can also be used as ladders to ascend by contemplation to the Divine Presence (Casewit 2017, p. 146). Applying Ibn Masarra's *iʿtibār* to the genre of the Commentary on the Divine Names is one of the innovative aspects of Ibn Barrajān's thought. According to his doctrine, the effects of the Names in creation can be apprehended by human beings and used in the contemplative ascent.

Along with the notion of *al-Ḥaqq al-makhlūq bi-hi*, that of Universal Servant (*al-ʿabd al-kullī*), is probably the one that had the greatest influence on the work of Ibn ʿArabī. Ibn Barrajān's definition of Universal Servant as the one "who possesses the *nūr al-mubīn*, which is the essence of the [prophetic] news (*inbāʾ*) and the Divine Revelation (*waḥy*)" is close to what Tustarī or Tirmidhī refer to as muḥammadan light or muḥammadan reality (Īḍāḥ, Mahmut Paşa 3, fol. 254b in González Costa 2013, p. 234). Although Ibn ʿArabī usually uses the expression Perfect Man (*insān kāmil*) for that same notion, in some passages we can find it cited in the manner of Ibn Barrajān:

> Perfect Man (*insān kullī*) is more perfect than the cosmos in its totality, since he is a transcript of the cosmos letter by letter, and he adds to it the fact that his reality does not accept shrinking (...). Shrinking only takes place in relation to a precedent elevation, but the Universal Servant (*al- ʿabd al-kullī*) has no elevation in his servanthood. (Ibn ʿArabī 1972, vol. 2, p. 615 in Chittick 1989, p. 371)

It is, in point of fact, his condition of total servant—for he serves God in an all-encompassing way, through all His Names—that elevates him to the rank of perfect man. Other points in common between both mystics and Ibn Masarra have been highlighted by Küçük (Küçük 2013b), such as the concept of light and wisdom, the metaphor of knowledge of the internal realities of the human being as anatomical dissection, as well as order and correspondence between the three levels of reality: universe, Qur'an and human being[55].

From the partial comparison of elements of Andalusian mysticism outlined here, it can, however, be concluded in general terms, building on the line of argument pursued by previous researchers, that there exists an explicit recognition by Ibn ʿArabī of the influence of certain doctrines of Ibn Masarra and Ibn Barrajān on his thinking, along with a tacit presence of other elements (with names that may differ) originating from these authors. Ibn Arabi's texts also suggest a desire to distance himself or remain silent with regard to questions with which he is familiar, but which could draw him towards philosophy or theurgy. If the spiritual authority of Ibn Masarra and Ibn Barrajān over Ibn ʿArabī is clear, the extent of the intellectual heritage with which Ibn ʿArabī engages in dialogue,

55 Man as microcosms is the interpreter of the verses and the letters of the Book of Existence by means of an ascendant path of meditation, reflexion and transposition (*iʿtibār*) and through an analogical reading of both the Qur'an and the signs of the Cosmos as a divine discourse that aims to uplift the reader to this higher reading (*al-tilāwat al-ʿulyā*) to contemplate the cosmic and inner dimension of the Qur'an. (González Costa 2009; Gril 2007).

rejects or re-adapts his own cosmological and hermeneutical system can still lead to new developments, especially in light of the recent "discovery" of the figure of Ibn Barrajān and his extensive work. This comparison of mystics at local level must also be understood, as has been the case in recent years, within a broader framework, as a result of the influence of the earlier mysticism of the Islamic East upon that of the West.

Funding: This research has been carried out within the framework of the I+D+i research project "Género y santidad: experiencia religiosa y papel social a través de las vidas de mujeres santas en el Norte de Marruecos (Tánger, Tetuán)" (PID2019-104300GB/I00 MINECO-FEDER).

Institutional Review Board Statement: Not applicable.

Informed Consent Statement: Not applicable.

Conflicts of Interest: The authors declare no conflict of interest.

References

Addas, Claude. 1992. Andalusī Mysticism and the Rise of Ibn 'Arabī. In *The Legacy of Muslim Spain*. Edited by Salma Khadra Jayyusi. Leiden: Brill, pp. 909–33.

Addas, Claude. 1993. *Quest for the Red Sulphur. The Life of Ibn ʿArabī.* Translated by Peter Kingsley. Cambridge: The Islamic Texts Society.

Al-Ḥallāj, Ḥussayn ibn Manṣūr. 1936. *Akhbār al- Ḥallāj: texte ancien relatif à la predication et au supplice du mystique musulman al-Ḥosayn b. Manṣour al- Ḥallāj*. Edited by Louis Massignon and Paul Kraus. Translated by Louis Massignon, and Paul Kraus. Paris: Éditions Larose.

Aladdin, Bakri, Pablo Beneito, Jorge Lirola Delgado, Gracia López-Anguita, Estela Navarro i Ortiz, and Salvador Peña. 2009. Ibn al- ʿArabī al-Ḥātimī/al-Ṭāʾī, Muḥyī l-Dīn. In *Biblioteca de Al-Andalus*. Edited by J. Lirola and J. M. Puerta Vílchez. Almería: Fundación Ibn Tufayl, vol. 2, pp. 158–332.

Al-Affifi, Abul Ela. 1964. Points of Correspondence between Ibnul 'Arabi and Ikhwan as-Safa'. In *The Mystical Philosophy of Muhyid Din-Ibnul Arabi*. Lahore: Ashraf Press, pp. 185–88.

Al-kutub al-sitta. 2000. *Supervised by Ṣāliḥ b. ʿAbd al-ʿAzīz Muḥammad b. Ibrāhīm*. Riyad: Dār al-Salām li-l-nashr wa-tawzīʿ.

Al-Nābulusī, ʿAbd al-Gānī. 1995. *Al-wujūd al-ḥaqq*. Édition critique du texte arabe du texte arabe et présentation par Bakri Aladdin. Damascus: IFEAD.

Altmann, Alexander. 1967. The Ladder of Ascension. In *Studies in Mysticism and Religion Presented to Gershom G. Scholem on His Seventieth Birthday by Pupils, Colleagues and Friends*. Edited by E. E. Urbach, W. Werblowski and C. Wirszubski. Jerusalem: Magnes, pp. 1–32.

Asín Palacios, Miguel. 1914. *Abenmasarra y su escuela: Orígenes de la filosofía hispano-musulmana*. Madrid: Maestre.

Badawi. 1955. Autobibliografía de Ibn ʿArabi. *Al-Andalus* 2: 107–28.

Bellver, Jos. 2013. Al-Ghazālī of al-Andalus: Ibn Barrajān, Mahdism and the Emergence of Learned Sufism on the Iberian Peninsula. *Journal of the American Oriental Society* 133: 659–81. [CrossRef]

Brown, Vahid J. 2006. Andalusī Mysticism: A recontextualization. *Journal of Islamic Philosophy* 2: 69–101. [CrossRef]

Casewit, Yousef. 2012–13. A Reconsideration of the Life and Works of Ibn Barrajān. *Al-Abhath* 60–61: 111–42.

Casewit, Yousef. 2014. The Forgotten Mystic: Ibn Barrajān (d. 536/1141) and the Andalusian Muʿtabirūn. Ph.D. dissertation, Yale University, New Haven, CT, USA.

Casewit, Yousef. 2016. A Muslim Scholar of the Bible. Prooftexts from Genesis and Matthew in the Qur'an Commentary of Ibn Barrajān of Seville (d. 536/1141). *Journal of Qur 'anic Studies* 18: 1–48. [CrossRef]

Casewit, Yousef. 2017. *The Mystics of Al-Andalus. Ibn Barrajān and Islamic Thought in the Twelfth Century*. Cambridge: Cambridge University Press.

Chittick, William. 1989. *The Sufi Path of Knowledge. Ibn al-ʿArabī's Metaphysics of Imagination*. Albany: State University of New York Press.

Chittick, William. 1998. *The Self-Disclosure of God: Principles of Ibn al-ʿArabî 's Cosmology*. Albany: State University of New York Press.

Chittick, William. 2020. Ibn 'Arabî. In *The Stanford Encyclopedia of Philosophy*. Spring 2020 Edition. Edited by Edward N. Zalta. Available online: https://plato.stanford.edu/archives/spr2020/entries/ibn-arabi/ (accessed on 5 September 2020).

Cornell, Vincent. 1996. *The Way of Abū Madyan. Doctrinal and Poetic Works of Abū Madyan Shuʿayb ibn al-Ḥusayn al-Anṣārī (c. 509/1115–16-594/1198)*. Cambridge: The Islamic Texts Society.

De Callataÿ, Godefroid. 2014. Philosophy and Batinism in Al-Andalus: Ibn Masarra's *Risalat al-i'tibār* and the *Rasā'il Ikhwān al-ṣafā'*. *Jerusalem Studies in Arabic and Islam* 41: 261–312.

De Callataÿ, Godefroid. 2014–15. From Ibn Masarra to Ibn ʿArabī: References, Shibboleths and Other Subtle Allusions to the *Rasāʾil Ikhwān al-Ṣafāʾ* in the Literature of al-Andalus. In *Labor Limae. Studi in onore di Carmela Baffioni. Studi Maġrebini*. Edited by A. Straface, C. De Angelo and A. Manzo. Naples: Universitá degli Studi di Napoli "L 'Orientale", pp. 217–67.

Ebstein, Michael, and Sara Sviri. 2011. The so-called *Risālat al-Ḥurūf* (*Epistle on Letters*) adscribed to Sahl al-Tustarī and Letter Mysticism in Al-Andalus. *Journal Asiatique* 299: 213–70.

Ebstein, Michael. 2014. *Mysticism and Philosophy in al-Andalus. Ibn Masarra, Ibn al-ʻArabī and the Ismāʻīlī Tradition*. Leiden: Brill.

Ebstein, Michael. 2020. Classifications of Knowledge in Classical Islamic Mysticism: From Eastern Sufi Sources to the Writings of Muḥyī l-Dīn Ibn al-Arabī. *Studia Islamica* 115: 33–64. [CrossRef]

Elmore, Gerald. 2001. Shaykh ʻAbd al-ʻAzīz al-Mahdawī, Ibn al-ʻArabī's mentor. *Journal of the American Oriental Society* 121: 593–613. [CrossRef]

Ferhat, Halima. 2005. L´organisation des soufies et ses limites à l´epoque almohade. In *Los almohades: Problemas y perspectivas*. Edited by Maribel Fierro and Francisco García Fits. Madrid: CSIC-Casa Velázquez, pp. 685–704.

Fierro, Maribel, and Francisco García Fitz, eds. 2005. *Los almohades: Problemas y perspectivas*. Madrid: CSIC-Casa Velázquez.

Fierro, Maribel. 1996. Bāṭinism in Al-Andalus. Maslama b. Qāsim al-Qurṭubī (d. 353/964), Author of the *Rutbat al-Ḥākim* and the *Ghāyat al-Ḥākim* (*Picatrix*). *Studia Islamica* 84: 87–112. [CrossRef]

Fierro, Maribel. 2016. *The Almohad Revolution. Politics and Religion in the Islamic West during the Twelfth-Thirteenth Centuries*. New York: Routledge.

Garrido Clemente, Pilar. 2008a. Sobre la morada de las cinco columnas mencionadas en la obra Futūḥāt makkiyya de Ibn al-ʻArabī. In *El viaje interior entre Oriente y Occidente*. Edited by P. Beneito and P. Garrido. Madrid: Mandala, pp. 87–93.

Garrido Clemente, Pilar. 2008b. Textos relativos al trono en la obra de Ibn Masarra, contrastados con las doctrinas que Ibn ʻArabī e Ibn Ḥazm le atribuyen. In *El viaje interior entre Oriente y Occidente*. Edited by P. Beneito and P. Garrido. Madrid: Mandala, pp. 141–47.

Garrido Clemente, Pilar. 2008c. Traducción anotada de la *Risālat al-iʻtibār* de Ibn Masarra de Córdoba. *Estudios Humanísticos-Filología* 30: 139–63. [CrossRef]

González Costa, Amina, and Gracia López-Anguita, eds. 2009. *Historia del sufismo en Al-Andalus: Maestros sufíes de Al-Andalus y el Magreb*. Córdoba: Almuzara.

González Costa, Amina. 2009. Un ejemplo de la hermenéutica sufí del Corán en al-Andalus. El comentario coránico Īḍāḥ al-ḥikma de Ibn Barrajān (m. 536–1141). In *Historia del Sufismo en al-Andalus y el Magreb*. Edited by Amina González Costa and Gracia López-Anguita. Córdoba: Almuzara, pp. 43–67.

González Costa, Amina. 2013. Estudio y edición (primera mitad) del Īḍāḥ al-Ḥikma, comentario coránico del sufí Ibn Barrajān de Sevilla (m. 536/1141). Ph.D. dissertation, University of Seville, Sevilla, Spain.

Gril, Denis. 2004. The Science of Letters. In *The Meccan Revelations*. Edited by Michel Chodkiewicz. Translated by Cyrille Chodkiewicz, and Denis Gril. New York: Pir Press, vol. II, pp. 105–220.

Gril, Denis. 2007. L´interprétation par transposition symbolique (*iʻtibār*), selon Ibn Barraǧân et Ibn ʻArabî. In *Symbolisme et Herméneutique dans la pensée d ´Ibn ʻArabī*. Edited by Aladdin Bakri. Damasco: IFPO, pp. 147–62.

Ḥakīm, Suʻād. 1981. *Al-Muʻjam al-ṣūfī. Al-Ḥikma fī ḥudūd al-kalima*. Beirut: Dandara.

Ibn al-ʻArīf. 1993. *Miftāḥ al-saʻāda wa-taḥqīq ṭarīq al-irāda*. Edited by ʻAbd al-Laīf Dandash. Beirut: Dār al-garb al-islāmī.

Ibn ʻArabī. 1911. *Al-futūḥāt al-makkīya*. 4 vols. Cairo: Bulaq.

Ibn ʻArabī. 1919. ʻUqlat al-Mustawfiz. Al-tadbīrāt al-ilāhīya. Kitāb inshā᾽ al-dawā᾽ir. In *Kleinere Schriften des Ibn al-ʻArabī*. Edited by H. S. Nyberg. Leiden: Brill.

Ibn ʻArabī. 1946. *Fuṣūṣ al-ḥikam*. Edited by Abū l-ʻAla al-ʻAffīfī. Beirut: Dār al-kutub al-ʻarabī.

Ibn ʻArabī. 1948. *Rasā᾽il ibn al-ʻArabī*. Hyderabad: Deccan.

Ibn ʻArabī. 1971. *Sufis of Andalusia: the Rūḥ al-Quds and the Dhurrat al-fākhira*. Translated into English by R. W. J. Austin. London: Allen and Unwin.

Ibn ʻArabī. 1972. *Al-futūḥāt al-makkīya*. 14 vols. Edited by Osman Yahya. Cairo: Al-hay᾽at al-miṣrīya al-ʻāmma li-l-kitāb.

Ibn ʻArabī. 1997. *El secreto de los Nombres de Dios*. Edited and Translated into Spanish by Pablo Beneito. Murcia: Editora Regional.

Ibn ʻArabī. 1999a. *Al-futūḥāt al-makkīya*. 9 vols. Edited by Aḥmad Shams al-Dīn. Beirut: Dār al-kutub al-ʻilmīya.

Ibn ʻArabī. 1999b. *Kitāb ʻAnqā᾽ Mugrib. Islamic Sainthood in the Fullness of Time: Ibn Arabi´s Book The Fabulous Gryphon*. Edited by Gerald Elmore. Translated by Gerald Elmore. Leiden: Brill.

Ibn ʻArabī. 2001a. *Rasā᾽il Ibn Arabi (Kitāb al-mīm, pp. 83–91)*. Beirut: Dār al-kutub al-ʻilmīya.

Ibn ʻArabī. 2001b. *Mashāhid al-asrār al-qudsiyya wa-maṭāliʻ. Contemplation of the Holy Mysteries and the Rising of the Divine Lights*. Translated by Cecilia Twinch, and Pablo Beneito. Oxford: Anqa Publishing.

Ibn ʻArabī. 2002. *Kitāb al-Mīm wa'l-Wāw wa'l-Nūn*. Edited by C. -A. Gilis. Beirut: Albouraq.

Ibn ʻArabī. 2002–03. *Iṣṭilāḥāt al-ṣūfiyya*. Edited by Abd al-Raḥīm Māridīnī Damascus. Beirut: Dār al-Maḥabba-Dār Āyya.

Ibn ʻArabī. 2006. *Risālat ittiḥād al-kawnī. The Universal Tree and the Four Birds*. Edited by Denis Gril. and Translated into English by Angela Jaffrey. Oxford: Anqa Publishing.

Ibn ʻArabī. 2009. *Los engarces de las sabidurías. Fuṣūṣ al-Ḥikam*. Translated into Spanish by Andrés Guijarro. Madrid: Edaf.

Ibn ʻArabī. 2019. *Fī maʻrifat kīmīyā᾽ al-saʻāda. The Alchemy of Human Happiness*. Translated into English by Stephen Hirtenstein. Oxford: Anqa.

Ibn Barrajān. 2000. *Sharḥ asmā᾽ Allāh al-Ḥusnà. Comentario sobre los nombres más bellos de Dios*. Edition and Introductory study by Purificación de la Torre. Madrid: CSIC-AECI.

Ibn Barrajān. 2015. *Īḍāḥ al-Ḥikma bi-aḥkām al-ʻibra. A Qur᾽ān Commentary of Ibn Barrajān of Seville (d. 536/1141)*. Edited by Gerhard Böwering and Yousef Casewit. Leiden-Boston: Brill.

Ibn Masarra. 2007a. *Risālat al-iʿtibār*. Critical edition by Pilar Garrido Clemente. *Miscelánea de Esudios Árabes e Islámicos* 56: 81–104.

Ibn Masarra. 2007b. *Risāla Jawāṣṣ al-ḥurūf*. Critical edition by Pilar Garrido Clemente. *Al-Andalus-Magreb* 14: 51–89.

Kenny, Joseph. 2002. Ibn-Masarra: His *Risāla al-iʿtibār*. *Orita: Ibadan Journal of Religious Studies* 34: 1–26.

Knysh, Alexander D. 1999. *Ibn ʿArabi in the Later Islamic Tradition. The Making of a Polemical Image in Medieval Islam*. New York: State University of New York Press.

Küçük. 2013a. Light Upon Light in Andalusian Sufism: Abū al-Ḥakam Ibn Barrajān (d. 536/1141) and Muḥyī l-Dīn Ibn al-ʿArabī (d. 638/1240) as Developer of His Hermeneutics. Part I: Ibn Barrajān's Life and Works. *ZDMG* 163: 87–116.

Küçük. 2013b. Part II: Ibn Barrajān's Views and Legacy. *ZDMG* 162: 383–409.

López-Anguita, Gracia, ed. 2018. *Ibn ʿArabī y su época*. Seville: University of Seville.

López-Anguita, Gracia. 2014. Aproximación a la angelología en la mística islámica. In *El cielo en el Islam*. Edited by Fátima Roldán. Sevilla: University of Sevilla-University of Huelva, pp. 207–27.

Lory, Pierre. 1989. *Alchimie et mystique en Terre d 'Islam*. Lagrasse: Verdier.

Lory, Pierre. 2004. *La science des lettres dans l 'Islam*. Paris: Dervy.

Lory, Pierre. 2006. Ibn Masarra. In *Diccionario crítico del esoterismo*. Edited by J. Servier. Translated into Spanish by F. J. González. Madrid: Akal, vol. 1, pp. 834–37.

Nakamura, K. 1994. Imām Ghazālī´s Cosmology Reconsidered with Special Reference to the Concept of Jabarūt. *Studia Islamica* 80: 29–46. [CrossRef]

Nwiya, Paul. 1970. *Exégèse coranique et langage mystique. Nouvel essai sur le lexique technique des mystiques musulmanes*. Beirut: Dar el-Machreq.

Pacheco Paniagua, Juan Antonio. 2007. Ibn Arabi, 353–371: Número y Razón. In *Simbolisme et herméneutique dans la pensée d 'Ibn Arabi*. Edited by B. Aladdin. Damasco: IFPO, pp. 99–111.

Pacheco Paniagua, Juan Antonio. 2012. Ibn Arabi and Aristotelian Logic. *Ishraq: Islamic Philosophy Yearbook* 3: 1–19.

Pacheco Paniagua, Juan Antonio. 2017. *Filosofía y pensamiento espiritual en al-Andalus*. Córdoba: Almuzara.

Pacheco Paniagua, Juan Antonio. 2019. *Ibn Arabi. El Maestro Sublime*. Cordoba: Almuzara.

Qashānī, ʿAbd al-Razzāq. 1992. *Al-iṣṭlāḥāt al-ṣūfīya*. Edited by Shāhin. Cairo: Dār al-Manār.

Qashānī, ʿAbd al-Razzāq. 2005. *Laṭā ʾif al-iʿlām fī ishārāt ahl-ilhām*. Edited by al-Ṣā ʾiḥ. El Cairo: Maktaba al-thaqafa al-dīnīya.

Qunawī, Ṣadr al-Dīn. 1983. *Mir ʾat al-ʿārifīn. Reflection of the Awakened*. Edited by S. Ḥassan ʿAskarī. Translated by S. Ḥassan ʿAskarī. London: Zahra Trust.

Rosenthal, V. Franz. 1988. Ibn ʿArabī between "Philosophy" and "Mysticism": "Sūfism and Philosophy are Neighbors and visit each Other". *Fa-inna al-taṣawwuf wa-t-tafalsuf wayatazāwarānī Oriens* 31: 1–35.

Shafik, Ahmad. 2012. Filosofía y mística de Ibn al-ʿArīf: Su *Miftāḥ al-saʿāda*. *Anales del Seminario de Historia de la Filosofía* 29: 433–48.

Stroumsa, Sarah, and Sara Sviri. 2009. The Beginnings of Mystical Philosophy in Al-Andalus: Ibn Masarra and his *Epistle on Contemplation*. *Jerusalem Studies in Arabic and Islam* 36: 201–53.

Stroumsa, Sarah. 2006. Ibn Masarra and the Beginnings of Mystical Thought in al- Andalus. In *Mystical Approaches to God: Judaism, Christianity and Islam*. Edited by P. Schäfer. Munich: Historisches Kolleg, Oldenbourg, pp. 97–112.

Stroumsa, Sarah. 2016. Ibn Masarra´s (d. 931) Third Book. In *The Oxford Handbook of Islamic Philosophy*. Edited by Khaled el-Rouayheb and Sabine Schmidt. Oxford: Oxford University Press, pp. 83–100.

Ṭāhir Ḥasanayn, Ḥāmid. 1985. Al-walāya wa-l-nubuwa ʿinda Muḥyī l-Dīn ibn ʿArabī. Taḥqīq wa-dirāsa li-naṣṣ lam yasbiq našrahu. Sainthood and Prophecy: A Study and Edition of an Epistolary Manuscript by Ibn ʿArabī. *Alif* 5: 7–38.

Tornero Poveda, Emilio. 1985. Nota sobre el pensamiento de Abenmasarra. *Al- Qanṭara* 6: 503–6.

Tornero Poveda, Emilio. 1993. Noticia sobre la publicación de obras inéditas de Ibn Masarra. *Al- Qanṭara* 14: 47–64.

Valsan, Michel. 2016. La science propre à Jésus (*al-ʿilm al-ʿīsawī*). *Fûtûḥāt*, chapitre 20. In *L 'Islam et la fonction de René Guénon*. Nuits-Saint-Georges: Science sacrée, pp. 135–46.

Zine, Mohammed Chaouki. 2004. L´apport de Ġazālī aux fondaments mystiques et philosophiques de la connaissance et l´objection d´Ibn ʿArabī à la question de la vision de Dieu. *Studia Islamica* 98–99: 131–56.

religions

MDPI

Article

Qurrat al-ʿAyn, the Maiden of the Kaʿba: On the *Themenophany* Inspiring Ibn ʿArabī's *Tarjumān*

Pablo Beneito

Departamento de Traducción e Interpretación, Universidad de Murcia, 30001 Murcia, Spain; pbeneito@um.es

Abstract: Qurrat al-ʿAyn is the name of the enigmatic Maiden who appeared alongside Ibn ʿArabī when he was inspired to recite the four verses that open *The Interpreter of Desires*, as he was wandering around the Kaʿba. In this article, through the analysis of the passage in which she is mentioned, the identity of the Maiden is explored from various perspectives typical of the author's theo-anthropo-cosmovision, characterised by his concept of theophany (*tajallī*) or divine self-revelation, resorting especially to both the analysis of the lexical inter-reference in the roots of the Arabic terms used by Ibn ʿArabī in his *Tarjumān al-ashwāq*, as well as the study of the symbolism characteristic of the Arabic alphanumeric system. Furthermore, the article proposes that the kaleidoscopic structure of this collection of odes, studied here for the first time, is the result of a *themenophany* of the Kaʿba: the *Tarjumān* has been "inspired" by/on the Kaʿba itself, so that in a sense it is a *bibliophany* of the so-called House of God, to whose geometry—four corners, six faces, seven ritual turns, eight vertices—its structural conception corresponds. The symbolism of Arabic geomancy in relation to the structure of the *Tarjumān* is also considered.

Keywords: Ibn ʿArabī; theophany; Qurrat al-ʿAyn; *Tarjumān al-ashwāq*; *Interpreter of Desires*; science of letters; *abjad*; Sufism; Arabic geomancy

check for updates

Citation: Beneito, Pablo. 2021. Qurrat al-ʿAyn, the Maiden of the Kaʿba: On the *Themenophany* Inspiring Ibn ʿArabī's *Tarjumān*. *Religions* 12: 158. https://doi.org/10.3390/rel12030158

Academic Editor: Cristobal Serran-Pagan Y Fuentes

Received: 26 January 2021
Accepted: 23 February 2021
Published: 27 February 2021

1. Introduction

The work of the Andalusian Muḥyī l-Dīn Ibn ʿArabī (Murcia 1165–Damascus 1240 CE), widely recognized as the greatest exponent of the sciences of Sufism and Islamic esotericism, can be considered the culminating expression of thought in al-Andalus and, without comparison with any other in the literary scope, the most significant contribution of the Andalusian cultural environment to the legacy of the Arab language and the universal Islamic culture. His most famous, translated, and commented collection of odes, the *Tarjumān al-ashwāq*, part of his enormous poetic production, of particularly lyrical inspiration, reveals that Ibn ʿArabī is also among the greatest poets in the history of literature, especially in the field of mystical poetry. Among other aspects, the rich plurality of expressive registers in his writing, in inspired interaction, also makes his poetry the culminating expression of its genre in the Arabic language.

In this brief study[1], which analyses various technical aspects of the preface to *The Interpreter of Desires*—in particular the only episode in which the personal figure of Qurrat al-ʿAyn is mentioned as such in the writings of Ibn ʿArabī—I am obliged to assume a certain familiarity of the reader with Ibn ʿArabī's thought, as well as with his hermeneutical procedures, which include both (1) the use of the alphanumeric calculation system[2] as

[1] This article is based on the text presented at the MIAS Symposium *Counsel my people*, celebrated at the Wolfson College, Oxford, October 2019.

[2] Known in Arabic, among other denominations, as *ḥisāb al-jummal*, the language of arithmosophy is very significantly used and transmitted by the author, a master par excellence in this contemplative art, in many of his writings, especially in Chapter II of *al-Futūḥāt al-makkiyya*. See Winkel's translation (Ibn ʿArabī 2018c, pp. 161–286). On the science of letters see the general studies by Pierre Lory (2004) and Denis Gril (2004). On the use of the *ḥisāb al-jummal*, see for example, *Patterns of contemplation* (Beneito and Hirtenstein 2021) and, among the many works by ʿAbd al-Bāqī Miftāḥ, his *Mafātīḥ fuṣūṣ al-ḥikam* (Miftāḥ 1997, p. 62). As a tool for counting, including a table of letter values, see also the Abjad Calculator (https://www.abjadcalc.com, accessed on 10 January 2021) which follows Ibn ʿArabī's main *abjad* rules (although it never considers *shadda*, 'reduplication', and always counts *tāʾ marbūṭa* as a *hāʾ*).

a symbolic reference frame that refers to a structural geometry in the background of the texts, and (2) the use of associations characteristic of lexical inter-reference (*ishtiqāq* or morphosemantic derivation), so significant when it comes to understanding the framework of his writings (Beneito 2006, p. 25 ff.) as well as (3) the constant practice of intertextuality, principally with the written references of the Koran and the Hadith and with his own work, although also, of course, with the works of other Sufis or Arab authors.

1.1. The Passage from Ibn ʿArabī's Tarjumān on Qurrat al-ʿAyn

I will comment in particular on the passage, relating to the female figure of Qurrat al-ʿAyn, that appears both in the original short preface of the *Tarjumān*[3], where there is not a single mention of the young maiden al-Niẓām—who is generally considered to be the inspiration for this collection—and in the longer preface to the extensive commentary on the work, the *Dhakhāʾir al-aʿlāq* (Ibn ʿArabī 1995, pp. 179–81), which includes the whole of the original text of the *Tarjumān*. To introduce the context, I include first, with minimum comments, a new translation of the brief passage whose parts I will study later in detail. In that passage, Ibn ʿArabī says,

> And part of it [i.e., of the composition of the collection] is a conversation in the course of an episode that happened during the circumambulation [of the Kaʿba]. I was circumambulating one night around the House [of God], when my [spiritual] moment (*waqt*) became propitious and shook me up a state I already knew. I then left the paved space to [get away from] people and continued to walk around on the sand. Then some verses presented themselves to me and I began to recite them, making them audible to myself and to whoever might have been with me, if anyone could have been there. And [the verses] are:
>
> Would that I were aware whether they knew what heart they possessed!
>
> And would that my heart knew what mountain-pass they threaded!
>
> Dost thou deem them safe or dost thou deem them dead?
>
> Lovers lose their way in love and become entangled"[4].
>
> [After quoting the enigmatic verses, the passage continues]
>
> I felt nothing but the touch, between my shoulders, of the palm of a hand softer than silk. When I turned around, [I found that] there was a maiden from among the daughters of Rūm[5]. I have never seen a more beautiful face, nor [heard] sweeter language, nor more penetrating glosses, nor more subtle meanings, nor allusions so delicate, nor conversations so graceful. She is ahead of [all] the people of her time in grace, courtesy, beauty and knowledge. Then she said: 'My Lord, how hast thou said [when declaiming ...]'? And I answered [repeating the first verse] ... "

[Then follows the young Maiden's commentary and questioning, verse by verse. At the end, Ibn ʿArabī ends the story of the encounter in these terms]:

> I then asked her: 'Cousin, what is your name?' and she said: 'Qurrat al-ʿAyn [Pleasure of the Eye]'. To which I replied: '[The pleasure is] mine'. (Ibn ʿArabī [1955] 2003, pp. 11–12)

This concludes the account of the meeting with Qurrat al-ʿAyn, whose personal name, as has been said, is not mentioned by such a personal name, to my knowledge, in any other

3 I differ from Nicholson's thesis (Ibn ʿArabī 1911a, pp. 3–6), who considered this second, more extensive preface, in which Niẓām appears explicitly mentioned, as the original one. I consider, on the contrary, that the analysis of the manuscript copies (in particular ms. Ragib Pasha 1453/181b–202b and ms. Manisa 6596/81b–90a) shows that the short version—without mentioning Niẓām, the poet's beloved friend—is the original preface of the collection, conceived of as a 'section' (*juzʾ*) of a wider inclusive *dīwān* before it became, after the addition of the author's own commentary, a complete independent book by itself. On more details and references of these textual issues, see (Beneito 2022, *sub voce*).

4 I quote here the translation by Nicholson (Ibn ʿArabī 1911a, p. 48), from whose editing and interpretation I only slightly differ in a couple of secondary terms. My Spanish version in *El compás de la inspiración* (Beneito 2022) will be accompanied by an in-depth study of the poem.

5 On the meaning of this expression, see below Section 2.4.

passage of the author's works. This is followed by Ibn ʿArabī's own commentary on the four verses (so that these are repeated three times in their entirety in the final text, perhaps in correspondence with the three axes or dimensions of the Temple's cube and the human cubic/spherical constitution[6] as conceived by Ibn ʿArabī). It is understood that the author is thus responding to the questions raised by Qurrat al-ʿAyn, whose objections and demands—which provide a teaching proper to an inspired master of interpretation—would be from this perspective the true spiritual motif of the composition of Ibn ʿArabī's subsequent commentary, above the secondary request of his companions because of the objections raised by a certain contemporary *faqīh* teaching in Aleppo[7]. Qurrat al-ʿAyn questions the verses which, according to my understanding, she herself has actually inspired, thus requiring Ibn ʿArabī's subsequent commentary which she herself also inspires.

1.2. The Dating of the Tarjumān in Relation to Its Structure

One aspect of Ibn ʿArabī's hermeneutic features—commonly neglected—is the symbolic significance—perceived as providential—of dates relating to the composition of his writings. From this perspective, we note that the year in which Ibn ʿArabī arrives in Mecca, as explained by the author at the very beginning of the original preface, is 598 H., a figure which by reduction to units, according to the so-called minor calculation system (598 = 5 + 9 + 8 = 22 = 2 + 2)—is equivalent to 4, corresponding to the four verses, the four corners of the Kaʿba, and the numerical value of the first letter of the title of the work, the *tāʾ* of *tarjumān*, because these four verses constitute, in a way, the interpretative key of the 598 verses of the book. So, 598 is in correspondence with the same year 598 of the author's arrival in Mecca, a correlation that has not been observed before but is fundamental to understanding the structure of the collection.

On the other hand, Ibn ʿArabī mentions that the meeting with Qurrat al-ʿAyn and, therefore, the gestation of the poems from its matrix of four verses, took place in 604[8], a figure which corresponds to the number of the 60 odes and the 600 verses of the final edition of the *Tarjumān al-ashwāq* (a title with a total numerical value of 6), contained thus in synthesis in the four verses of the poem that generated the work. On the other hand, 604 is equivalent (6 + 4 = 10) to 1, the value of both the word *Tarjumān* (= 4 + 2 + 3 + 4 + 1 + 5 = 19/1 + 9 = 10 = 1) and the word *ashwāq* (without article, 1 + 1 + 6 + 1 + 1 = 10/1), the two terms of the book's title. We shall see later another approach to the structure 60+4.

There is one more very significant date concerning the composition of the *Tarjumān* only referred to in some early copies. The text of the manuscript copy Ragip Pasha 1453 (as well as the copy Manisa 6596) begins directly with the original preface (contained in all successive recensions), in the first person, after an initial brief mention of the author, saying,

> I asked God Most High for inspiration (*istakhartu*) and I have gathered in this section (*juzʾ*) [of my poetic production] which I have called *The Interpreter of*

6 On the question of the six directions (*jihāt*) and the three dimensions (*abʿād*) of the human constitution, see (Ibn ʿArabī 2017, vol. 6, pp. 300–1). On the six-fold character of the heart (Ibn ʿArabī 1911b, vol. 3, p. 305) and the illumination Ibn ʿArabī experienced in Fez in 593/1197, of which he says 'I had no sense of direction, as if I had become completely spherical', see (Ibn ʿArabī 1911b, vol. 2, p. 486; Hirtenstein 2010, p. 40).

7 See the references on this episode as translated by M. Gloton (Ibn ʿArabī 1996a, p. 51).

8 The episode (*ḥikāya*) with Qurrat al-ʿAyn is dated by the author at the very end of the original text of the *Tarjumān* (see ms. Ragib Pasha 1453, fol. 202b, l. 1), where he states that it happened *precisely* (*khāṣṣatan*) the year 604. This copy of the *Tarjumān* is dated ten years later the fifth of Rajab of the year 614 h. in Malatia (ll. 4–5) and it contains a certificate (*samāʿ*) of Ibn ʿArabī's direct audition and approval of the text which reads as follows: 'Says Muḥammad b. ʿAlī b. Muḥammad Ibn al-ʿArabī al-Ṭāʾī: 'The *faqīh*, *imām* and most complete scholar (*al-ʿālim al-akmal*) ʿImād al-Dīn Ḥibr b. ʿAlī b. ʿAlī al-Barmakī has read in my presence this *juzʾ*, entitled *Tarjumān al-ashwāq*, composed by me (*min inshāʾī*), while I listened to him in the course of a single session (*majlis wāḥid*) and I have given him permission to transmit from me (*al-ḥadīthʿannī*) the entirety of my transmissions (*riwāyātī*) and my own compositions (*muṣannafātī*) according to the conditions customary among the people of this purpose (*bayna ahl hādhā l-shaʾn*) [it is interesting to note that this last *nūn* appears as a complete circle with the dot in the centre, symbolising the transmission of both exoteric and esoteric knowledge, this last corresponding to the otherwise invisible part of the circle in the common writing of the *nūn*] and I have formulated my [general] authorisation (*talaffaẓtu la-hu bi-l-ijāza*) to him on the third of Rajab of 614' (ms. Ragib Pasha 1453, fol. 202b, ll. 6–10). Because it was not reproduced in other copies, the particular date of the meeting with Qurrat al-ʿAyn in 604 has not been mentioned by any scholar previously. On other related events of the year 604, see the appendix on chronology in Addas (1993). I will comment more on the significance and details of this and other dates in Beneito (2022, *sub voce*).

> *desires* the verses I have composed, in the style proper to erotic and amatory lyric poetry (*ghazal / nasīb*), in the city of Mecca—with the good omen (*tayammun*) and the blessing conferred by the nobility of this place, whose elevation God has so exalted—from what was inspired to my in the [holy] months of Rajab, Shaʿbān and Ramaḍān—and exclusively in that period (*lā ghayr*)—of the year 611 ... (ms. Ragib Pasha 1453, fol. 181b, ll. 8–12)[9]

As we see, in this opening, the author confers great importance to the fact that the *Tarjumān* has been inspired and composed in Makkah. The significance of those precise months and the year 611, among other aspects of this paragraph, will be analysed elsewhere in relation to the structure of the work (Beneito 2022). For now, the most relevant aspect of this initial fragment is its connection to the imagery of the Kaʿba, since 6 + 1 + 1 equal the number of angles of the Temple (in correspondence with the 62 odes of the book, counting the two in the preface), another significant symbolic perspective.

1.3. *The Kaʿba, Heart of Existence, and the Human Heart*

In order to understand some of Ibn ʿArabī's symbolical procedures, let us consider one of the main symbols implied in the *Tarjumān*: the transitive relationship of the Kaʿba and the heart. Ibn ʿArabī summarises some of the main aspects of the significance of the Temple in a few lines from *Futūḥāt*, part of a section in rhymed prose and poetry on the secret of the Kaʿba[10]. It says,

> This Kaʿba is the heart of existence (*qalb al-wujūd*) and My Throne is for this heart a delimited body. Neither of these [neither My heavens nor My earth] encompasses Me, nor do I give notice of Me through what I have referred to [in prophetic revelation]. But My house, by virtue of which your heart 'contains Me'—the purpose [of creation]—is deposited in your perceptible body, so that those who circumambulate your heart are the secrets (*asrār*), which are in the abode of your bodies when they circumambulate these stones. The circumambulators who circle [and carry] Our Throne that surrounds you [on the spiritual plane] are like those who circle around you in the world of [perceptible] tracing (*ʿālam al-takhṭīṭ*). Just as, with respect to you, the degree of the [composite] body is lower than that of your simple heart (*basīṭ*), so [is the degree of] the Kaʿba with respect to the [lower degree of the] Throne that encompasses everything (*al-ʿarsh al-muḥīṭ*). (Ibn ʿArabī 2017, vol. I, pp. 203–6)

Note the symbolic parallelism—between Throne, body, and circle, on the one hand, and Kaʿba, heart, and centre point, on the other—which, in intertextual reference to the divine hadith—where God speaks in the first person, saying 'neither My heavens nor My earth can encompass Me, but the heart of My faithful servant encompass Me' (Hirtenstein 2010, p. 28)—expresses this complex passage. Its final sentence reverses the order of the four terms suggesting a mirror relationship: (1) Kaʿba: heart of existence (centre of the sphere); (2) Throne: delimited body (as an inclusive sphere); (3) Heart: Kaʿba deposited in the body of the servant and centre of the sphere; (4) perceivable (*mashhūd*) Human Constitution, sphere that contains the heart (House of divinity) as the divine Throne surrounds the Kaʿba.

With regard to divinity, the Kaʿba, House of God, is superior in degree to the inclusive sphere of the Throne, just as the non-dimensional point or centre is superior to the circumference. In parallel, with respect to the perfect human being, a microcosmic synthesis, the Kaʿba of his heart is, as a non-dimensional centre, superior in degree to the sphere of his

[9] The same passage and date are reproduced in the ms. Manisa 6596 (*Tarjumān*, fols. 78a–91a, not dated), copied from the copy of a close disciple of Ṣadr al-Dīn Qūnawī, Ibn ʿArabī's adoptive son. An annotation on the cover page, after the title, reads, 'I have written this *risāla* and those that follow it from the copy of the virtuous Bahāʾ al-Dīn b. Ḥāmid b. ʿUthmān ... one of the disciples of the *shaykh* and most complete guide (*min talāmīdh al-shayḫ al-imām al-akmal*...) Ṣadr al-Dīn Qūnawī' (fol. 78b, ll. 7–14). For the passage containing the date of composition 611 h., see fol. 79a, l. 6.

[10] The passage rhymes in *-ūd -humā* and *-īṭ* (three rhymes of numerical value 10 = 1), as well as *-ār* (of value 3).

body. Ibn ʿArabī uses the term 'world of tracing' (*ʿālam al-takhṭīṭ*)[11] meaning the domain of relationships, the lines that link all the points of manifestation in geometry, writing, sound waves, rays of light ... It is the domain of the First Intellect (*al-ʿaql al-awwal*, where *ʿaql* = 2), that is, the first creation which is the letter *bāʾ* (=2) with which the revelation begins (whether in the Torah or the Koran) and which implies the *dualitude* of the created with respect to the undifferentiated Unity of the unfathomable, non-dual principle, beyond any trace. The root of *ʿaql* does indeed mean 'to bind', as in the expression 'to tie up', and necessarily implies the dualitude of two: the point and the circle, the two points connected by a line. The drown world of layout or 'design' (*takhṭīṭ*, a term connected to *khaṭṭ*, 'line', 'writing') is the exterior and apparent (*ẓāhir*), that is to say, any domain that can be witnessed (*shuhūd*) with respect to absolute concealment (*ghayb muṭlaq*), or the visible world (*ʿālam al-shahāda*) with respect to the interior, intelligible world of the Order (*ʿālam al-amr*)[12].

The Kaʿba is to the human heart as the human constitution is to the divine Throne, which contains all existence in every 'perceptible' domain. However, only the heart contains the mystery of the divine Presence. As a fruit of symbolic transitivity, Ibn ʿArabī understands here the hadith in that sense: the heart, as the House of God, contains the divine Throne—that is, the entire existence as a manifestation of God in the likeness of the theomorphic human constitution (proper to man created in the image of God) in dimensional creation. Temple and Heart are thus the symbol and abode of the unlimited, non-dimensional Mystery; they are the veil that reveals the unconditioned, the point of all possibility that points, at the centre of the circle, to its hidden principle, to pure undifferentiated Unity. As Ibn ʿArabī explains,

> When God created your body, He placed within it a Kaʿba, which is your heart. He made this temple of the heart the noblest of houses in the person of faith (*muʾmin*). He informed us that the heavens, in which there is the Frequented House (*al-bayt al-maʿmūr*), and the earth, in which there is the [physical] Kaʿba, do not encompass Him and are too confined for Him, but He is encompassed by this heart in the constitution of the believing human. What is meant here by 'encompassing' is knowledge of God. (Hirtenstein 2010, p. 27; translated from Ibn ʿArabī 1911b, vol. 3, p. 250)

As a significant example of symbolic condensation, this correlation of the cubic Temple, the centre of its sphere, and the human heart as Temple and centre, is fundamental to understand the dynamic of symbolic relationships, the imaginal experiences, and the theophanic manifestations that Ibn ʿArabī reports in the passage from *Tarjumān* that we are studying here, in particular, and in his writings in general.

Represented in Arabic as a square, the dot itself can be seen—from a three-dimensional perspective—as a six-sided cube. Such is for Ibn ʿArabī the shape of the heart with six faces and eight angles, but only four corners in the plane of the seven ritual turns, corresponding to the seven essential attributes in their circular unfolding[13].

Within a sphere, 4, 6, 7, and 8 are thus fundamental figures of the symbol of the House, the Heart, the Throne, and the human constitution, as can be seen in many of the author's texts, such as *Tāj al-rasāʾil* (Ibn ʿArabī 2018a).

1.4. On the Meaning of the Expression Qurrat al-ʿayn

Qurrat al-ʿAyn, the very name of the enigmatic maiden who appeared when Ibn ʿArabī was inspired by the four verses preceding the 60 poems[14] of *The Interpreter of Desires*

---

11 *Futūḥāt* contains only two mentions of the term *ʿālam al-takhṭīṭ* (Ibn ʿArabī 2017, vol. I, pp. 205 and 366). See Winkel's commentaries in (Ibn ʿArabī 2018c, p. 153).

12 On this matter see also (Beneito and Hirtenstein 2021).

13 On the seven attributes and this correspondence, see (Ibn ʿArabī 2018c, pp. 148–51).

14 The translators, without exception, have numbered 61 odes following Nicholson, but it seems clear that the four-line poem in the preface has to be considered separately by virtue of its specificity. There are 60 odes that together symbolise the hexad (six faces of the Kaʿba and the heart, corresponding to the total value of the title of the work) represented by the letter *ṣād* = 60 (see Ibn ʿArabī 2018c, pp. 235–41) and the letter *wāw* = 6 (*ibid.* pp. 247–48), symbol of the Perfect Human Being.

as he wandered around Kaʿba is, in addition to a well-known expression of the Prophet Muhammad—in the hadith relating to the three things he loved most in this world, where he refers to the 'freshness of the eye' that results from the practice of prayer (Ibn ʿArabī 1946, pp. 218 and 225)—a term that frequently occurs in the works of Ibn ʿArabī, who uses it in a technical way alluding in various contexts to different expressive aspects of its symbolic polyvalence.

Thus, for example, in 'Theophany 82' of his *Kitāb al-Tajalliyāt*, the divine speech is addressed to Ibn ʿArabī himself calling him Qurrat ʿAynī ('Freshness of My Eye') and the commentary on the work notes as an explanation: 'You are the one through whom I [God] see everything (*unẓur fī kulli shay'*)', adding that the meaning of the expression here is 'the place [of the gaze] of the [divine] eye' (Ibn ʿArabī 1988, p. 467).

In the *Futūḥāt al-makkiyya*, as we shall see more in detail in a following second part of this article, Ibn ʿArabī calls the Kaʿba itself by the name Qurrat al-ʿAyn (Ibn ʿArabī 2017, vol. IV, p. 102). This is an eminent example of symbolic transitivity and transjectivity (the permeable transfer of the interrelations of object, subject, condition, act, attribute, essence, etc.), fundamental aspects of the science of symbols in Ibn ʿArabī and, in particular, of its highest and most original expression in the imaginal domain, the science of letters and numbers[15].

The Akbarian conception of *tajallī*, divine self-revelation or theophanic irradiation, is the hermeneutic key that allows the articulation of the relationships that arise by virtue of this symbolic transitivity in the thought and poetry of Ibn ʿArabī. His work constantly refers to a polyhedral, kaleidoscopic cosmovision in which the transjective manifestations have to be interpreted as much in relation to God as in relation to Man, the Book, the Universe, the Temple, the Names[16], or the Angel. Therefore, in order to differentiate different relationships and perspectives of this transitivity, I propose throughout the text a series of neologisms that are necessary to specify different modalities and interrelationships of theophanic manifestation.

It should be understood that for Ibn ʿArabī, in a universally inclusive sense, all existence is a theophany. However, human beings are veiled in the domain of ordinary distinctive perception, so that they only perceive the theophanic nature of Reality when it is revealed to their inner vision through unveiling. Then the theophanic dimension of manifestation appears in the domain of creative Imagination. Thus, the perception of a particular theophany is linked to a particular moment of inspired unveiling.

Let us therefore briefly summarise the general meanings of the terms in the Arabic name of the Maiden:

> The feminine term *qurra*, from the root *q-r-r*, means 'comfort', 'relief', 'freshness', 'consolation', 'solace' and, in that sense, 'pleasure', 'joy'. By virtue of the original basic meaning of its lexical root, which, as can be deduced, metaphorically links cold – in a sense of relief from the desert heat - with solidity (as in reference to the solidity of ice), it also has the meanings of *qarār*: 'stillness', 'permanence', 'rest', 'residence', 'dwelling', 'stability'.

On the other hand, the word *ʿayn*, which is also feminine, in addition to being the name of the letter so called, with a numerical value of 70 or 7, also means 'eye', 'source', 'entity', 'identity', 'disc' (for example, in the common expression *ʿayn al-shams*, 'the solar disc').

Thus the conjunction of the two terms can mean various interconnected realities that reveal multiple aspects of the mysterious Maiden.

Ibn ʿArabī often refers to the idea that there is only one unique *ʿayn* (*al-ʿayn al-wāḥida*), although on the plane of plurality there are an infinity of *aʿyān ṯābita*, eternal exemplars, latent realities, immutable entities, timeless prototypes hidden in divine knowledge... that

[15] For a discussion on symbolic transitivity and transjectivity, see (Beneito 2022). On the relevance of the science of letters and numbers as *al-miftāḥal-awwal*, 'the First [hermeneutic] Key', see (Ibn ʿArabī 2017, vol. 1, p. 282).

[16] As an example of transjectivity in the context of a commentary on the divine Name al-Shahīd, see (Ibn ʿArabī 1996b, p. 345).

are shaped into an endless number of concrete eye-entities on the plane of manifestation when the existentiating light illuminates the corresponding latent eye-entities that constitute their spiritual support. Symbolically, every entity is, in a sense, the expression of a circle, of the sphere of a perceiving eye, and therefore of the universal sphere which encompasses all existence.

There is another dimension in which the letter *ʿayn*, its numerical value, and the numerical value of the letters in its name are highly significant in relation to *The Interpreter of Desires*. The structure of the work, as in other cases of Akbarian writings, is kaleidoscopic—that is, it responds simultaneously to several complementary structural conceptions. Without going into detail, what is important here is to understand that the book as a whole is symbolically a circle (on the two-dimensional plane) or a sphere (on the three-dimensional plane), that is, an *ʿayn*, an eye-entity that 'interprets desires': an allusion to the seven ritual cycles around the Kaʿba that correspond to the seven essential attributes. In this sense, Qurrat al-ʿAyn herself embodies this value 7 of the circumambulation. On the other hand, the numerical value of the name *ʿayn* (70 + 10 + 50 = 130 = 13/1 + 3) is equivalent to 4, the four corners of the Kaʿba. In this sense, Qurrat al-ʿAyn *is* the Kaʿba.

Therefore, the four inspiring verses of the poem are equivalent to a complete turn or circumambulation around the four corners of the Kaʿba. Moreover, considering its preceding numerical components as a tetraktys, 4 is equal to 1 + 2 + 3 + 4 = 10, so that the initial tetrad is equivalent to a ten, after which follow 60 additional poems, i.e., six t10s. In a symbolic reading, each ten corresponds to one of the seven ritual turns (since the 10, as we have indicated, is contained in the four) and expresses the culmination of a cycle. If we consider 61 poems (6 + 1) we have the value seven of the letter *ʿayn*. If we consider seven 10s (for the value 10 of the four verses), then we have its value 70. Thus, symbolically, the *Tarjumān* is Qurrat al-ʿAyn herself or vice versa: more precisely, the *Tarjumān* is the *bibliophany* of Qurrat al-ʿAyn; Qurrat al-ʿAyn is the *gynaecophany* of the *Tarjumān*.

Naturally, this is only one possibility of interpretation, since the 60 odes themselves constitute a complete symbol: 60 is the numerical value of the letter *ṣād*—a letter that Ibn ʿArabī considers graphically circular and to which he dedicates the longest and most personal section of the chapter on letters in the *Futūḥāt* (Ibn ʿArabī 2018c, pp. 235–41). The name of the letter *ṣād* (60 + 1 + 4) is equivalent to 65/6 + 5 = 11 = *Huwa*, the divine Ipseity, a fundamental symbol in the writings of Ibn ʿArabī. The number 60 corresponds to the six faces of the Temple's cube, the heart (*qalb* = 6), the title's expression *Tarjumān al-ashwāq* (= 6) and the concept of Perfect Human Being (*al-insān al-kāmil* = 6), which is ultimately the very Interpreter of desires.

Before we proceed to comment on the passage, it seems fundamental to know that, in the Preface to the *Dhakhāʾir*, special reference is made to two hadiths that provide a written foundation for the imaginal transitivity of epiphanies. There is a significant expression in the proem of the *Tarjumān*, referring to the Prophet, where he is called *al-mutajallā ilay-hi bi-aḥsani l-ṣūra* ('he to whom God reveals himself in the most beautiful form'). This refers to the hadith of the lordly theophany in the best of forms[17]. On the other hand, the expression *al-makhṣūṣ bi-l-kamāl al-kullī wa-tanzīl al-diḥyī*—which only the Lebanese edition of the *Dhakhāʾir* (Ibn ʿArabī 1995, p. 171) reproduces correctly—refers to the hadith according to which Gabriel appeared to the Prophet in the form of his young and beautiful contemporary Diḥya[18], thus as a *diḥyī* angelophany which is also, therefore, an anthropophany, like the appearance of Qurrat al-ʿAyn under the image—which the text

17 This hadith (al-Tirmidhī, *Jāmiʿ*, Tafsīr al-Qurʾān, Book 47, Hadith 3541; al-Tabrīzī, *Mishkāt al-maṣābīḥ* 725, Book 4, Hadith 154) says in Arabic,

"أَتَانِي اللَّيْلَةَ رَبِّي تَبَارَكَ وَتَعَالَى فِي أَحْسَنِ صُورَةٍ ... فَوَضَعَ يَدَهُ بَيْنَ كَتِفَيَّ حَتَّى وَجَدْتُ بَرْدَهَا بَيْنَ ثَدْيَيَّ أَوْ قَالَ فِي نَحْرِي فَعَلِمْتُ مَا فِي السَّمَوَاتِ وَمَا فِي الأَرْضِ".

The expression 'in the best form' (*fī aḥsan ṣūra*) also resonates with the expression *fī aḥsan taqwīm* in Q 95: 4, referring to the creation of human being in the best form.

18 According to well-known transmissions (see for example Muslim, *Ṣaḥīḥ*, Īmān, 167; al-Tirmidhī, *Jāmiʿ*, Manāqib, Book 49, Hadith 4010; al-Nasāʾī, *Sunan* 47/7, 4991), the angel Gabriel appeared to the Prophet under the appearance of Diḥya b. Khalīfa al-Kalbī, one of his companions (*ṣaḥāba*) whose face, according to the sources, stood out for its beauty.

does not make explicit, but suggests indirectly in the final preface of the *Dhakhāʾir*—of the young and learned lady called al-Niẓām[19], whose company Ibn ʿArabī frequented in Mecca.

In the introduction to the most extensive preface of the *Dhakhāʾir*, the initial salutation to the Prophet—where Ibn ʿArabī calls him (in affinity with the terms of his epistolary entitled *Tāj al-rasāʾil*) by his personal name Muḥammad b. ʿAbd Allāh (Ibn ʿArabī 2018a, p. 252), thus referring to his inclusive epiphanic condition (because the name Allāh contains all names)[20] —is key to understanding the poem, and says of him

> God bless the one to whom He reveals Himself in the most beautiful form (*ṣūra*) [. . . the one], distinguished with universal perfection and the descent [of the angel of revelation] in the image of Diḥya (*al-tanzīl al-diḥyī*) . . .

This is a fundamental reference that announces the manifestation of al-Niẓām and/or Qurrat al-ʿAyn as a *diḥyī* gynaecophany—that is, the possibility that the themenophany of the Kaʿba may take the imaginal form of al-Niẓām[21], just as Gabriel adopted for Muhammad the appearance of the young Diḥya.

2. Progressive Commentary on the Episode with Qurrat al-ʿAyn

Let us now proceed by steps, after these preliminaries, to the study of the passage quoted at the beginning.

2.1. Around the Four Corners of the House of God

As the author states,

> And part of it [i.e., the composition of the *Tarjumān*] is a conversation (*ḥikāya*) [in the course of an episode] that took place (*jarat*)[22] during the circumambulation (*ṭawāf*) [of the Kaʿba]. I was circumambulating one night around the House [of God], when my [spiritual] moment (*waqtī*) became propitious (*ṭāba*) and shook me up (*hazza-nī*) a state I already knew . . .

The term *ḥikāya* used here means both a 'conversation' and a 'story' or 'episode' and even, in a sense relevant to understanding the scope of the situation, a 'transmission'. While the word *ḥikāya[t]* is equivalent to 16 = 7, in line with the seven ritual turns and the value 7 of the final letters of the rhyme of the poem below, the term *ṭawāf* is equivalent to 24 = 6 (without the article), in line with the six faces of the Kaʿba and the value of the term *qalb*, 'heart', or to 28 = 1 (with article), in accordance with the 28 letters of the alphabet and with the result of multiplying the seven ritual turns by the four corners of the temple, as well as with the value 1 of the final letter (*alif*) of the poem's rhyme.

Symbolically, an essential matter here is to understand that the *ṭawāf* implies the circulation of the square or the drawing of a circle around its central point. On the other hand, the whole episode takes place during the night, which in the context specially refers to the intimacy of interiority. That interior night, the poet circumambulates the House (*bayt*), a term that in Arabic also means 'verse' and is numerically equivalent to 7 (2+1+4). Then the personal spiritual instant, the *kairos* of the contemplative, became propitious (*ṭāba*, from the root of *ṭīb*, 'goodness', 'perfume') and an inner state with which Ibn ʿArabī was familiar moved him: a radical qualitative change takes place, a decisive movement from the exteriority of common perception to the interiority of imaginal vision.

19 On this character and her signification, see (Ibn ʿArabī 1996a, pp. 19, 29–34), where Gloton's very rich commentary does not differentiate al-Niẓām from Qurrat al-ʿAyn.

20 See (Ibn ʿArabī 1996b, pp. 30–32).

21 Note that the name of this beautiful woman, epitome of virtues, means 'harmony', 'order (of the cosmos)', and 'poetry' (*naẓm*). Niẓām is equivalent to 18 (= al-Ṣamad, thus 1 + 8 = 9) in the western (or *gharbī*) *abjad* system favoured by Ibn ʿArabī and to 19 (= *wujūd* = *wāḥid* = 1 + 9 = 1) in the eastern (*sharqī*) system, but with the article, more significantly (as this is how she is mentioned in the prologue), al-Niẓām is equivalent to 22/2 + 2 = 4 (corners of the Kaʿba), in western system, or 23 (= 5, value of the letter and pronoun *hā'* which symbolises the *huwiyya* or Divine Identity throughout the book), in eastern system.

22 The verb *jarat* is in the same form and lexical root of the term *jāriya* (its active participle) which will later be commented on in relation to Qurrat al-ʿAyn as *jāriya*.

The use of the expression *hazza-nī* alludes to the divine imperative addressed to Mary in the Quran: 'And move (*huzzī*) towards you the trunk of the palm tree ... ' (Q 19:25). This is followed by the instruction ' ... and refresh your eyes with joy (*qarrī ʿaynan*)' (Q 19:26), where the expression used, with the same terms of the expression *qurrat al-ʿayn*[23], may also be understood as 'refresh yourself [as an eye-entity] with joy'. The verb *hazza* also means 'making someone rejoice', so that the whole sentence implies goodness and joy.

2.2. Circumambulating on the Sand: The Imaginal Geomancy of the Interpreter

And he goes on to say:,

> I then left the paved space (*balāṭ*) to [get away from] people (*li-ajli l-nās*) and continued to walk around on the sand (*al-raml*).

This mention of the sandy ground, not tiled and therefore, so to speak, unconditioned, is more significant than it may seem at first sight. The expression offers a certain ambivalence: while it is understood in the first instance that he 'came out of the tiled area by [moving away from] the people', it can also be understood that he 'came out of the pavement [added to the ground around the temple] in order to [facilitate the movement of] people', so that he simply came out of the space of plurality and limited perception, a sense which in the context of the story is also revealing. Ibn ʿArabī, on entering that spiritual state which shakes him up, moves to the original naked land of inspiration. The term *al-raml* is equivalent to 13 = 4 (1 + 3 + 2 + 4 + 3, including the article), in line with the four verses of the poem below[24], or to 9 (2 + 4 + 3, excluding the article), corresponding to the value nine of the extended rhyme *-kū* of the poem (written *kāf-wāw-alif* = 2 + 6 + 1 = 9). Although no translator or previous study has pointed this out before, this connection between the sand and the four verses is extremely significant. The verses are composed precisely in the meter called *al-ramal*, a term that in common writing without vowels is written in exactly the same way as sand (رمل) and therefore has the same numerical value. Technically, the word *ramal* also designates the light steps of the pilgrim during the first three ritual turns of the Kaʿba. However, there is also a more complex symbolism here, related to the fact that the term *rammāl* not only means 'the one who works with sand', but also a 'geomancer', which came often to imply an 'astrologer geomancer' practising as well the science of letters (Melvin-Koushki 2020, p. 790).

Thus, the inspirational four-lines poem as well as the previous hermeneutic ode in the preface (*Kulla-mā adhkuru-hu* ...), with 16 lines, also in the same meter *ramal*, are associated with the sand (*raml*). As we shall see, there seems to be a significant confluence of the use of this poetic rhythm and the particular number of verses of these two poems in the preface of the book, which would indicate a symbolic relationship between the verses number and the practice of geomancy. These poems describe, respectively, 40 images relating to epiphanies of the Beloved (in 16 verses, with a rhyme in *mīm*, of value 40, corresponding to the 40 images) and four questions (in four verses, connected to four degrees). Those numbers directly correspond to the four lines and 16 positions or figures in the patterns used in Islamic geomancy, a science which is precisely called *ʿilm al-raml*, 'science of [the interpretation of] the sand'. When Ibn ʿArabī specifies that his encounter with Qurrat al-ʿAyn took place in the sand—which is also the 'earth'—he is suggesting that the *Tarjumān* could be understood as a geomantic—that is, symbolic—interpretation, in a poetic understanding, of both the Vast Earth of Imagination and the Kaʿba *herself*. Effectively, the House, in consideration of its four corners surrounded by sand, the seven ritual turns (16/1 + 6 = 7) performed by Ibn ʿArabī on the sand—and, perhaps, the sum of the eight angles (upper and lower) of the visible or terrestrial Kaʿba and the 8 of the invisible, celestial or interior Kaʿba (Beneito 2006, p. 38), so that 8 + 8 = 16—combines

[23] There are six other related expressions in the Quran (see Q 20:40, 25:74, 28:9, 28:13, 32:17, and 33:51).

[24] It is also interesting to note that all the hemistiches of the four verses end in *wāw-alif* (= 6 + 1), except the third which ends with the word *darā*, with the same value (4 + 2 + 1), in correspondence with the seven circumambulations.

symbolically main numerical relations associated with this science of Arabic and Islamic roots.

Matthew Melvin-Koushki summarises in a paragraph the fundamental terms (denominations and elements) of geomancy or 'science of sand/earth' that are needed here to show the intimate symbolic correlation that Ibn ʿArabī allusively establishes between geomancy and the composition of the *Tarjumān* in the passage we are analysing:

> The Latin term *geomantia* imprecisely translates the Arabic *'ilm al-raml*, the "science of sand"; like other Arabic terms for the art (*khaṭṭ al-raml, ḍarb, ṭarq*), this refers to its original procedure of drawing 16 random series of lines in the sand or dirt to generate the first four tetragrams of a geomantic Reading [. . .] (Melvin-Koushki 2020, p. 788)[25]

Let us consider those Arabic alternative terms for geomancy in relation to the images we find in the *Tarjumān*. Note that while Ibn ʿArabī circumambulates over the sand, he describes the line of the circle. The practice of *khaṭṭ al-raml* precisely means 'drawing lines on the sand'). There, he refers to the 'blow' (*ḍarba*) of a hand by using the word *ḍarb* that also designates geomancy. Even the term *ṭarq* can be understood here, either as a 'blow' or as a reference to the path (*ṭarīq*)—both the circular way of circumambulation and 'the path to the heart' (*al-ṭarīq ilā l-qalb*)—,a term with which Ibn ʿArabī explains the meaning of the word *shiʿb* ('mountain path') which occurs in the second verse of the inspirational four-line poem in *ramal*. Thus, we see that all the various names for this art of 'sand' are symbolically present or evoked in the passage that describes the meeting with Qurrat al-ʿAyn.

The *Tarjumān al-ashwāq*, so deeply inspired by its style and motifs in the old poetry connected to the desert landscapes of Arabia, often refers to the image of the loving poet following traces in the sand in search of the beloved: thus the geomancer's search becomes a metaphor of the lover's journey through desert and mountains, which is itself a metaphor of the *viator* (*sālik*), the pilgrim traveling towards the Real, interpreting His signs in the horizons and within his own self as an *hermenaut* (hermeneutic traveller) of the worlds. Through this transparency of perspectives, topography—in particular that of the Hijaz and the sacred places connected to the Islamic pilgrimage (*ḥajj*) -, becomes in the *Tarjumān* the imaginal geography of the spiritual journey.

Let us remember that describing one of the special privileges he had been granted, the Prophet of Islam said, 'The [entire] earth was made a place of worship for me' (*juʿilat lī al-arḍ masjidan*) [26] (Hirtenstein 2010, p. 20). This universal mosque of the entire Earth is thus the referent for contemplative geomancy.

We find that each figure or position in the pattern of 16 figures of the science of sand consists in turn of four elements (lines or dots), so that there are a total of 64 signs between dots and lines[27], precisely in correspondence with the 60 odes of the *Tarjumān* and its four initial foundational verses (60 + 4) or with the 60 poems and the four poetic pieces (the two

---

25 On the same page we read, 'As with I Ching trigrams, the four lines of a geomantic figure (*shakl*) are generated by the odd (*fard*) or even (*zawj*) result of each line, creating a binary code represented as either one dot (*nuqṭa*) or two dots respectively—hence the science's alternative name of *'ilm al-nuqṭa* or *'ilm al-niqāṭ*, whence its close association with lettrism (*'oilm al-ḥurūf*), coeval Arabic twin to Hebrew kabbalah. This binary code is then deployed according to set procedures to capture the flux patterns of the four elemental energies (fire, air, water, earth) as a means to divine past, present and future events, and indeed the status of every thing or being in the sublunar realm'. This theme will be further developed in our book *El compás de la inspiración*.

26 Muslim, *Ṣaḥīḥ*, Masājīd, 3.

27 Melvin-Koushki (2020, p. 789) adds, 'The number of possible combinations of figures in a geomantic tableau is 16^4, or 65,536 in all. Each of the 16 geomantic figures acquired a full suite of specific elemental, astrological, calendrical, numerical, lettrist, humoral, physiognomical and other correspondences; the first 12 houses of the geomantic chart were likewise mapped onto the 12 planetary houses, and occasionally constructed in the form of a horoscope. Detailed information can thus be derived from the figures and their relationships about virtually any aspect of human experience, whether physical, mental or spiritual, whether past, present or future'. This has to be related to the verse in the ode 11 of the *Tarjumān* where Ibn ʿArabī refers to himself, symbolically, as a *munajjim* or 'astrologer' (Ibn ʿArabī [1955] 2003, p. 46). Note that geomancy is a very inclusive science: 'Arabo-Persian geomancy in its mature form is predicated on the deployment of cycles (sg. *dāʾira*), or specific orders of the 16 figures (sg. *taskīn*), to reveal with precision such categories of data as the following: numbers, letters, days, months, years, astral bodies and divisions, body parts, physical and facial characteristics, minerals, precious stones, plants and plant products, animals and animal products, birds, fruits, tastes, colors, places, directions, regions, topographies, genders, social classes, nations, weapons, diseases, etc.' (Melvin-Koushki 2020, p. 790).

preliminary odes and the two isolated verses of the final preface), as well as with the year 604, the date of the meeting with Qurrat al-ʿAyn already mentioned[28].

Bearing in mind that both poems are in *ramal*, the number 64 also results from multiplying the 16 verses of the first poem, corresponding to the 16 figures of the geomantic diagram, by the four verses of the inspirational poem, the matrices of the book, corresponding to the four elements of each geomantic figure, such that 16 × 4 = 64.

It should be understood that the expression Qurrat al-ʿAyn ('Stability of the Sphere', or 'of the 'eye' or 'of the disc' as a sphere) is equivalent to seven (*qur[r]at*) of four (*ʿayn*, without the article) in correspondence with 16 (1 + 6 = 7) figures for four elements. With another variant of calculation (counting *hāʾ* for the *tāʾ*), Qurrat al-ʿAyn is directly equivalent to 16, in precise correspondence with the value of ʿAyn al-Shams wa-l-Bahāʾ[29], 'Disc of the Sun and [Source] of Splendour', the alternative nickname of al-Niẓām in the preface to the *Dhakhāʾir* (Ibn ʿArabī 1995, p. 173). Therefore, we note that al-Niẓām herself is dubbed ʿAyn, in clear correlation with the name Qurrat al-ʿAyn, indicating that, in a sense, the two are one and the same entity.

A question arises here: did the appellation Qurrat al-ʿAyn hide the personal name of Niẓām in the first recension of the preface before the commentary was added together with her name in the second recension? The fact is that, in support of this idea, the word *niẓām* appears six times in the poems of the *Tarjumān*[30].

2.3. The Arrival of the Four Verses

He goes on to say,

> Then some verses (*abyāt*) presented themselves to me (*ḥaḍarat-nī*) and I began to declaim them, making them audible to myself (*nafs-ī*) and to whoever might have been with me (*man yalī-nī*), if (*law*) anyone (*aḥad*) could have been there (*hunāka*).

When the author says, before he realised the presence of Qurrat al-ʿAyn, 'some verses presented themselves to me', he implies that the verses appeared to him, in the domain of imaginal perception, endowed with an autonomous entity: they are inspired to him as living imaginal presences. The word *abyāt* (= 9, as in the value of the complete extensive rhyme of the four-line poem[31], which is also the value of *raml*), is a plural of the same word *bayt*, 'house', previously used in the text, so that the term 'verses' is associated with the Kaʿba as House: the four verses are dwellings (the four corners) of the House (Kaʿba) and are, by the numerical value of the word and the value 9 of the extended rhyme in the poem, an expression of all the figures from one to nine that make up the matrix of all language.

' ... and then I began to recite them ... ', i.e., the four verses of the four verses poem *Layta shiʿrī* which are quoted below. The verb used (*anshada*) means 'to recite' or 'to sing', but it also has the prior meanings of 'to seek' and 'to pursue' (as when the poet seeks the beloved by following her trail) and, in the first form, 'to beg' (for example, to ask of God, since the search is itself a supplication). Although this verb is entirely common to introduce the quotation of a poem, in the deeply conscious language of Ibn ʿArabī the four verses are thus presented, in the process of recitation, as a search in the course of the ritual turning around the four corners of the temple. These four verses include four questions—even if the translation does not fully reflect the interrogative style—as if they were a plea for an answer.

28 As for the number of verses in the *Dhakhāʾir*, it is interesting to note that in the 1995 Lebanese edition, the preliminary poem incorporates three verses that only appear in a single manuscript. If these three added lines are counted together with the addition of two lines to the original ode 10 and the two isolated verses at the beginning of the preface to the *Dhakhāʾir*, the final total of the verses in the work (597—that is, the initial 598, minus one omitted in the final version—plus 7) would be equivalent to 604, precisely the year of the meeting with Qurrat al-ʿAyn.

29 Counting the western values (1 of *shīn* and 3 of *sīn*), *ʿAyn al-shams wa-l-bahāʾ* = 7 + 1 + 5 / 1 + 3 + 1 + 4 + 3 / 6 + 1 + 3 + 2 + 5 + 1 = 13 + 12 + 18 = 4 + 3 + 9 = 16 = 7. See the mention of this nickname in Ibn ʿArabī ([1955] 2003, p. 8).

30 The term *niẓām* can be found six times in the following five odes: 4/verse 2, 19/16, 28/14 (twice), 45/14 and 54/4. The term *qurra* is not used in the poems. The term *ʿayn* appears four times in the poems (in 24/9, 27/11, 44/2 and 44/6), i.e., six times in total in the *Tarjumān* if we count the only two mentions of the word in the preface (in the names Qurrat al-ʿAyn and ʿAyn al-Shams). The related word *ʿīn* (in plural), with different vowels but the same writing, appears twice in 8/5 and 12/5.

31 The full rhyme –not just the *rāwī* or main rhyming letter- is composed graphically of the letters *kāf-wāw-alif* (= 2 + 6 + 1 = 9).

'... making them audible to myself ... ', that is, to 'my own soul' (*nafs* = 5 + 8 + 3 = 16 = 7 personal attributes of the self in correspondence with the ritual turns)[32], so that it can be understood that they were declaimed inwardly, and to 'whoever is [or might have been] with me' (*man yalī-nī* = [4 + 5] + [1 + 3 + 1 + 5 + 1] = 20)'. Note that the verb used is from the lexical root of *walāya*, the relationship of proximity to God. The expression may be understood as external spatial proximity 'whoever is next to me', but it can also be understood as an allusion to a co-presence within.

'... if someone could had been there': the author uses here the conditional particle *law*, which expresses unreality and impossibility, instead of *in*—which would express the possibility of realization—, with the sense therefore of 'if there had been someone (who was not...)'. This leads us to interpret that 'there' was only the imaginal presence of his own anima or *nafs* (a feminine word corresponding to the figure of Qurrat al-ʿAyn) and there was no place for any otherness apart from his spiritual *alter ego*. Is Qurrat al-ʿAyn the name of an imaginal presence of the author's 'anima'?

From the point of view of arithmosophy, we may interpret that Ibn ʿArabī was reciting the verses internally to Muḥammad (= 20, the author's own name, that is, to 'himself') and to the Kaʿba (al-Kaʿba, with final graphic *h* = 20). It should also be considered that the sum of 20 (*nafsī* in the eastern system) + 20 (*man yalī-nī*), equal to 4(0), is equivalent to the four corners and the number of the verses, corresponding to the terms 'there' (*hunāka* = 13 = 4) and 'someone' [lit. 'one'] (*aḥad* = 13 = 4) which repeat the value 4 and the underlying unity (4 = 10). Thus, the epiphanies of the Temple and the Poem are perceived as spiritual presences and mediations in the realm of a single *nafs* and not as 'someone else' added to the outside of the scene: it is about the otherness lived within one's own unity[33].

Two verses from Ibn ʿArabī's major *Dīwān* particularly support this understanding. He says concerning *qurrat al-ʿayn*: 'The "freshness of the eye" is nothing other than my own eye-entity (*ʿaynī*), for all passionate love is between me and me' (Ibn ʿArabī 2018b, p. 389)[34]. And in another poem, he explains,

> The most joyful day for me is a day in which I see the light of my own eye-entity (*ʿaynī*): / this is the eye of the heart, a full moon, a freshness of the eye (*qurrat ʿayn*) for every eye-entity. / My beloved, God did not separate (*farraqa*) in between your breaths (*anfāsu-kum*) and me. (Ibn ʿArabī 2018b, p. 151)[35]

So Qurrat al-ʿAyn may also be understood as Ibn ʿArabī's own eye-entity, i.e., as his very self, which is the eye-entity of the heart and a joyful freshness for all eye-entities.

Then follow the four enigmatic verses[36] of the passage. I will only add now that the four questions contained in the verses correspond to the four degrees of love that Ibn ʿArabī defines later in his commentary on them. The poem is fundamentally about the High Dwelling of Perplexity (*ḥayra*)—a station implying the reconciliation of opposites, the integrating—but ambivalent—experience of the essential unity in contrast to the multiple

32 In the eastern system *nafs-ī* = 5 + 8 + 6 [+ 1 of the pronoun] = 19 (= 1, without pronoun) or 20, value of the name al-Kaʿba(h), with article, and of the name Muḥammad.

33 '... if (*law* =9) had been (*kāna* =9) there (*hunāka* = 13 = 4) someone [one] (*aḥad* = 13 = 4)', equivalent to saying, 'if the 9 (rhyme) were there 4 (verses)' or 'if multiplicity were there only unity'. The poem is thus in the tone of unreality in consonance with this particle *law* that also appears in the third hemistich of the poem with the same meaning, that is, the non-realisation of the answer to those four questions of the poet that necessarily add to the perplexity of lovers.

34 In the context, it is relevant to note that this is the first verse of a seven lines poem on the *ʿayn*. The letter *ʿayn* has a value 7 corresponding to the seven verses and to the value of the rhyme in *-aynī* (1 + 5 + 1 = 7). The poem can also be found in ms. Leiden Or 2687, fol. 54b. In Arabic, the verse reads as follows:

ما قُرَة الْعَيْن غَيْر عَيْني فَبَيْني كان الْهَوَى وبَيْني

35 In this short poem, fully translated here, the third final verse is almost the same that concludes the previous poem whose first verse I just quoted. Only a pronoun changes: 'God did not separate (*farraqa*) in between your breaths (*anfāsu-hu*) and me' (Ibn ʿArabī 2018b, p. 389). Note that the word *anfās* is significantly connected to the word *nufūs* (souls). In the continuation of this article, to appear soon, the figure of the Fatā inspiring *Futūḥāt makkiyya* (Ibn ʿArabī 2018c, chp. 1, pp. 131–58) will also be considered at the light of this perspective.

36 I do not proceed to analyse these verses now, as their detailed study, together with Qurrat al-ʿAyn's questions, Ibn ʿArabī's later explanation and Abdullah Bosnevi's late and revealing commentary, will be the subject of a large section in my book *El compás de la inspiración*.

theophany and of its language par excellence, Poetry—the origin of all manifested things (Addas 2016, p. 153)—as the domain of symbols.

2.4. The Subtle Touch of a Hand and the Appearance of the Maiden (jāriya)

Then, after this poetic insemination, the passage continues:

> I felt (*ashʿur*) nothing but the touch[37], between my shoulders (*katifay-ya*), of the palm of a hand (*kaff*)[38] softer (*alyan*) than silk (*al-khazz*)[39]. When I turned around, [I found that] there was a maiden (*jāriya*) from among the daughters of Rūm.

The entire passage implies the arrival of a subtle but palpable presence that is *imaginalised*. Just when the poem (*shiʿr*) arrives - which begins with the term *shiʿrī* ('my perception of the veiled ... ')—the poet says with a verb of the same form and root: 'I felt... (*ashʿur*) ... a touch ... '—in correspondence with the shake that affected him—and here he turns—it can also be understood that he does so inwards—and the Maiden is 'there', characterized—as we will see in the text—by six aspects and four attributes.

Let us recapitulate by considering the hadith implicitly associated with the episode: when Ibn ʿArabī relates that he felt that 'the palm of one hand (*kaff* = 10/1) softer (*alyan*) than silk touched him between his two shoulders (*bayna katifay-ya*, where *katifay-ya* = 6 as the expression *kaffu-hu* in the text)' he is alluding to the aforementioned hadith which refers that the Prophet contemplated his Lord 'in the most beautiful of forms (*fī aḥsan ṣūra*)', the same literal expression of the *Dhakhāʾir*'s proem, and that this lordly manifestation put his hand between his shoulder blades (*fa-waḍaʿa kaffa-hu bayna katifay-ya*). Thus, Ibn ʿArabī felt a slight impact at the centre of his neck's base, just as the Prophet felt the hand of the angel. When he says before that the state was already known to him, he might have been referring to this Prophetic precedent. He then refers to the 'maiden (*jāriya*) from among the daughters of Rūm', that is, a young woman who came from lands previously 'Roman', that is, Greek, Byzantine—then connected to philosophy and associated with Jesus and Mary. We suggest that this Maiden is to be understood as the themenophany of the Kaʿba, as a lordly epiphany 'in the most beautiful of forms', in this case in an imaginal *ḍiḥyī* mode of the human form of al-Niẓām (ʿAyn al-Shams ...), whose attributes of beauty are described in the Preface (final version of the *Dakhāʾir*) in very similar terms to those describing Qurrat al-ʿAyn.

If we analyse the alphanumeric value of the key expressions in the passage, we see that the word 'maiden' (*jāriya* = 3 + 1 + 2 + 1 + 5)[40] is equivalent to 12 (= 3), while *min banāt al-Rūm* (4 + 5 / + 2 + 5 + 1 + 4 / + 1 + 3 + 2 + 6 + 4 = 9 + 12 + 16 = 37 = 3 + 7) is equivalent to 10 (allusion to plural unity), so that together (12 + 10 = 22 / 2 + 2 = 4), they correspond to the four corners of the Kaʿba and the four verses of the poem *Layta shiʿrī*. According to this perspective, this poem has to be understood, by the nature of its inspiration and its condition of literary matrix of all the further development of the *Tarjumān*, as a *poemophany*, i.e., an epiphany in the form of a poem. We also see, on the other hand, that, without the article, Rūm has the same value 12 (= 1 + 2 = 3) as *jāriya*, which could allude, among other things, to the triad of relations that an epiphany entails: love-lover-beloved, reflected for example in the verse of the *Tarjumān* that says, 'My Beloved is one and triple' (Ibn ʿArabī [1955] 2003, 11, p. 46). On the other hand the value 7 of al-Rūm

37 As in other cases, I try as far as possible to maintain the literalness and syntax of the expression, because although translating 'I felt the touch of a hand ... ' would be more fluid, the author's expressions contain subtleties that would be lost by altering the structure of the sentence. Here denial introduces us into a climate of ambivalence and rules out any other perception: 'I felt nothing but the touch of a hand ... ', that is, everything else vanished in that sensation of the spiritual instant.

38 The successive translators of the work have not indicated that here there is a very significant allusion to the hadith in which the Prophet says, ' ... and [the angel] placed the palm of his hand between my shoulders (*fa-waḍaʿa kaffa-hu bayna katifay-ya*)'. See above note 17.

39 The word *khazz* (= 4; with the article = 8, as *ḍarba* = 8), which in addition to 'silk' also means in contrast 'to prick' or 'to wound', resonates here, as an alliteration and play on related words, with the verb *hazz* (to shake) used before. The intensity of the blow, the shudder and the change of state is, nonetheless, as soft as silk and, at the same time, as intense as an open wound when pricked.

40 In the major system 3 + 1 + 200 + 10 + 5 = 219. However, if alternatively the *tāʾ marbūṭa* is counted with value *tāʾ* = 400 (instead of the graphic value 5 of *hāʾ*), then *jāriya(t)* (3 + 1 + 200 + 10 + 400 = 614 / 6 + 1 + 4) is equivalent to 11, corresponding to the name Muḥammad (= 92 / 9 + 2 = 11) and to the divine name Huwa.

with the article (12 + 4 = 16 / 1 + 6 = 7) is particularly revealing, as it comes to mean that the Maiden is a daughter of the 7, which is equivalent to saying 'daughter of the 7 ritual turns to the four corners', that is to say, the circulation of the square, or 'daughter of the seven abodes or celestial Kaʿba-s in the successive spheres of ascension.

Furthermore, the first four figures of Arab geomancy are called 'mothers', and the next four derived from these are called daughters (*banāt*). As we see in this passage, Qurrat al-ʿAyn is said to be '[one] of the daughters of Rūm (*min banāt al-Rūm*)', where the term al-Rūm is precisely equivalent to 16, the number of the geomancy figures. These two initial tetrads of geomancy can also be correlated with the four upper and the four lower angles of the Kaʿba. In any case, we find here another possible symbolic correlation between the episode of inspiration of the *Tarjumān* and the science of sand[41].

Note that the word *jāriya*, 'girl', 'maiden', 'servant', is first of all an active participle that also means 'current', 'in progress', 'actual', 'in circulation'. This term or others from the same lexical root are used by Ibn ʿArabī in similar contexts, referring to both the Kaʿba and the Ark of creation (*markab*) or divine Throne (Ibn ʿArabī 2018a, p. 249). In a sense, *jāriya* meaning 'servant' should be understood in connection to the notion of the inclusive name ʿAbd Allāh, since the Maiden is, certainly, a Servant of God. In another sense, Qurrat al-ʿAyn, the interlocutor of the *Tarjumān*, is presented here as a feminine epiphany or *gynoecophany* of the cosmic Throne—connected to the eight carriers of the Throne—that is to say, as a *thronophany* or *cosmophany* in one sense and, in another sense, as a *themenophany* of the temple of the Kaʿba (with eight vertices) in affinity with the epiphanic figure of the Fatā (8 + 4 + 1 = 13/1 + 3 = 4, with article al-Fatā = 8), the Knight or anthropophany that inspired *al-Futūḥāt al-makiyya* (The openings revealed in Mecca) in the very same enclosure of the Kaʿba (Ibn ʿArabī 2018c, chapter 1).

The 'Greek' origin of the young Maiden may also refer to an intense whiteness of her appearance, that is, either to her luminosity as a *luminophany* (in fact the term *tajallī*, because of its lexical root, implies the luminous dimension of an irradiation), or to a *chromatophany* of white brilliance. Subtly, this description seems to refer to a "feminine", luminous and 'circular' epiphany (which is manifested in the circumambulation as *jāriya*, 'circulating') of the Kaʿba. In fact, given that in the first brief preface that appears in the copies of the original *Tarjumān*[42], prior to the addition of the later commentary, there is not even a mention of al-Niẓām, it is possible to understand that somehow the work is originally 'dedicated', even if this is not explicitly formulated, to Qurrat al-ʿAyn. Nonetheless, it certainly seems that Qurrat al-ʿAyn, as suggested, adopted the imaginal form of al-Niẓām, since both—in this sense only one—receive a very similar praise from the author. In this sense, by virtue of the transparency of figures, Qurrat al-ʿAyn would be a *diḥyī* gynecophany with the appearance of al-Niẓām.

We may consider that just as the *Futūḥāt* is finally dedicated to al-Mahdawī, Ibn ʿArabī's companion, although it was inspired by the spiritual Fatā, in a similar way the *Tarjumān / Dhakhāʾir* is subsequently dedicated to al-Niẓām, although it was conceived under the inspirational presence of Qurrat al-ʿAyn, in conjunction with her as a gynecophany.

Once we conceive of the possibility that the imaginal Maiden took the form of al-Niẓām, just as Gabriel in some visions of the Prophet took the appearance of Ḍiḥya al-Kalbī, we may understand better the subsequent explicit dedication to al-Niẓām in the *Dhakhāʾir*. From this perspective, the collection of odes is dedicated to Niẓām/Qurrat al-ʿAyn (ʿAyn al-shams wa-l-bahāʾ). The correspondence between the total number of verses in the work (598 [5 + 9 + 8 = 22 / 2 + 2 = 4] in the original version) and the year in which the author met al-Niẓām on his arrival in Mecca reinforce the idea of an original implicit dedication to Niẓām herself. Both maidens also share the condition of not being Arabs (one from Rūm,

41 In the geomantic tableau or shield chart (*takht*), 'from right to left, the first four figures in the top row are termed Mothers (*ummahāt*), which are combined to produce the second four in the same row, termed Daughters (*banāt*); the four figures the Mothers and Daughters produce in the next row are termed Nieces (*ḥafīdāt, mutawallidāt*) . . . ' (Melvin-Koushki 2020, p. 789).

42 See, for example, ms. Manisa 6596, fol. 78b.

the other Persian or Iraqi), i.e., coming from another contrasting land that symbolises a heritage of ancestral wisdom.

2.5. *Six Aspects and Four Attributes of the Maiden*

He goes on to say,

> I have never seen a more beautiful face, nor [heard] sweeter language, nor more penetrating glosses, nor more subtle meanings, nor allusions so delicate, nor conversations so graceful. She is ahead of [all] the people of her time in grace, courtesy, beauty and knowledge. Then she said [to me]: 'My Lord, how hast thou said [when declaiming . . .]'? And I answered [repeating the first verse . . .].

As we have seen, once the Maiden appears, the poet enumerates precisely six aspects of Qurrat al-ʿAyn's virtues, corresponding to the six faces of the cube, and then mentions four further attributes, in correlation with the four corners of the Kaʿba and the four degrees of love commented on at the end of the section. The precise geometry of these descriptive attributes shows us once again that, in one sense, Qurrat al-ʿAyn *is* the very Kaʿba itself.

Then Qurrat al-ʿAyn comments verse by verse in a critical tone that shows her mastery revealing, in an ambivalent way, certain keys to the poem. The Maiden goes so far as to exclaim aloud (*ṣāḥat*) with amazement: 'How strange that you . . . !', as she questions the poet. At the end of her observations on the fourth verse, Ibn ʿArabī addresses the Maiden for the last time in the story.

2.6. *On the Name and Ascent of the Maiden and the Imaginal Earth*

> I asked her: 'Cousin (*yā Bint al-khāla*), what is your name?' and she said: 'Qurrat al-ʿAyn (Pleasure of the Eye)'. [To which] I replied: '[The pleasure is] mine (*lī*)'.

At the very end of their conversation about the verses, the poet transcribes this final dialogue in which he addresses Qurrat al-ʿAyn in familiar terms. Their very brief exchange is of extraordinary subtlety. In a sense the last expression is a gallantry of the most refined courtesy: when she says that she is called 'Pleasure of the eye', the poet replies, with the elegance of only two letters, that the pleasure is his, saying that the Maiden's name evokes the same joyful repose he feels on meeting her. In another sense, he is literally affirming (as is the first sense of the preposition *li-*) '[You are] mine', meaning that she is 'his' Qurrat al-ʿAyn. That is, she is an inner presence of his own being, or more, his own *ʿayn* (= 13/1 + 3 = 4 = *lī*), his own essence manifesting itself in the form of a spiritual female counterpart, the epiphany of the inner Kaʿba of his heart.

After this significant conversation, linked to the inspiration of these four verses that originated the collection of odes called *Tarjumān al-ashwāq*, Ibn ʿArabī explains that he contemplated in her female presence "what nobody has previously described about the four subtle knowledges" (*laṭāʾif al-maʿārif al-arbaʿ*)[43]. Then, in the *Dhakhāʾir*, he immediately begins after this explanation his own "Commentary on the four verses" which he concludes in parallel by speaking about the four modes of love which, evidently, correspond to the four subtle modes of knowledge and to the four verses of the poem, and thus to four moments of a circumambulation that passes by the four corners of the Kaʿba.

Before asking her name, Ibn ʿArabī addresses Qurrat al-ʿAyn as Bint al-khāla[44], 'Daughter of the maternal aunt', where *khāla(h)* = 15/1 + 5 = 6, so that the expression would mean 'daughter of the 6', or 'daughter of the Kaʿba' as a cubic temple with six faces, or 'daughter of the heart'. On the other hand, the lexical root of *khāla* (*kh-w-l*) is directly related to *kh-y-l*, the root of the word 'Imagination' (*khayāl*). By virtue of this lexical inter-reference, 'the maternal aunt' is an allusion to the domain of the creative Imagination: in this case the way

[43] On these four types of love (*ḥubb, wadd, hawā* and *ʿishq*) see Gloton's full translation of the *Dhakhāʾir* (Ibn ʿArabī 1996a, pp. 60–61), the most valuable available version of the text.

[44] *Yā bint al-khāla(h)* = [1 + 1 = 2] + [2 + 5 + 4 = 11] + [1 + 3 / + 6 + 1 + 3 + 5 = 19] = 5, value of the letter *hāʾ* of the *huwiyya* - as in the calculation of *khāla(t)*, counting *tāʾ* = 14 = 5—which, according to ode 41 (Ibn ʿArabī [1955] 2003, p. 161) is the only object of the poet's search. On other numerical values of this expression and the polyvalence of its symbolism, see Beneito (2022, *sub voce*).

the poet addresses that presence would mean 'daughter of the Imaginal realm', precisely because she is an epiphany in the presence of active Imagination or even the eye/entity of Imagination itself.

In this sense, we have to remember that Islamic tradition calls the Palm Tree 'Adam's sister' because *she* was created with the leftover clay from his creation, and she is thus the aunt of humanity. From the leftover clay from the creation of the Palm Tree, the Vast Earth of Reality or Imaginal World was then created (Beneito 2001, pp. 88–91). Thus 'maternal cousin' comes to mean—since implicitly the Palm Tree (as well as the Earth of Imagination) is also Eve's sister and, therefore, aunt in the maternal line (*khāla*)—'daughter of the Palm' or 'daughter of the Imaginal World'.

3. Qurrat al-ʿAyn according to Bosnevi: The Perfect Mirror of Muhammadian Sainthood

I did not find previously any relevant commentary on Qurrat al-ʿAyn in any other work or study, but recently, working on Bosnevi's book *Qurrat ʿayn al-shuhūd wa-mirʾāt ʿarāʾis maʿānī al-ghayb wa-l-jūd*—which includes a commentary on Ibn ʿArabī's *Tāʾiyya*—(Bosnevi 2015) together with Stephen Hirtenstein, we came across and translated possibly the most significant passages ever written on the figure of Qurrat al-ʿAyn that inspires the very title of his book. Although we treat the matter more extensively in our forthcoming book *Patterns of contemplation*, it seems necessary to mention here some references that are particularly meaningful in the context of this article. In his introduction, Bosnevi (d. 1644 CE) places particular emphasis on attributing his inspiration to the same source as Ibn ʿArabī, namely Qurrat al-ʿAyn (Bosnevi 2015, pp. 86–87).

For him, the Maiden is 'the image of the Muhammadian Sainthood and the all-inclusiveness of the divine Names, who is disclosed in the places of manifestation of the perfect ones among the people of Rūm', or in other words, she is 'the place of disclosure (*maḥall*) of the divine sciences and the sublime spiritual places of contemplation'. Thus, Ibn ʿArabī, according to Bosnevi's words, 'contemplated in her mirror the form of Muhammadian Sainthood and the form of the divine sciences which he reflected in his own writings . . . ' (Bosnevi 2015, p. 86).

Later on, Bosnevi explains more about the very important principle of 'the Complete Mirror' (*al-mirʾāt al-tāmma*), the ultimate nature of Qurrat al-ʿAyn, saying that 'the order of revelation (*tajallī*) never ceases to manifest [. . .] to the universal mission of Muhammad and the total unitive form, which is the Complete Mirror [that is] in total correspondence to the divine Form and the essential Unity of Uniqueness' (Bosnevi 2015, p. 55). As we can observe in his text, he is not considering any other approach to Qurrat al-ʿAyn: she is the Perfect Mirror of the divine disclosure. Note that, significantly, in the previous passage the consonantal writing of the word 'mirror' (*mirʾā*)—a feminine term—is in Arabic exactly the same (with the only variation of an auxiliary *mādda* on the *alif* instead of a *hamza*) as the word 'woman' (*marʾa*)', which seems to be an implicit allusion in this passage: Qurrat al-ʿAyn would be, from this perspective, both the Complete Mirror (المرآة التامة) and the Complete Woman (المرأة التامة) as a female expression of perfect receptivity, in correspondence with Ibn ʿArabī's conception in the *Fuṣūṣ al-ḥikam* (Ibn ʿArabī 1946, p. 217), according to which contemplating God in women constitutes the most perfect contemplation.

Bosnevi interprets the name of the Maiden as 'the sign of her appearance' and he explains that 'Qurrat al-ʿAyn' has a numerical value of 1030 (100 + 200 + 200 + 400 + 70 + 10 + 50, without counting the article), which was precisely the date of the first draft (*taswīd*) of his commentary (Bosnevi 2015, p. 55), i.e., the culmination of the period of inspiration of this writing.

Showing his familiarity with arithmosophy and the imaginal transparency of figures when reading Ibn ʿArabī, he adds that the year 1031 is the numerical value of the 'sign of her manifestation', the word *khāla*(*t*) we have seen referring to Qurrat al-ʿAyn's 'maternal aunt', which he counts as 600 + 1 + 30 + 400 explaining that the resulting number 1031 is precisely the date of the completion and writing-out (*tabyīḍ*) of his own treatise. And he adds, 'It is because of this correspondence (*munāsaba*) [of the two signs and the two dates] that I have entitled it with her name [Qurrat al-ʿAyn] . . . ' (Bosnevi 2015, p. 87). In a sense, through these detailed explanations Bosnavi seems to suggest that his book—which includes the composition of his own *tā'iyya* (a poem rhyming in *tāʾ* as the other longer one he extensively comments in this work)—is directly connected to the inspiration of the Maiden.

4. Conclusions

With Bosnevi's significant testimony on her condition of perfect mirror of the divine form, we conclude that Qurrat al-ʿAyn is an imaginal presence, a themenophany of the Kaʿba in the image of al-Niẓām, a cosmophany of the Universal Complete Mirror in Ibn ʿArabī's heart. As the matrix of the *Tarjumān* she is also a poemophany (in the form of the four verses) and she can even be seen as a bibliophany in the form of the *Tarjumān's* collection of odes.

We have shown that the kaleidoscopic structure of the *Tarjumān* corresponds, on the one hand, to the symbolism of the Kaʿba, where the numbers 4, 6, and 7 are particularly important in relation to its conception and its correlation with the values and meanings of the Arabic letters *ʿayn* and *ṣād*, as circular/spherical expressions, and on the other hand, to the symbolism of Arabic geomancy, understood as the spiritual science of interpreting the Imaginal Earth.

Through the article, which is also a reflection on Ibn ʿArabī's hermeneutical procedures applied to his own writing, the symbolic significance of the dates, or the number of verses of the poems, or their prosody, or the numerical value of their rhymes, or lexical inter-reference among words, among other features, has also been emphasized, showing that a text by the author can't be wholly understood without taking into account all the keys that he uses himself in his hermeneutic treatment of scriptural or traditional texts. Studying these procedures in his writings is also fundamental in order to understand later key authors from all Islamic lands and periods, such as the Andalusian Ibn Sabʿīn al-Riqūṭī, the Emir ʿAbd al-Qādir al-Jazāʾirī or ʿAbd al-Ghanī al-Nābulusī from Damascus, just to mention some examples among hundreds of main Sufi figures inspired by Ibn ʿArabī's teachings until our own time.

This article will soon be followed, as a continuation, by another one entitled 'The emissaries to the Kaʿba: on the structure of Ibn ʿArabī's *Tāj al-rasāʾil*', exploring again—from new texts and perspectives—main terms studied here, such as *qurrat al-ʿayn* or *jāriya*, as well as other epiphany figures of spiritual mediation, such as the Fatā or 'the emissaries of the Names servants' (rusul ʿabīd al-asmāʾ, Ibn ʿArabī 2018a, p. 251), mainly in the *Dīwān al-maʿārif* (also known as *al-Dīwān al-kabīr*) and two other major works by Ibn ʿArabī, *al-Futūḥāt al-makkiyya* and *Tāj al-rasāʾil*, whose respective structures and modalities of inspiration will be considered from similar perspectives to what we have presented here with regard to the *Tarjumān*.

Funding: This research received no external funding.

Conflicts of Interest: The author declares no conflict of interest.

References

Addas, Claude. 1993. *Quest for the Red Sulphur. The Life of Ibn ʿArabī*. Translated by Peter Kingsley. Cambridge: The Islamic Texts Society.

Addas, Claude. 2016. El lenguaje poético y su razón de ser. *El Azufre Rojo: Revista de Estudios sobre Ibn Arabi* 3: 147–59.

Beneito, Pablo. 2001. A Summary of the Life of the Prophet by Ibn ʿArabī and the Miracle of the Palm Tree of Seville. *Journal of the Muhyiddin Ibn Arabi Society* 30: 73–103.

Beneito, Pablo. 2006. The Ark of Creation: the *markab* Motif in Sufism. *Journal of the Muhyiddin Ibn Arabi Society* 40: 21–57.

Beneito, Pablo, and Stephen Hirtenstein. 2021. *Patterns of Contemplation: On the Blessing-Prayer of Effusion upon the Reality of Muhammad*. Oxford: Anqa, Forthcoming.

Beneito, Pablo. 2022. *El compás de la inspiración en Ibn Arabi: estructura y claves alfanuméricas de El Intérprete de los deseos*. Madrid: Mandala, Forthcoming.

Bosnevi, Abdullah. 2015. *Qurrat ʿayn al-shuhūd wa-ʿarāʾis maʿānī al-ghayb wa-l-jūd*. Edited by Amīn Yūsuf ʿAwdah. Beirut: ʿĀlam al-Kutub al-Ḥadīth.
Gril, Denis. 2004. The Science of Letters. In *The Meccan Revelations*. Edited by Michel Chodkiewicz. Translated by Cyrille Chodkiewicz, and Denis Gril. New York: Pir Press, vol. II, pp. 105–220.
Hirtenstein, Stephen. 2010. The Mystic's Kaʿba—The Cubic Wisdom of the Heart according to Ibn ʿArabī. *Journal of the Muhyiddin Ibn Arabi Society* 48: 19–43.
Ibn, ʿArabī. 1911a. *Tarjumán al-ashwáq: A Collection of Mystical Odes*. Edited and translated by R. A. Nicholson. London: Royal Asiatic Society.
Ibn, ʿArabī. 1911b. *Al-Futūḥāt al-makkiyya*. 4 vols. Cairo: Bulaq.
Ibn, ʿArabī. 1946. *Fuṣūṣ al-ḥikam*. Edited by Abū l-ʿAlāʾ ʿAfīfī. Beirut: Dār al-Kitāb al-ʿArabī.
Ibn, ʿArabī. 1988. *al-Tajalliyāt al-ilāhiyya*. Edited by Osman Yahia. Tehran: Iran University Press.
Ibn, ʿArabī. 1995. *Dhakhāʾir al-aʿlāq*. Edited by Muḥammad ʿAlam al-Dīn al-Shaqīrī. Minya: Ein/Minya University.
Ibn, ʿArabī. 1996a. *L'interprète des désirs*. Translation into French by Maurice Gloton. Paris: Albin Michel.
Ibn, ʿArabī. 1996b. *El secreto de los Nombres de Dios (Kashf al-maʿnā ʿan sirr asmāʾ Allāh al-ḥusnā)*. Arabic edition and translation into Spanish by Pablo Beneito. Murcia: Editora Regional de Murcia.
Ibn, ʿArabī. 2003. *Tarjumān al-ashwāq*, 3rd ed. Beirut: Dār Ṣādir. First published 1955.
Ibn, ʿArabī. 2017. *Al-Futūḥāt al-makkiyya*. 13 vols. Edited by ʿAbd al-ʿAzīz Sulṭān al-Manṣūb. Cairo: al-Majlis al-Aʿlā li-l-Thaqāfa.
Ibn, ʿArabī. 2018a. *Rasāʾil Ibn al-ʿArabī (Tāj al-rasāʾil wa-minhāj al-wasāʾil)*. Cairo: al-Quds, Part II (*al-juzʾ al-thānī*). pp. 243–302.
Ibn, ʿArabī. 2018b. *Al-Dīwān al-kabīr*. Edited by Abdelilah Ben Arfa. Beirut: Dār al-Ādāb, vol. 4.
Ibn, ʿArabī. 2018c. *The Openings Revealed in Makkah (al-Futūḥāt al-Makkīyah)*. Books 1 & 2. Translated by Eric Winkel. New York: Pir Press.
Lory, Pierre. 2004. *La Science des lettres en Islam*. Paris: Dervy.
Melvin-Koushki, Matthew. 2020. Geomancy in the Medieval Islamic World. In *Prognostication in the Medieval World: A Handbook*. Edited by Hans Christian Lehner, Klaus Herbers and Mattias Heiduk. Berlin: De Gruyter, vol. II, pp. 788–93.
Miftāḥ, ʿAbd al-Bāqī. 1997. *Mafātīḥ fuṣūṣ al-ḥikam li-Ibn ʿArabī*. Marrakesh: Dār al-Qubba al-Zarqāʾ li-l-Nashr.

religions

MDPI

Article

The Ontology, Arrangement, and Appearance of Paradise in Castilian Kabbalah in Light of Contemporary Islamic Traditions from al-Andalus

Avishai Bar-Asher

Department of Jewish Thought, The Hebrew University of Jerusalem, Jerusalem 9190501, Israel; avishai.bar-asher@mail.huji.ac.il

Received: 18 September 2020; Accepted: 20 October 2020; Published: 26 October 2020

Abstract: This study is a comparative analysis of the appearances of the lower and upper Paradise, their divisions, and the journeys to and within them, which appear in mystical Jewish and Islamic sources in medieval Iberia. Ibn al-'Arabī's vast output on the Gardens of divine reward and their divisions generated a number of instructive comparisons to the eschatological and theosophical writing about the same subject in early Spanish Kabbalah. Although there is no direct historical evidence that kabbalists knew of such Arabic works from the region Catalonia or Andalusia, there are commonalities in fundamental imagery and in ontological and exegetical assumptions that resulted from an internalization of similar patterns of thought. It is quite reasonable to assume that these literary corpora, both products of the thirteenth century, were shaped by common sources from earlier visionary literature. The prevalence of translations of religious writing about ascents on high, produced in Castile in the later thirteenth century, can help explain the sudden appearance of visionary literature on Paradise and its divisions in the writings of Jewish esotericists of the same region. These findings therefore enrich our knowledge of the literary, intellectual, and creative background against which these kabbalists were working when they chose to depict Paradise in the way that they did, at the time that they did.

Keywords: the book of Zohar; Moses de León; Castilian kabbalah; Paradise; the Garden of Eden; eschatology; Ibn al-'Arabī; al-Futūḥāt al-Makiyyah; hekhalot; comparative mysticism

1. Introduction

In the history of Jewish literary depiction of Paradise and Hell, the second half of the thirteenth century proved a formative period. The Castilian kabbalists of the time conjured richly vivid descriptions of the aery pavilions of Paradise and the infernal rings of Hell. The most well known of these were attributed pseudepigraphically to the rabbinic sages of Antiquity and were written in a pseudo-Aramaic dialect. They were consolidated in the Zohar, the masterwork of medieval Kabbalah that would, in time, become canonical. Less famous, but also anonymous, Hebrew texts on this subject circulated during the same period. These included the treatise Seder Gan 'Eden, and shorter visionary passages that purported to be from an ancient "Book of Enoch", part of the ancient apocrypha associated with that biblical figure. A close reading of these works reveals their crucial role in the formation of an eschatology of the soul within the burgeoning theosophical Kabbalah of thirteenth-century Iberia. Ultimately, all these writings can be traced to a single Castilian kabbalist, R. Moses b. Shemtob de Léon, whose name is also inextricably bound up with the initial dissemination—and even composition—of the Zohar.

In these writings, the loci of divine retribution are treated within broader discussions on the fate of the human soul and on the emanated Godhead, both of which were influenced by the Neoplatonism of the Arab philosophers. In these two contexts, these spaces of ultimate reward are said to correspond to aspects of the divine soul and, in turn, to emanations or hypostases—the *sefirot*—of the Godhead and its dynamic flux. The divine plane of existence therefore meets the earthly one in the eschatological Paradise, an intermediate realm that allows for passage from one side to the other. In this biplanar view of all existence, it comes as no surprise that Paradise itself is split into two halves of identical design: there is an earthly Paradise with a latitude and longitude, and a heavenly Paradise of Edenic castles in the sky. Owing to the mostly midrashic or exegetical nature of medieval kabbalistic literature, this notion of a multidimensional Paradise had to be grounded in Scriptural exegesis. In the endeavor to endow Scripture with layers of meaning that reach beyond the immediate historical-contextual one, Genesis 3 was read using allegorical, figurative, typological, and symbolic strategies in order to interpret the Garden of Eden as a spiritual realm, and even an aspect of the Godhead.

Previous scholarship has suggested that this outpouring of speculative kabbalistic writing on Paradise was a delayed amplification of a faint echo of the late antique visionary literature (the *Hekhalot* and *Merkavah* corpus), with additional input from pre-kabbalistic trends in western European Jewish esotericism. The unique characteristics of these texts, however, not only render this an oversimplification, but underscore the necessity of properly situating them historically and culturally. I have shown at length elsewhere that we can better grasp the genesis and features of this corpus in light of the scientific and geographic writing of medieval Christians and Muslims about the nature and location of the earthly Garden, which in Jewish hands often turned mythological or symbolic; the rich literature of adventures in terrae incognitae, with their exotic climes and fantastic creatures, such as the ever-popular romance of Alexander the Great; the medieval Jewish visionary texts describing the in-body visitation of rabbinic sages to the Garden of Eden and Gehenna; and more (Bar Asher 2019).

In this study, I set out to demonstrate the importance of another, oft-neglected context in the scholarly study of Spanish Kabbalah, namely, the neighboring Islamic cultural milieu of al-Andalus. Thinkers there had a deep well to draw from in delineating the sights and sounds of Paradise and cataloging its divisions and degrees: the Quran, the hadith literature, and the more contemporary legends, especially concerning Muhammad's night journey to Jerusalem and thence to the celestial Garden (*al-'isrā' wa-'l-mi'rāj* [Quran 17:1]). Informed by medieval Islamic theology and philosophy, which absorbed neo-Aristotelian and Neoplatonic conceptions of the soul's immortality, the Andalusians tended to spiritualize the Garden and assign it an eschatological role.

It is my contention here that the writings of one prolific Andalusian in particular, Muḥyī al-Dīn Ibn al-'Arabī (1165–1240), who died in Damascus in 1240, can throw new light on these kabbalistic treatments of Paradise. In his *Meccan Revelations* (*al-Futūḥāt al-Makiyyah*), Ibn al-'Arabī distinguishes between a physical Garden for the body and a spiritual one for the soul, which accords with Neoplatonic ontology and the Platonic dualism of body and soul. The earthly Garden is not a place to enjoy mere pleasures of the flesh but to bask in bliss of the spirit, to experience spiritual ecstasies that culminate in beholding the divine light—and even God Himself—in a celestial gathering.

This study consists of four parts. First, I briefly survey the state of scholarship on the historical background of medieval Jewish mystical writing about Paradise; the influence of Islamic sources on speculative writing about Paradise in Western Europe; and the possible connections between Ibn al-'Arabī's works and kabbalistic literature. In the following section, I enumerate the main elements of the terrestrial and celestial Paradises in Castilian Kabbalah of the second half of the thirteenth century, and of Ibn al-'Arabī's comprehensive and complex writing on the paradisiacal Gardens, particularly in his *Meccan Revelations*. After setting this stage, I dedicate the next part of this study to the many and major similarities between the two. Finally, I explore potential literary contacts that would account for these convergences, given the absence of any historical evidence that medieval Jewish esotericists of the thirteenth-century Iberian Peninsula read the Arabic writings of contemporary Islamic mystics and esotericists. These are meant to pave the way for further scholarly consideration of possible literary contacts between the kabbalists of northern Spain and the Andalusian mystics to the south.

2. Literature Review

The Zoharic and related literature depicts Paradise as consisting of stations, gradations, or levels which, recruiting an ancient term, are often called *hekhalot* (lit. "palaces"). Gershom Scholem surmised that the rich kabbalistic descriptions of these *hekhalot* originated with the first kabbalists, who to his historiographical thinking were active in late twelfth-century Provence and owed some debt to the contemporary mysticism of the "German Pietists" (Scholem 1934; 1945; Dan 1968). Recent scholarship, however, has questioned Scholem's broader historiographical assumptions and theories, with ramifications for his narrower speculation on topics such as this as well (Bar Asher 2019). Isaiah Tishby pushed back the source of inspiration for the Zohar's *hekhalot* (and angelology) to the *Hekhalot* literature of late Antiquity, while also speculating that another unknown source was used (Tishby 1989; cf. Dan 1993; Elior 2010; Dan 2017). However, a careful reading of the pertinent Zoharic material reveals stark differences from the ancient visionary literature, and quite limited conceptual, terminological, or literary borrowing (Idel 2005; Bar Asher 2019). It is possible that a number of works on the periphery of the *Hekhalot* literature, grouped by their kinship in genre and dating, might have helped frame the kabbalistic imagining of Paradise, and these are *Seder Rabbah di-Vereshit* (Schäfer 2004), *Massekhet Hekhalot / Ma'aseh Merkavah* (Jellinek 1853; Farber 1987), and others (Busi 1996). Along the same lines, scholars have proposed linking the Zoharic portrayal of Paradise to medieval elaborations of a Talmudic legend, in which a very much alive sage, R. Joshua b. Levi, enters the Garden of Eden (Ginzberg 1913; cf. Gaster 1893). Recent studies have looked at how Franco-German versions of this legend developed in connection with Christian accounts of journeys to the next world and with the intensifying Jewish valorization of martyrdom (Shepkaru 2002; Kushelevsky 2010). One scholar has further proposed that this legend was a critical source given a kabbalistic spin in the Zohar (Perry 2010). In trying to account for Iberian phenomena, this explanation privileges the literary output of western Christendom, mediated through Midrash produced specifically in the cultural sphere of Ashkenaz.

The present study argues that to better grasp the depiction of Paradise in Castilian Kabbalah we should look not to western Christendom but to the western Islamicate. Already a century ago Miguel Asín Palacios called attention to Islamic influence on the speculative depictions of Paradise in Western Europe, more specifically on those found in Dante Alighieri's *Divine Comedy* and its sources (Asín Palacios 1919). To account for similarities between the epic poem and Islamic literary traditions about Muhammad's miraculous night journey, he hypothesized that none other than Ibn al-'Arabī was the conduit by which Dante or his Christian sources learned of those traditions. Shared motifs include the geometric design of Paradise as eight concentric circles, the nearly endless subdivisions, and the moral yardsticks dictating one's assigned place (see also Gardiner 1989). In spite of Asín Palacios' insistence on the closeness between Dante's vision and Islamic tradition, the closer one examines the two the more one finds resemblances too general to be meaningful or no resemblances at all. Indeed, for over a century, scholars have taken many opportunities to pick apart his assumptions and findings (Cerulli 1949; Silverstein 1952; Gabrieli 1953; Cantarino 1965; Corti 2001; Ziolkowski 2007). Among other things, they have criticized his overemphasis on the contribution of Islamic sources to Dante's *Divine Comedy* at the expense of earlier Christian visionary literature (Silverstein 1952; Himmelfarb 1983).

Even if Ibn al-'Arabī cannot be fingered as Dante's cultural ambassador, that does not mean he left no literary footprint in medieval Christendom. It has been surmised that copies of his works arrived in Europe from the East and were available to Christians in the second half of the thirteenth century in Catalonia (such as Ramon Llull) (Carra de Vaux 1923; Albarracín 2016). Of even greater relevance to this study, Ariel Bension, following on the work of Asín Palacios, claimed that Ibn al-'Arabī himself was influenced by ancient Jewish visionary literature, Aggadah, and Geonic literature, and that although the Zohar's authors were familiar with this material, they preferred "the distinctive Spanish garments in which Ibn Arabi clothed the ideas he took from them" (Bension 1932, p. 48). Although his speculation has no firm basis, and again the similarities—at best—belong to general form rather than to specific content, it has provoked scholars into rethinking what we know about these two bodies of thought and their possible interrelationship. Today, we fortunately have more recent studies that point to the need to reassess the relationship between Ibn al-'Arabī's writings and medieval theosophical Kabbalah (Sviri 1996; McGaha 1997; Pedaya 2002; Wolfson 2008; Ebstein 2014), as well as those that pinpoint parallels between his oeuvre and the Zohar (Kiener 1982; Wolfson 1990). It is my hope that the present study, working in this same direction, further demonstrates the necessity and fruitfulness of reading kabbalistic literature with Ibn al-'Arabī in mind.

3. Depicting Paradise in Castilian Kabbalah and in the Mystical Writings of Ibn al-'Arabī

The mostly anonymous Castilian kabbalistic texts from the period in question, set in a distinctly poetic and midrashic style, immerse the reader in a lush earthly Garden with borders and bowers, and provide the celestial schematics of the many-citadeled heavenly Garden. Although the powers of description and the specifics vary from one composition to the next, the tie that binds them all is the fundamental conceptualization of Paradise as split between two planes, with a Garden of Eden below and a Garden of Eden above. In these accounts, the two are structured in ways that resonate with the complex configuration of the emanated Godhead and of the heavenly hosts.

The concept of a split-level Paradise preceded the Castilian kabbalists in the works of the Catalonian kabbalist R. Moses Nahmanides (1194–1270). He formulated the theory of mirrored Gardens in his discussions of divine recompense, the most extensive of which can be found in the concluding chapters of his *Torat ha-Adam* (Nahmanides 1964, pp. 264–311). It is almost certain that his thinking was shaped by medieval Jewish Neoplatonism found in works like *Kitāb ma'ānī al-nafs* (late 11th cent.), which lays out a two-stage theory of divine retribution under the influence of Arabic sources. Another key element in Nahmanides' account is the concept of the soul donning a "garment" after the body dies, which sprang from a robust theory about the divine source of the human soul and its postmortem fate (Scholem 1955; Cohen-Aloro 1987; Wolfson 1990; Perani 1996; Bar Asher 2013). The Zohar and de Léon's Hebrew writings greatly elaborate this concept of a spiritual "garment" that allows the soul to pass from one Paradise to the next, with souls putting on and removing "garments" continuously for a person's entire lifetime, that is to say, even in this world. Passage between Gardens is also rooted in a singular theosophical exegesis of the Tree of Life in Genesis 3, which is identified as the axis mundi by which the human soul moves between ontological planes (Yisraeli 2010).

The lengthy Aramaic sections devoted to Paradise that are scattered throughout the Zohar share a basic architectural plan of seven *hekhalot* or gradations arranged hierarchically, which correspond to the seven ascending potencies on the sefirotic tree of kabbalistic theosophy (Zohar I:38a–39b; II: 244b–262b). Other shorter literary units appear throughout the Zohar's homilies on the Torah, as well as in the corpus known as *Midrash ha-Ne'elam*, printed alongside the body of the Zohar or as a separate work. Similar depictions appear in the Hebrew writings of Moses de Léon, in the passages on the "courtyards of the World to Come" in his *Sefer ha-Rimmon* (Moses de Leon 1988), in fragments of "the Book of Enoch" in *Mishkan ha-'Edut* (Moses de León 2013; Reeves and Reed 2018), and in *Seder Gan 'Eden* attributed to the rabbinic sage R. Eliezer b. Hyrcanus (Jellinek 1855; Scholem 1990; Moisés de León 2007). I have characterized, categorized, and chronologized these texts elsewhere, where I emphasize their common foundations and record their differences (Bar Asher 2019).

Visually speaking, there is much worthy of note in de Léon's writings. First, there is a kind of travelogue of a celestial journey to the firmament atop the Garden, with movement between the four rings of heaven, each of which is located at a different cardinal endpoint, is colored by a separate hue, and is under the charge of an appointed angel who calls out letters of the Tetragrammaton. Another unique and new image is of Paradise constructed with three rings of concentric walls, with different groups of people inhabiting the intermural and adjoining courtyards. The walled-off spaces beginning with the curtain wall and ending in the innermost sanctum represent an ascending hierarchy of reward and of the spiritualization of the soul. The groups enumerated by de Léon as denizens of these precincts include "the pious of the nations", "those martyred by the regime", "innocent schoolchildren", the profoundly righteous who did some wrong, the penitent, and more.

He also links Jewish religious praxis and his theosophical views with the populations designated for the ranked courtyards of Paradise. We are even told of the Messiah's future encounters with some of these groups, as in the dramatic account of the Messiah leaving the *hekhal* called "the bird's nest", garbed in apocalyptic vengeance, to exact recompense from slayers of the righteous (Liebes 1993; Wolfson 1994).

Beside the visual aspects, these works have particular ideas in common: paradise having preexisted Creation, the two Paradises possessing the same form or model, and an axis mundi joining the two that allows for bidirectional movement. Moreover, the supernal Paradise and its multiple dimensions symbolize or otherwise represent the *sefirot* of the Godhead as conceived by theosophical Kabbalah, which also correspond to the earthly Paradise with its brick-and-mortar subdivisions. Furthermore, the divisions of the Garden, the population assignments, and the movement from one arena to another express a static or dynamic state of the *sefirot* reverberating through lower levels of existence. Theurgic shifts within Paradise, whether engendered by movement between the two planes or by the apocalyptic deeds of the Messiah, play a central role in kabbalistic literature and set it apart from all other Jewish eschatological or visionary writing. We should also mention the narratives and stories included in the Zohar's homilies on the Torah that include treks and mystical ascents to the lower Paradise or to the precincts of Heaven (Idel 1982; Meroz 2000; Wolski and Carmeli 2007). Most of these have a mythological cast, where an individual or fellowship on an expedition is turned into clothed spirits wandering from a physical place to the spirit-realm of divine forms.

Ibn al-'Arabī's intense preoccupation with eschatology and the afterlife is evident from a number of his works, which exerted immense influence on subsequent Islamic thought (Chittick 1988; Chodkiewicz 1986). In his voluminous *Meccan Revelations*, he works the many names and descriptions of Paradise in the Quran and the hadith tradition into a very creative and complex architecture of divine reward. An analysis of the relevant material shows that the book contains two adaptations of a hadith tradition, according to which Paradise is divided into seven or eight Gardens (*jannāt*) given names from Quranic verses (el-Ṣaleḥ 1971; Gardet 1960; Porter 2007; Abdel Haleem 2017). The underlying hadith is attributed to 'Abd Allāh Ibn 'Abbās, companion of Muhammad. One widespread version enumerates seven gardens (the ordering changes among tradents), whereas the version recorded by Al-Kisā'ī (al-Kisa'ī [1922] 1923, p. 17) includes an eighth (Tamari 1999; Lange 2017).

Ibn al-'Arabī inserts this reworked hadith about seven or eight Gardens into a framework of more basic divisions. First of all, he posits an ontological and epistemological bifurcation of paradisiacal reality: there is a sensible (*maḥsūsa*) and "coarse" Paradise, and an abstract (*ma'nawiyya*) and "rarefied" Paradise, which are conceived of in the same way that body and soul are integrated into a single entity (Ibn al-'Arabī 2010, p. 65). In addition, Paradise has a certain vitality, which is why it is called the "the abode of the living" (*al-darr al-ḥayawān*). In this connection, Ibn al-'Arabī invokes a hadith about four of Muhammad's loyal followers for whom the Garden yearned. In like fashion, he writes about four mystical-spiritual conditions (*aḥwāl*) that transmute the Garden in the presence of four types of people fit to enter: Muhammad, "the folk of the conditions", the faithful, and the heretics, including those who reject the existence of a sensible Garden. It bears noting here that in his *Bezels of Wisdom* (*Fuṣūṣ al-ḥikam*), Ibn al-'Arabī interprets Quran 89:30, in which Allah invites the soul into his Garden, in mystical terms. Entry to the Garden is compared to man entering his own soul, whereby he gains two kinds of knowledge: "knowledge of Him through yourself, and knowledge of Him through yourself, with respect to Him, not to you" (Ibn 'Araḥī 1946, p. 7; Abrahamov 2015, p. 61).

On top of this biplanar model, Ibn al-'Arabī further divides Paradise into three loci of reward, all termed "Gardens" and housing different groups in accordance with their degree: (1) the Garden of Divine Specification (*jannat ikhtiṣāṣ*) "entered by children who have not reached the limit of deeds ... from the time of birth until they complete six years. [...] the mad who have not had the rational faculty, the folk of cognitive *tawḥīd*, the folk of the gaps [between prophets]"; (2) the Garden of Inheritance (*jannat mīrāth*) "attained by all those I have mentioned as entering the Garden and by the faithful. It is the places that would have been designated for the folk of the Fire had they entered [the Garden]";

and (3) the Garden of Deeds (*jannat al-a'māl*), "within which people dwell in keeping with their deeds", achievements, and mystical attainments (Chittick 1998, p. 399). The third, the Garden of Deeds, is further subdivided into degrees and divisions, and it is here that we find the seven- or eightfold division of walled gardens, the highest of which is termed the Garden of Eden (*jannat 'adn*). It is the description of this Garden that is among the theosophical pinnacles of the rich account of Paradise in Ibn al-'Arabī, because located in this supernal garden is the citadel that houses the Dune (*al-katīb*) by which people will behold the "Truth", meaning Allah (Ibn al-'Arabī 2010, pp. 65, 371).

Another main theme is the relationship between the observance of Sharia and the sites of specific recompense in these Gardens. A central hadith, repeatedly cited by Ibn al-'Arabī from the collection of al-Bukhārī, states that the Garden has eight gates (Lange 2016) that correspond to the eight central directives of Islam (prayer, charity, fasting, jihād, and more). Entrants to Paradise are directed to the gate that fits their deeds: devotees of prayers are assigned to the gate of prayer, devotees of jihād to the gate of jihād, and so on (Ṣaḥīḥ al-Bukhārī 2001, vol. 10). In Ibn al-'Arabī's spiritual conception of the Garden, the layout of the eight gates integrates with a mystical conception of the Garden and of entering it, whereby the commandment kept becomes a kind of mystical key that unlocks one of the Garden's gates. The eight gates further parallel the eight limbs on which the commandments devolve (*al-taklīf*), and being among the devotees of a particular commandment allows one to enter through the gate that corresponds to the limb that performed that commandment (Ibn al-'Arabī 2010, p. 371). In this context, he elaborates on the mystical states of human activity attained by each limb at the same time and by existing simultaneously in two places. He characteristically illustrates this by drawing on his own experiences of mystical illumination and split consciousness.

4. Convergences between the Paradisiacal Accounts of Castilian Kabbalah and of Ibn al-'Arabī

There is no historical evidence whatsoever that thirteenth-century Jews living in northern Spain had any familiarity—even secondhand—with the work of Ibn al-'Arabī. Yet, there is no denying the profound similarities found in the Zohar, the Hebrew writings of Moses de Léon, and the other anonymous Hebrew compositions traced to his pen, which I enumerate below. I will return to the challenging question of how to explain this in the next section.

First, all these works exegetically and ontologically split Paradise in two. The twofold Paradise of Ibn al-'Arabī is quite close to the duality formulated by Nahmanides and developed by subsequent kabbalists; at the very least, all base themselves on the same Neoplatonic worldview. The same is true of the distinction made by the later kabbalists in assigning physical delights to one place and spiritual ecstasies to another, all the while presenting the spiritual and physical within a unified paradigm. In kabbalistic theosophy, this facilitated linking the sensual earthly Garden and its various sites with the supernal Garden and its *hekhalot* or precincts, which then correspond to aspects of the Godhead.

Second, this ontological divide gave rise to an interesting conception in which the soul is dominant in Paradise and the body persists as a kind of tagalong. In this context, it is fitting to reproduce a citation from an unknown work of Ibn al-'Arabī by the later Egyptian scholar and mystic 'Abd al-Wahhab al-Sha'rānī, one of the most important expositors of *Meccan Revelations*. Initially, he says that "the bodies of the folk of the Garden are covered and wrapped in their spirits, so that the spirits are the body's garment", and then he attributes the following to Ibn al-'Arabī: "some of the revelationists erred in saying that spirits are not gathered in without bodies, as they discerned from the transformation of the folk of the Garden according to their will" (al-Sha rānī 1959, p. 670). The opinion he rejects maintains that on the day of resurrection only the spirits or souls will be gathered in. This might have been the position of the Brethren of Purity (Ikhwān al-Ṣafā'), who believed that the Garden is a spiritual realm and that the resurrection is when souls are gathered in (Rasā'il Ikhwān al-Ṣafā' 1957, p. 41). The opposing view affirmed by Ibn al-'Arabī, according to which bodies persist while enclothed by shapeshifting spirits, calls to mind the notions of wearing a spiritual "garment" developed in multiple directions by thirteenth-century kabbalists. While there is a self-evident difference in the details of this eschatological ontology, I think they are adopting a similar approach to a problem that was inherent to their shared worldview.

Third, the two corpuses divide Paradise into three separate sections arranged hierarchically, which mostly aligns with distinct groups of people. Here even the likeness in the particulars is intriguing. Ibn al-'Arabī places adherents of other religions and the "folk of the gaps" in the Garden of Divine Specification, and de Léon likewise puts "the pious of the nations" and converts in the outermost courtyard. Ibn al-'Arabī mentions "children who have not reached the limit of deeds" and de Léon includes "innocent schoolchildren", a motif further developed in the fantastical figure of R. Gaddi'el Na'ar of Seder Gan 'Eden.

Fourth, the kabbalistic works divide the innermost circle of Paradise into seven divisions, which correspond to individual or group attainments in religious observance (punctilious observance of the commandments, acts of kindness, Torah study, etc.). This resembles Ibn al-'Arabī's descriptions of Paradise, where one of eight religious directives meticulously observed takes on a spiritual-eschatological significance and is molded into a key that unlocks a gate of Paradise.

Fifth, Ibn al-'Arabī places the Garden called "Eden" (*jannat 'adn*) at a sublime rung of existence because it is theosophically linked to his belief that there the "Truth", Allah, is revealed to his servants through the Dune. This begs comparison to the theosophic idea of the supernal Eden in thirteenth-century Kabbalah. Furthermore, Nahmanides wrote of an "apprehension" that "is called Eden, which is called 'the bundle of the living'" (Nahmanides 1964, p. 297), which is the divine repose of souls, and Ibn al-'Arabī wrote about the divine nature of Eden and even used the name given it by Allah, "the abode of the living."

Finally, these works share another motif, the wondrous tree planted in the Garden: for the theosophical kabbalists, this is the Tree of Life (*eṣ ha-ḥayyim*), and for the tradition used by Ibn al-'Arabī, this is the Ṭūbā Tree (*šajarat ṭūbā*). Asín Palacios, the reader will recall, sought to link the blueprints of Paradise in the works of Ibn al-'Arabī and Dante. To that end, he argued that the eschatological *Candida Rosa* of the Divine Comedy is parallel to the supernal Ṭūbā Tree of Islamic tradition (Asín Palacios 1919). The basis for this parallelism is ultimately the precedence given to the geometric architecture of Paradise, which supposedly is reflective of a cosmology to which both authors subscribed. I propose that we not lose sight of the stress Ibn al-'Arabī placed on the moral hierarchy of the Garden and its eschatological, and perhaps even theosophical, import. In his account, the architecture of the Gardens and their subdivisions are founded on the special status of Muhammad's *umma*, the absolute superiority of the prophet, and the matchlessness of Allah's messenger among all creations. It seems to me that comparing the Ṭūbā Tree to the eschatological Tree of Life of the theosophical kabbalists is more productive than the rather formal parallelism with Dante's *Candida Rosa*. More specifically, these kabbalists envision the Tree of Life as the source of the human soul, and Ibn al-'Arabī writes: "Know that the Ṭūbā Tree among the Gardens is like a man with respect to his children; in the same way Allah planted it by hand and made it, He blew into it from his spirit, in the same way he did with Mary, into whom He blew from His spirit, which is how Jesus could resurrect the dead and cure the blind and leprous" (Ibn al-'Arabī 2010, p. 371). In this way, the Ṭūbā Tree, like the Tree of Life, has a similarly etiological role with respect to the human soul emerging from God

5. Possible Literary Contacts between Castilian Kabbalah and Ibn al-'Arabī's Writings

So far as we know, it can be said about Castilian kabbalists and Ibn al-'Arabī's writings that never the twain did meet. Yet, we have observed many striking similarities between the works. To try to account for this, I will explore their joint philosophical heritage, touch on the possibility of an early medieval transfer of ideas, and argue that the Castilian kabbalists knew of a book with similar content to Ibn al-'Arabī's.

Both bodies of writing examined above exhibit similar ontological and psychological assumptions. Their biplanar conception of Paradise is squarely grounded in the ontology of medieval Neoplatonism. The view of reality as a continuum that gradually shifts from the "coarse" to the "rarefied" comes from the same body of thought, and is expressed in their writing through the bidirectional movement between the two Gardens and the theories of an "enclothed" soul. The underlying psychology assumes the Platonic dualism of body and soul, which entails distinct eschatological fates for each of the two, and so we end up with more than one Paradise. Additionally, Ibn al-'Arabī's philosophico-mystical writing and that of the Castilian kabbalists both tie the plane of the Godhead to these lower ontological-cosmological dimensions of reality. In other words, the idea of multiple planes of existence, with their respective pleasure Gardens, is integrated with a complex doctrine of emanation, in which there is a long chain of unified being. What is perhaps most noteworthy is that in all of this Castilian Kabbalah and Ibn al-'Arabī's thought are much closer to each other than either of them are to other medieval Jewish or Muslim thought, including the rationalist Mu'tazilah schools and their Jewish elaborations and the Aristotelian Islamic philosophy of al-Farabi, Avicenna, or Maimonides. When one considers the fact that the kabbalists were mainly interpreting their own canonical literature (Scripture, Talmud, and Midrash) and Ibn al-'Arabī his (the Quran, its early interpretation, and hadith literature), the affinity between what resulted from each is remarkable. Several scholars have formulated the theory, which has yet to be fully worked out, that Iberian Jewish thinkers, kabbalists included, were familiar with Neoplatonic writings, especially of the Ismaili variety, in Arabic (Pines 1980; Goldreich 1987; Ebstein and Weiss 2015; Krinis 2016). Although no study has investigated kabbalistic use of specific Arabic texts in the areas of divine retribution and eschatology, such a study is expected to yield rich results, given the high concentration of similarities found in our phenomenological and terminological comparison of a relatively limited sample.

In searching for avenues of (mutual) influence, we should not set our temporal parameters too rigidly but be open to finding it centuries before any of our authors were alive. One study has compared conceptions of Paradise in the hadith literature and in Midrash, even going so far as to claim that the hadith can be traced to texts like a midrash about R. Joshua b. Levi's entry to the Garden of Eden (Tamari 1999). Another study attempts to draw parallels between late Shiite traditions on the *mi'rāj* and ideas found in the Midrash, Talmud, and *Hekhalot* literature (Halperin 1995). Neither of the two passes historical criticism, especially due to the late dating of many of the texts involved, but together they highlight the need to reexamine early medieval literary contacts between the two faiths. With respect to the depiction of Paradise in medieval Kabbalah, pursuing these kinds of literary leads would counterbalance the preoccupation with finding precedents in Franco-German Midrash or *Hekhalot* writings that supposedly absorbed Christian ideas. For example, when it comes to the groupings in Paradise dictated by moral considerations, Castilian Kabbalah has more in common with *Meccan Revelations* than it does with any texts from Franco-Germany, where the focus on morality is said to have been the product of confronting Christian ideals like asceticism.

Another line of thinking, in which the historically plausible connection can be followed all the way from one end to the other, is that Jewish esotericists encountered Islamic literature on Muhammad's ascent in their mother tongue. *Kitāb al-Mi'rāj* is the reconstructed Arabic title of a work translated into Castilian in 1264 at the court of King Alfonso X (the Wise). This work thickly portrays the various waystations of the celestial journeys taken by Muhammad and the archangel Gabriel. There are visions from the heavens, Paradise, and Gehenna with their diverse precincts and inhabitants, interspersed with motifs from the Quran and the hadith literature (Tottoli 2017). The parts about Muhammad's visits to Paradise (Cerulli 1949) contain traces of many traditions that appear in the writings of Ibn al-'Arabī, who himself wrote extensively about the *mi'rāj* and celestial journeys (Morris[1987] 1988). *Kitāb al-Mi'rāj* also presents a basic layout of seven hierarchical Gardens with further subdivisions, and many Islamic traditions about Paradise take on new form, such as the entire vision dedicated to the Ṭūbā Tree, which details its massive dimensions, its exquisite fruit, and more. The account presented therein, it should be noted, only contains a latent distinction between the twofold Paradise motive ("paradise of action" vs. "paradise of presence"). Significantly, the translation of this Arabic work into the Castilian vernacular was executed by a Jewish physician named Don Abraham, who was employed by the king's court as a translator (Heullant-Donat and de Beaulieu 1991; Gil 1985). Furthermore, it has been suggested that the lost Arabic source was influenced partly by ancient Jewish and Christian ideas that later appeared in the visionary apocrypha or even *Hekhalot* literature (Weill 1991). It is not beyond the realm of possibility that the reintroduction of kabbalists to this ancient material via this Arabic work in translation inspired them to reimagine Paradise in all its glory, and even led them to absorb its ideas and motifs (Idel 2005; Liebes 2011). In fact, beyond the general similarity in conceptualization and imagery, the very publication of visionary literature on Paradise based on Muhammad's ascent, in the same time and place as the kabbalists discussed here, can help explain the sudden emergence of a new kabbalistic genre of Hebrew and Aramaic pseudepigraphy, in which elaborate visions of Paradise were attributed to venerated figures of Jewish tradition.

6. Conclusions

This study is a comparative analysis of the appearances of the lower and upper Paradise, their divisions, and the journeys to and within them, which appear in mystical Jewish and Islamic sources in medieval Iberia. Ibn al-'Arabī's vast output on the Gardens of divine reward and their divisions generated a number of instructive comparisons to the eschatological and theosophical writing about the same subject in early Spanish Kabbalah. Although there is no direct historical evidence that kabbalists knew of such Arabic works from Catalonia or Andalusia, there are commonalities in fundamental imagery and in ontological and exegetical assumptions that resulted from an internalization of similar patterns of thought. It is quite reasonable to assume that these literary corpora, both products of the thirteenth century, were shaped by common sources from earlier visionary literature. The prevalence of translations of religious writing about ascents on high, produced in Castile in the later thirteenth century, can help explain the sudden appearance of visionary literature on Paradise and its divisions in the writings of Jewish esotericists of the same region. These findings therefore enrich our knowledge of the literary, intellectual, and creative background against which these kabbalists were working when they chose to depict Paradise in the way that they did, at the time that they did. One hopes that the relatively uncharted territory which this study has begun to map will become well trodden and comprehensively surveyed.

Funding: This research is supported by the Israel Science Foundation (grant number 727/19).

Conflicts of Interest: I declare no conflict of interest as author of this article.

References

Abdel Haleem, Muhammad. 2017. Quranic Paradise: How to Get to Paradise and What to Expect There. In *Roads to Paradise: Eschatology and Concepts of the Hereafter in Islam*. Leiden and Boston: Brill, vol. 1.

Abrahamov, Binyamin. 2015. *Ibn al-'Arabī's Fuṣūṣ al-Ḥikam: An Annotated Translation of 'The Bezels of Wisdom'*. London and New York: Routledge.

Albarracín, Francisco Martínez. 2016. Ramón Llull: Arte y mística. Imágenes, memoria y dignidades (Para una comparación entre las místicas del amor de Ibn ʿArabī y Ramón Llull). *El Azufre Rojo* 3: 222–40.

al-Kisa'ī, Muhammad ibn 'Abd Allah. 1923. Muhammad ibn 'Abd Allah. In *Qisas Al-Anbiya*. Edited by Isaac Eisenberg and E. J. Leiden. Chicago, IL: Kazi Publications. First published 1922.

al-Shaʿrānī, ʿAbd al-Wahhāb ibn Aḥmad. 1959. *al-Yawāqit wa al-Jawāhir fi Bayān Aqā'id al-Akābir*. Cairo: Mustafā al-Bābī al-Halabi, vol. 2.

Asín Palacios, Miguel. 1919. *La Escatología Musulmana en la Divina Comedia: Discurso Leído en el Acto de su Recepción*. Madrid: Impr. de E. Maestre.

Bar Asher, Avishai. 2013. Kabbalistic Interpretations of the Secret of the Garment in the 16th Century. *Da'at* 76: 191–213. (In Hebrew).

Bar Asher, Avishai. 2019. *Journeys of the Soul: Concepts and Imageries of Paradise in Medieval Kabbalah*. Jerusalem: Magnes Press. (In Hebrew)

Bension, Ariel. 1932. *The Zohar in Muslim and Christian Spain*. London: Routledge.

Busi, Giulio. 1996. L'architettura mistica dell'Eden nella tradizione ebraica. *Annali di Storia dell'Esegesi* 13: 465–75.

Cantarino, Vicente. 1965. Dante and Islam: History and Analysis of a Controversy. In *A Dante Symposium in Commemoration of the 700th Anniversary of the Poet's Birth*. Edited by William De Sua and Gino Rizzo Chapel. Hill: University of North Carolina Press, pp. 175–98.

Carra de Vaux, Bernard. 1923. *Les penseurs de l'Islam*. Paris: Librairie orientaliste Paul Geuthner.

Cerulli, Enrico. 1949. *Il 'Libro della Scala' e la questione della fonti arabo-spagnole della Divina Commedia*. Vatican: Biblioteca Apostolica Vaticana.

Chittick, William C. 1988. Death and the World of Imagination: Ibn al-ʿArabī's Eschatology. *The Muslim World* 78: 51–82.

Chittick, William C. 1998. *The Self-Disclosure of God: Principles of Ibn al-'Arabī's Cosmology*. Albany: SUNY Press.

Chodkiewicz, Michel. 1986. *Le Sceau des Saints, Prophétie et Sainteté dans la doctrine d'Ibn 'Arabî*. Paris: Éditions Gallimard.

Cohen-Aloro, Dorit. 1987. *The Secret of the Garment in the Zohar*. Jerusalem: Research Projects of the Institute of Jewish Studies. (In Hebrew)

Corti, Maria. 2001. Dante e la cultura islamica. In *Per Correr Miglior Acque...: Bilanci e Prospettive Degli Studi Danteschi Alle Soglie del Nuovo Millennio, Atti Del Convegno (Verona-Ravenna, 25–29 October1999)*. Edited by Lucia Battaglia Ricci. Roma: Salerno, vol. 1–2, pp. 183–202.

Dan, Joseph. 1968. *The Esoteric Theology of Ashkenazi Hasidism*. Jerusalem: The Bialik Institute. (In Hebrew)

Dan, Joseph. 1993. The Ancient Hekhalot Mystical Texts in the Middle Ages: Traditoin, Source, Inspiration. *Bulletin of the John Rylands University Library of Manchester* 75: 83–96.

Dan, Joseph. 2017. From Hekhalot Rabbati to the Hekhalot of the Zohar: The Depersonalization of the Mysticism of the Divine Chariot. *Jewish Studies: Journal of The World Union of Jewish Studies* 52: 143–162.

Ebstein, Michael. 2014. *Mysticism and Philosophy in al-Andalus: Ibn Masarra, Ibn al-ʿArabī and the Ismāʿīlī Tradition*. Leiden: Brill.

Ebstein, Michael, and Tzahi Weiss. 2015. A Drama in Heaven: "Emanation on the Left" in Kabbalah and a Parallel Cosmogonic Myth in Ismāʿīlī Literature. *History of Religions* 55: 148–71. [CrossRef]

Elior, Rachel. 2010. Introduction. In *A Garden Eastward in Eden: Traditions of Paradise*. Edited by Rachel Elior. Jerusalem: Magnes Press. (In Hebrew)

el-Ṣaleḥ, Ṣoubḥî. 1971. *La vie Future selon le Coran*. Paris: Vrin.

Farber, Asi. 1987. The Concept of the Merkabah in Thirteenth-Century Jewish Esotericism: Sod ha-Egoz and its Development. Ph.D. dissertation, Hebrew University, Jerusalem, Israel. (In Hebrew).

Gabrieli, Francesco. 1953. New Light on Dante and Islam. *East and West* 4: 173–80. [CrossRef]

Gardet, Louis. 1960. Djanna. In *Encyclopedia of Islam*, 2nd ed. Edited by E. J. Leiden. Leiden: Brill, pp. 447–52.

Gardiner, Eileen. 1989. *Visions of Heaven & Hell Before Dante*. New York: Italica Press.

Gaster, Moses. 1893. Hebrew Visions of Hell and Paradise. *Journal of the Royal Asiatic Society* 25: 571–611.

Gil, José S. 1985. *La escuela de traductores de Toledo y sus colaboradores judíos*. Toledo: Instituto de investigaciones y estudios toledanos (CSIC).

Ginzberg, Louis. 1913. *The Legends of the Jews*. Philadelphia: Jewish Publication Society of America, vol. 1.

Goldreich, Amos. 1987. The Theology of the *Iyyun* Circle and a Possible Source of the Term "Ahdut Shava". *Jerusalem Studies in Jewish Thought* VI: 1441–56. (In Hebrew).

Halperin, David J. 1995. Hekhalot and Mi'raj: Observations on the Heavenly Journey in Judaism and Islam. In *Death, Ecstasy, and Other Worldly Journeys*. Edited by John J. Collins and Michael Fishbane. Albany: SUNY, pp. 269–88.

Heullant-Donat, Isabelle, and Marie Anne Polo de Beaulieu. 1991. Histoire d'une traduction. In *Le Livre de l'Échelle de Mahomet*. Paris: Le Livre de Poche.

Himmelfarb, Martha. 1983. *Tours to Hell: An Apocalyptic form in Jewish and Christian Literature*. Philadelphia: University of Pennsylvania Press.

Ibn 'Arabī. 1946. Muḥyī al-Dīn ibn al-ʿArabī. In *Fuṣūṣ al-Ḥikam*. Beirut: Dar al-Kitab al-ʿArabī.

Ibn al-'Arabī. 2010. Muḥyī al-Dīn ibn al-ʿArabī. In *al-Futûhât al-Makkiyya*. Tarīm: wizārat al-thaqāfa, vol. 12.

Idel, Moshe. 1982. The Journey to Paradise: The Jewish Transformations of a Greek Mythological Motif. *Jerusalem Studies in Jewish Folklore* 2: 9–16.

Idel, Moshe. 2005. *Ascensions on High in Jewish Mysticism: Pillars, Lines, Ladders*. Budapest and New York: Central European University Press.

Jellinek, Adolf. 1853. *Bet ha-Midrasch: Sammlung kleiner Midraschim und Vermischter Abhandlungen aus der ältern jüdischen Literatur*. Leipzig: Friedrich Nies, vol. 2.

Jellinek, Adolf. 1855. *Bet ha-Midrasch: Sammlung kleiner Midraschim und Vermischter Abhandlungen aus der ältern jüdischen Literatur*. Leipzig: Friedrich Nies, vol. 3.

Kiener, Ronald C. 1982. Ibn al-'Arabí and the Qabbalah: A Study of Thirteenth-Century Iberian Mysticism. *Studies in Mystical Literature* 2: 26–52.

Krinis, Ehud. 2016. Cyclical Time in the Ismāʿīlī Circle of Ikhwān al-ṣafā' (tenth century) and in Early Jewish Kabbalists Circles (thirteenth and fourteenth centuries). *Studia Islamica* 111: 20–108. [CrossRef]

Kushelevsky, Rella. 2010. *Penalty and Temptation Hebrew Tales in Ashkenaz Ms. Parma 2295 (de-Rossi 563)*. Jerusalem: Magnes Press. (In Hebrew)

Lange, Christian. 2016. *Paradise and Hell in Islamic Tradition*. Cambridge: Cambridge University Press.

Lange, Christian. 2017. The "Eight Gates of Paradise" Tradition in Islam. In *Roads to Paradise: Eschatology and Concepts of the Hereafter in Islam*. Leiden and Boston: Brill, vol. 1.

Liebes, Yehuda. 1993. The Messiah of the Zohar: On R. Simeon bar Yoḥai as a Messianic Figure. In *Studies in the Zohar*. Albany: State University of New York Press.

Liebes, Yehuda. 2011. *The Cult of the Dawn: The Attitude of the Zohar towards Idolatry*. Jerusalem: Karmel. (In Hebrew)

McGaha, Michael. 1997. The Sefer Ha-bahir and Andalusian Ṣūfism. *Medieval Encounters* 3: 20–57. [CrossRef]

Meroz, Ronit. 2000. Zoharic Narratives and Their Adaptations. *Hispania Judaica Bulletin* 3: 3–63.

Moisés de León. 2007. *Moisés de León: El Jardín del Edén*. Edited by Carlos del Valle Rodríguez, Norman Roth and Antonio Reguera Feo. León: Lobo Sapiens S.L.

Morris, James Winston. 1988. The Spiritual Ascension: Ibn 'Arabī and the *Mi'rāj*. *Journal of the American Oriental Society* 4: 629–52. 108: 63–7. First published 1987.

Moses de Leon. 1988. *The Book of the Pomegranate: Moses de Leon's Sefer ha-Rimon*. Edited by Elliot Wolfson. Providence: Brown Judaic Studies.

Moses de León. 2013. *R. Moses de León's Sefer Mishkan ha-Edut*. Critically edited introduced and annotated by Avishai Bar-Asher. Los Angeles: Cherub Press.

Nahmanides. 1964. *Kitvei Rabbenu Moshe ben Naḥman*. Edited by Charles B. Chavel. Jerusalem: Mossad Harav Kook.

Pedaya, Haviva. 2002. *Vision and Speech: Models of Revelatory Experience in Jewish Mysticism*. Los Angeles: Cherub Press. (In Hebrew)

Perani, Mauro. 1996. Il giardino dell'Eden in Nahmanide. *Annali di Storia dell'Esegesi* 13: 477–91.

Perry, Micha J. 2010. *Tradition and Change: The Transmission of Knowledge among medieval Jews*. Tel Aviv: Hakibbutz Hameucha. (In Hebrew)

Pines, Shlomo. 1980. Shīʿite Terms and Conceptions in Judah Halevi's *Kuzari*. *Jerusalem Studies in Arabic and Islam* 2: 165–251.

Porter, Yves. 2007. Paradis. In *Dictionnaire du Coran*. Paris: Robert Laffont, pp. 638–41.

Rasā'il Ikhwān al-Ṣafā'. 1957. *Rasā'il Ikhwān al-Ṣafā'*. Beirut: Dār Ṣādir, vol. 4.

Reeves, John C., and Annette Yoshiko Reed. 2018. *Enoch from Antiquity to the Middle Ages: Sources from Judaism, Christianity, and Islam*. Oxford: Oxford University Press, vol. 1.

Ṣaḥīḥ al-Bukhārī. 2001. *Abū ʿAbd Allāh Muḥammad ibn Ismāʿīl*. Edited by Muḥammad Zuhayr ibn Nāṣir. Beirut: Dār Tawq al-Najāh, vol. 10.

Schäfer, Peter. 2004. In Heaven as it is in Hell: The Cosmology of Seder Rabbah di-Bereshit. In *Heavenly Realms and Earthly Realities in late Antique Religions*. Edited by Raanan Boustan ' and Annette Yoshiko Reed. Cambridge: Cambridge University Press.

Scholem, Gershom. 1934. An Inquiry in the Kabbalah of R. Isaac ben Jacob Hacohen: III. R. Moses of Burgos, the disciple of R. Isaac. *Tarbiz* 3: 258–86. (In Hebrew).

Scholem, Gershom. 1945. Sideri de-Shimusha Rabba. *Tarbiz* 16: 196–209. (In Hebrew).

Scholem, Gershom. 1955. The Paradisic Garb of Souls aud the Origin of the Concept of *"Haluka de-Rabbanan"*. *Tarbiz* 24: 290–306.

Scholem, Gershom. 1990. The Sources of 'The Story of Rabbi Gadiel the Infant'. In *Devarim Bego 1*. Tel-Aviv: Am Oved, pp. 270–83.

Shepkaru, Shmuel. 2002. To Die for God: Martyrs' Heaven in Hebrew and Latin Crusade Narratives. *Speculum* 77: 311–41. [CrossRef]

Silverstein, Theodore. 1952. Dante and the Legend of the *Mi'rag*: The Problem of Islamic Influence on the Chrisitan Literature on the Other World. *Journal of Near Eastern Studies* 2: 89–110; 187–197. [CrossRef]

Sviri, Sara. 1996. Spiritual Trends in Pre-Kabbalistic Judeo-Spanish Literature: The Cases of Bahya ibn Paquda and Judah Halevi. *Donaire* 6: 78–84.

Tamari, Shmuel. 1999. *Iconotextual Studies in the Muslim Vision of Paradise*. Wiesbaden and Ramat Gan: Harrassowitz–Bar Ilan University.

Tishby, Isaiah. 1989. *The Wisdom of the Zohar: An Anthology of Texts*. London: The Littman Library of of Jewish Civilization, vol. 1.

Tottoli, Roberto. 2017. Muslim Eschatology and the Ascension of the Prophet Muḥammad: Describing Paradise in Miʿrāj Traditions and Literature. In *Roads to Paradise: Eschatology and Concepts of the Hereafter in Islam*. Leiden and Boston: Brill, vol. 1.

Weill, Isabelle. 1991. Le Livre de l'Échele et l'occident: judéo-crétien du XIIIe siècle. In *Le Livre de l'Échelle de Mahomet*. Paris: Le Livre de Poche.

Wolfson, Elliot R. 1990. The Secret of the Garment in Nahmanides. *Da'at* 24: xxv–xlix.

Wolfson, Elliot R. 1994. *Through a Speculum that Shines: Vision and Imagination in Medieval Jewish Mysticism*. Princeton: Princeton University Press.

Wolfson, Elliot R. 2008. Via *Negativa* in Maimonides and Its Impact on Thirteenth Century Kabbalah. *Maimonidean Studies* 5: 393–442.

Wolski, Nathan, and Merav Carmeli. 2007. Those Who Know Have Wings: Celestial Journeys with the Master of the Academy. *Kabbalah* 16: 83–114.

Yisraeli, Oded. 2010. The Tree of Life and Its Roots: A History of a Kabbalistic Symbol. In *A Garden Eastward in Eden: Traditions of Paradise*. Edited by Rachel Elior. Jerusalem: Magnes Press, pp. 202–21. (In Hebrew)

Ziolkowski, Jan M. 2007. Introduction. In *Dante and Islam*. Edited by Jan M. Ziolkowski. New York: Fordham University Press, pp. 1–34.

Publisher's Note: MDPI stays neutral with regard to jurisdictional claims in published maps and institutional affiliations.

religions

Article

The Divine Feminine Presence in Ibn 'Arabi and Moses de Leon

Julia Alonso

Vicerrectorado Investigación, Cátedra Internacional De Hermenéutica Crítica: Hercritia, National Distance Education University (UNED), 28015 Madrid, Spain; julia.alonso.dieguez@gmail.com

Abstract: This paper is an investigation of the divine feminine power as depicted in the texts of Hispanic mystics from Sufi, Hebrew, and Christian traditions. This work is intended to investigate the origin and subsequent development of a transcendent reconciliation of polarity, its diverse manifestations, and the attainment of a common goal, the quintessential of the Perfect Human Being. The architect of the encounter that leads to Union is "Sophia". She is the Secret. Only those who are able to discern Her own immeasurable dimension may contemplate the Lady who dwells in the sacred geometry of the abyss. Sophia is linked to the hermetic Word, She is allusive, clandestine, poetic, and pregnant with symbols, gnostic resonances, and musical murmurs that conduct the "traveler" through dwellings and stations towards an ancient Sophianic knowledge that leads to the "germinal vesicle", the "inner wine cellar", to the *Initium*, to the Motherland. She is the *Mater filius sapientae*, who through an alchemical transmutation becomes a song to the absent Sophia whose Presence can only be intuited. Present throughout the Creation, Sophia is the axis around which the poetics of the *Taryuman al-ashwaq* rotates and the kabbalistic Tree of Life is structured.

Keywords: Sophia; *Duende*; union; Eros; path; feminine power; heart; Presence; secret; theophany; Hieros Gamos

check for updates

Citation: Alonso, Julia. 2021. The Divine Feminine Presence in Ibn 'Arabi and Moses de Leon. *Religions* 12: 156. https://doi.org/10.3390/rel12030156

Academic Editor: Cristobal Serran-Pagan Y Fuentes

Received: 15 January 2021
Accepted: 19 February 2021
Published: 27 February 2021

1. Introduction

Those born in present-day Spain, whose Islamic and Jewish history has been concealed from us, have a responsibility to discover our long-neglected authors, the ones who wrote in other languages—Hebrew and Arabic. For centuries, their linguistic and cultural perceptions shaped the ways of life and interpretations of the world that, in conjunction, gave rise to that crucible of knowledge, the Andalusian world [andalusí], characterized by riches and originality. As Fernando Mora says, "It is an 'old vernacular disease' to think that our cultural background consists exclusively of texts written in Spanish" (Mora 2019, pp. 15–29). And it was the Spanish historian Sánchez Albornoz, in his book *Spain: a historical enigma*, who declared that Ibn al 'Arabi is "the personification of a Spaniard". (Mora 2019, pp. 15–29). The same can be said of Moses de Leon.

The purpose of this article is to address the subject of investigation within a broad historico-religious context whose written manifestation, as we know from Noah Kramer's translation of the Sumerian hymns, dates back to the bridal rite of (Ryan 2008, p. 7) Hieros Gamos[1]. To support the thesis of a common origin of the mystics linked to our authors, it is necessary to investigate the symbolic concurrences, exegesis,[2] and the process employed by each of these hermeneutical seekers whose object is to gain access to Supreme Knowledge, in which Sophia plays the primary role.

The present investigation reveals the exceedingly ancient Presence of a Feminine Power of Divinity, as displayed in diverse currents that shaped the many religious, spiritual,

[1] See: (Kramer 2001). *Sacred* sexual intercourse is thought to have been common in the Ancient Near East as a form of "Sacred Marriage" or *hieros gamos* between the kings of a Sumerian city-states and the High Priestesses of Inanna, the Sumerian goddess of love, fertility and warfare. See: (Vázquez 2007).

[2] The *Zohar* is an esoteric and mystical exegesis of the Torah, by means of the sacred value of its letters. Ibn 'Arabi, for his part, wrote an exegesis of his *Interpreter of Desires*, in which he elucidates the hidden meaning of the text.

and philosophical movements coexisting within Al-Andalus. Within that framework, despite a certain friction, Muslims and Hebrews were able to project their interpretation of spirituality and divine structure into Christian mystical theology, where it gave rise to a powerful and unique way of Knowledge whose originality consisted in the vision of a force emanating from the celestial order. This emanation results in a differentiated polarity that includes, yet transcends, the amorous entanglement that leads to an encounter with Unity. It is thus that a sort of irruption occurs within Hispanic mystical life, where the figure of Sophia linked to Love acquires enormous power and a singular means of expression. It is proposed to describe this original manifestation within the borders of Al-Andalus, focusing on the intimacy generated between the seeker and the Creative Force and on the ways in which this encounter is described in the three languages linked to the religions of the Book: Arabic, Hebrew, and Spanish. Underlying the Presence and manifestation of the Creative Force (known by various names, and to our authors as Ibn 'Arabi's, *Himma*, Moses de León's, *Shekhinah*, and St. John of the Cross', Wisdom), is the Science of Letters,[3] the knowledge of harmonic sounds, number, and sacred mathematics.

That said, the goal of this work will be to highlight the equilibrium of the Divine, the celestial hieros gamos that is projected into the human soul and into the Creation in a continuous circular flow, embracing differences and weaving them into a single manifestation that reflects, like a mirror, the unsayable.

The hypothesis is that diverse mystical paths suggest the existence of a universal entanglement that can be discovered by apparently divergent pathways, all leading to a common goal: intuitive, noetic knowledge. To those who undertake the voyage into uncertainty, Sophia grants this particular kind of knowledge in the form of Light. The current proposal will emphasize the original Unity, the power of the Cosmic Eros that guides the enamored soul toward an "experience" that takes place in dimensions of reality beyond time and space, incorporating harmonies and discrepancies, encounters and dis-encounters that emerge in an intermediary world between the spiritual and the material: the world of the soul.

It seems essential to investigate, beyond an anthropocentric and unidirectional vision, the relationship of the human being with Longing, with the Presence of a mysterious *Saudade*,[4] and definitively with the seductive figure of Sophia, who precipitates those who "suffer" from love-sickness into another dimension. This perspective will necessitate a search for cultural variations in varying contexts, perhaps even a sacred language that connects us to all the inhabitants of the universe, in an attempt to reveal the common indicators underlying the diverse interpretations and expressions and their hidden messages. Something is summoning us, challenging us, calling attention to the paths that coincide with this search, to the convergence of itineraries that guides us to the fulfillment of a lack. The present study will examine, in short, the common threads and concurrences that speak to us of a community formed around the sacred lineage of gnosis, which inhabits a very particular spiritual and contemplative cartography. Despite the diversity of perspectives and historical or cultural dissonances, it calls our attention to the sense of a universal gnostic communion. As the Murcian Sufi Ibn 'Arabi reminds us, the mystic souls of all historical epochs live out the same spiritual experience, though conditioned by the diversity of beliefs, by religious plurality. This diversity gives rise to what the *Sheikh* denominates as "the god created through the creeds", which is inevitably problematic, since according to a proverbial saying: "There are as many ways to God as there are human souls" (Chittick 2003, p. 7). It is precisely Ibn 'Arabi who emphasizes the diversity of opinions (Chittick

3 See: footnote 16.

4 See: (Borges 2008); De Sa ´Carneiro Mario, *No lado esquerdo da alma*, 2ª ed. Alma Azul, Coimbra 2004, p. 38; Pereira Dacosta Dalila, *Encontro na noite*, Lello & Irmão Editores, Porto 1973.
https://www.ler.letras.up.pt/uploads/ficheiros/17602.pdf (accessed on 12 December 2020).
Saudade is a Galician-Portuguese term and is therefore Iberian. There is no direct translation into English. It is an ardent desire or longing to return to the origin, which implies a deep sense of loss of an initial state of fullness in unity. This results in an emotional state of profound nostalgia or melancholia, predicated on the absence of a beloved person, place, object, or state of being.

2003, pp. 8–9) (*mas'ala jilāf*) that, according to him, has been established by the working of the divine *Wisdom* and *Compassion;* as he points out, "God himself is the source of all the diversity in the Cosmos" (Chittick 2003, pp. 8–9). Consequently, the Murcian master considers all creeds, even the discordant ones, as emanating from the same god, since they, as well as the variety of methodological perspectives, all contribute something vital. It requires only a certain effort to distance ourselves from our habitual epistemological framework in order to situate ourselves on another plane on which, according to the professor Teresa Oñate y Zubia, there occurs the coupling of differences, of masculine/feminine, light/darkness, above/below, dense/subtle.

When considering the mystical phenomenon within the geographical boundaries of Spain, Federico García Lorca's writings on *duende* are indispensable. It was the great Andalusian poet and dramatist who delineated the particular characteristics of the creators and seers native to this terrain in terms of their connection with *duende*, an overwhelming, indescribable, and irresistible Presence. This *duende* is nothing less than the relevant, magical divine feminine power that emerges in the texts of these Spanish mystics, leading through Sophia and her Wisdom to an intimate and theosophical encounter.

Thirteenth-century Spain experienced a very particular outburst of the mystical phenomenon, and any attempt to decipher the key points of the spiritual paths that channeled it demands the closest observation of the poetic and hermeneutic word, bearer of life yet foreign to rational discourse. A notable example would be the language of St. John of the Cross, where something that at first seems like nonsense turns out to be a cipher, a symbol, a key to unknown gateways and dimensions. Through this riddling use of language, the saint explored differentiated planes of reality: imaginal worlds, as in his Strange Isles[5] (Ínsulas Extrañas), or topographies of the frontiers where we find a phenomenology pregnant with sightings, revelations, locutions, visions, and other portents peculiar to the alchemical transmutation experienced by the Gnostic. These experiences foment the emergence of poetic language, allegory, and symbol, essential vehicles of a shared, alternative way of knowing linked to *noesis*.[6]

It is also necessary to consider other texts, far removed in time, in which we detect antecedents and fundamental parallelisms and where we confront reflections on Wisdom, that universal, millenary notion that penetrates the Hispanic theology of the Sufi, Hebrew, and Christian mystics. In examining the origins and later development of a certain sacred transcendent polarity, one discovers that complementary categories replace the oppositional ones, effecting, a conciliatory coupling that is projected through various manifestations and leads at last to the achievement of a goal common to all three "mystics of the Book": the Perfect Human Being. This Hispanic "symbolic nuptial process" is clearly opposed to the Rhenish "mystical theology of the essences", although some of its aspects—like those related to the Unsayable and Unthinkable, that Nothing that is Everything, the Divine Darkness of Dionysius the Areopagite may occasionally converge. Varying linguistic and cultural codes account for the disparity between particular approaches, but it is evident that these mystics are following parallel, not divergent, routes. Their discursive results, occasionally influenced by dogmas and theologies, arise from interpretations of their sacred texts.

The Hispanic mystical paths are diverse (centuries later the Sufis and the adherents of the Kabbalah were to contribute to the Christian mysticism of St. Teresa, St. John of the Cross, and Miguel de Molinos); however, all refer to biblical texts, sometimes amplified by millenary currents already evident in the Sumero-Akkadian hymns[7] whose defining characteristics were perpetuated in the *Song of Songs*. The metaphors, analogies, and double mean-

[5] Saint John of Cross: "strange isles" are analogous to Ibn 'Arabi's dimension of active imagination. It is in this intermediate state between the terrestrial and the celestial that visions, prophecies, and locutions occur.

[6] The Greek term νόησις refers to a "visión, intuition, intellection". *Noesis* is the mental activity ("nous") through which one accedes to a direct and unmediated knowledge of an object.

[7] See: Noah Kramer Samuel, *From the Poetry of Summer*, Berkeley, University of Berkeley Press, 1979, p. 71; Noah Kramer Samuel, *The Sacred Marriage Rite*, Bloomington, Indiana, University of Indiana Press, 1969; Graves Robert, *La Diosa Blanca*, Madrid, Alianza Editorial. 1996; p. 500.

ings of that supreme example of an erotico-spiritual text were in turn reflected both in the *Tarjumán al-Ashwáq* (*The Interpreter of Desires*) of the Murcian *sheikh*, Muhyiddin Muhammad Ibn al-Arabi (أبو عبد الله محمد بن علي بن محمد بن عربي الحاتمي الطائي, Murcia, 28 July 1165-Damascus, 16 November 1240), and in the *Zohar* attributed to the Spaniard Moses de León (?1240—Arévalo, 1305), known in Hebrew as Moshe ben Shem-Tov (ליאון-די טוב-שםבן משה). These texts describe the milestones of a spiritual journey with a common goal: the Union of the seemingly diverse, which is attained after an extremely personal itinerary that proceeds through dwellings, steps, stations, and wine cellars before arriving at Ibn 'Arabi's "Dwelling of Non-Dwelling", or at the *Zohar*'s "Opening of the Eyes", a station similar to the Seventh Wine Cellar of St. John of the Cross. In these mystics, this entire process is conveyed through an array of connections, attributes, categories and Names that contain a multitude of meanings in which it can be detect echoes of not only the *Song of Songs*[8] and other biblical texts but of the Sumero-Akkadian Hymns (3400 B.C.) whether anonymous or attributed to the Akkadian priestess Enheduanna (2285–2250 B. C), in which the transforming Feminine Presence of the Divine plays a central role.

2. Spanish Mystical Theology: A Mystical Theology with *Duende*

What is *duende*[9]? Only someone who is conscious of its Presence, which dwells in the heart and liberates the creative force that is inherent to it will be able to decipher *The Interpreter of Desires*, the *Zohar*, and *The Spiritual Canticle. Duende* seizes on the mystic writer and on the poet, emptying the heart of refletive thought. *Duende* is related to the Socratic *daimon*.[10] It pillages the soul and obliterates meaning, yet preserves the logic of reason even while transcending it, leading to the emergence of a "poetic rationality", as María Zambrano once maintained (*El Hombre y lo Divino, Filosofía y Poesía*). It is thus that Spanish philosophy, poetry, and symbolic texts invoke an obscure depth inhabited by the divine, the home of *duende*. Poetic Reason understands truth not as an adjustment to the facts of physical reality, but as revelation. Therefore, when mystics or poets communicate their revelation in writing, what occurs is the unveiling of a symbolic, transcendent truth from personal experience, destined to attain "the universal": the Unity of Being. As Zambrano says, poetic reason is a "half-awake thinking". It belongs to a liminal space in which the human being is neither awake nor asleep. That topography corresponds to the imaginal world, which the "pilgrim" enters by way of the kingdom of Intermediary Being.

As previously indicated, the supreme theorist of the Hispanic *duende* was Federico García Lorca (Farré 1998, p. 82), who outlined his theory in a conference he gave in Buenos Aires and Havana in 1933. We defer to him. Our great poet reflects very beautifully and poetically on the notion of *duende*, an idea linked to the Spain of a sensitive, poetic, impassioned, and overwhelming sentiment. It is normal, in reference to Lorca, that in referring to any splendid and sublime artistic creation one says, in our country, "it has a

8 On the *Song of Songs* and its relationship to the Sumerian hymns, see p. 9, ll. 318–26.

9 The Spanish word is close to untranslatable and has been adopted into the English lexicon. In reference to the notion of "Duende", the importance of this term lies in its summation of the particularity of all Hispanic art and thought. It is the "Duende," or Daimon, as masterfully described by Lorca, that inspires poetry, dance, painting, and philosophy. This concept is fundamental to an understanding of the consistency between the *Zohar* and the *Interpreter of Desires*, as the underlying principle of both texts is precisely this "magical and creative presence". Although it is unnecessary to expand throughout the article on the meaning of this term, it is indispensable to note that the spirit that pervades Duende is a key that unlocks the dimension of the active creative imagination where eroticism and the amorous mysticism meet.

10 Platón, *El Banquete*, Obras completas, Edic. Patricio de Azcárate, tomo 5, Madrid 1871, p. 293:

> "The Daimons occupy the intermediary space that separates heaven from earth; they are the link that unites the Great All with the human being. Since the Divinity never communicates directly with human beings, it is through the mediation of the Daimons (in Greek, δαίμων) that the divine Voice is transmitted. It is an intermediary between the subtle, or spiritual, and the material dimension. Socrates associates the Daimon with an inner voice."

There are many references to the Daimon in Plato's *Symposium*: "The divinity does not establish direct contact with men, but it is through this Daimon that every relationship and dialogue between gods and men occurs, both in sleeping and waking". "The proper function of the Daimon is to serve as an interpreter between gods and men, carrying the vows and homage of mortals from earth to heaven, and the commands and blessings of the gods from heaven to earth. This is why Love maintains harmony between the human sphere and the divine, and along with the other daimons, it is the link to the All."

lot of *duende*" (Lorca 2003). The author of *Blood Wedding* puts these words in the mouth of Manuel Torres, who, after hearing Manuel de Falla's interpretation of his *Nocturno de Generalife* [Nights in the Gardens of Spain], pronounces this magnificent sentence: "Anything that sounds dark has *duende*" (Lorca 2003). And listening and darkness always accompany *duende*, as they do those mystic states in which light is born out of darkness to the accompaniment of the harmonic sound of silence, the "sonorous silence" of St. John of the Cross. García Lorca (Lorca 2003) notes, in this sense, that "these dark sounds are the mystery, the roots sunk in the mire that we all know and we all ignore, but from which we receive everything that matters in art". He is referring to the hidden place that Gaston Bachelard (Bachelard 2000, p. 40) describes as "the sub-basements of the soul". That unknown place, that point within the heart, is the dwelling of every soul that allows itself to be penetrated in solitude by a "mysterious Power that everyone feels and that no philosopher can explain" (Lorca 2003).

This is the power that raises the indigent human from hearing to vision, and bestows on the pilgrim the eyes of the Creator, in a Return that deprives the sentient being of action and conceptual discourse. Thus, darkness, hearing, and mystery correspond to their etymological meanings: ἀ-μαυρόζ (invisible, somber, obscure, blind); ἀκούω (to pay attention, to hear a call, to be put to the test); and μυστήριον (arcane, secret, secret doctrine, secret cult). *Duende* assumes the function of a power receptive, obscure and mysterious, feminine, impassioned, committed, and creative. It remains remote from any action except those that provoke a rebirth deriving from an inner trauma, and in the process of rebirth, it destroys the self, which is annihilated by its seductive and demanding potency. It is a receptacle, a goblet whose contours are greater or lesser according to the content of the stored-up force and its insistent pulsations. Consequently, for the poet from Granada, "*Duende* is a power and not a labor, a struggle and not a thought". *Duende*, he says, "is not in the throat (from which sounds emerge): it rises inwardly from the soles of the feet" (Lorca 2003); and, as can be observed, it bursts forth in a delirium of song, poetry, and celestial harmony unrelated to earthly human concordances. Although it operates in a realm beyond the mundane, nevertheless, through an irradiation of the Good and the Beautiful, the Presence of this force alters the nature of the most ordinary experience. From this point of view, one can speak of a relationship with the transcendent in terms of love and passion.

The *duende* whose summons Lorca so eloquently describes resonates throughout Spanish mystical theology: vibrant, enigmatic, hermetic, and seemingly incoherent. For each individual, it is unique, inviting us to different paths, as many as there are human souls. *Duende* makes use of an obscure, audible, and mysterious desire, manifesting through signs engendered by Sophia. Anyone capable of capturing something of its excessive force trembles before it. In order to understand texts produced within this state, we need a symbolic hermeneutics whose goal is "festival", in the sense defined by Hans Georg Gadamer in *Truth and Method* and *The Beauty of the Word*. The sacred word is a milestone on the path, leaving coded messages; it sidles downward to the crypts of the soul along a spiral staircase that descends in a serpentine path towards the darkness where light is most concentrated. Undertaking this risky descent, the pilgrim discovers that it is, astonishingly, an ascent (see Kingsley 2016) toward an encounter within an "inner Castle" (Saint Thérèse of Avila.), which harbors a diamond, in a *Templum*, a palace of perfection, reached through states and stations or dwellings. In this displacement toward the heights of oneself, the *duende*, the Socratic *daimon*, impels the traveler who, like Odysseus, must pass the stormy seas in order to reach home: the point of origin, the Unity. It is not free from suffering, this voyage, because it requires a stripping away, a total abandonment, even of the ego itself, in order to reach, egoless, annihilated, the primordial I, the *Jungian* Self.

Duende knows the instant because it occurs within it. It emerges from an eternal un-time and interrupts chronological succession. Its Presence irrupts in the *Kairós*, an "opportune" time. Tearing the human being away from the known self, it suspends the quotidian, revealing subtle pathways that bifurcate in search of the bridegroom: God. Some of these pathways penetrate into an imaginal, intermediary world, the *barzakh*, in which, as Ibn 'Arabi (who, like Moses de León, was familiar with the Science of Letters) tells us, "the bodies are spiritualized and the souls materialized". Alternatively, they may travel through "Strange Isles" (as St. John of the Cross suggests) thanks to another "science of words", the sonorous play of alliteration; or they may end up conforming to an almost Riemannian geometry,[11] the Tree of Life, related to a sacred mathematics of the word, an alpha-numeric code known as Gematria,[12] as Moses de León insinuates.

Duende invariably arrives spontaneously, at just the right moment, when the ground has been prepared and the ego has learned to let go of itself, as the Portuguese poet Fernando Pessoa (1888–1935) informs us through the heteronymous master Alberto Caeiro (Pessoa 2001, p. 82). *Duende* cannot be invoked; it lays down no pathways; it simply bursts forth, transmitting to the subject who experiences it, in the words of Kierkegaard (1813–1855), "a searing atom of eternity". In this respect, Federico García Lorca (1898–1936) writes, "One only knows that the blood burns as if inflamed by an embrocation of broken glass, that *duende* exhausts, that it rejects a gentle, familiar geometry, that it disrupts styles" (Lorca 2003). The arrival of *duende* is closely related to death, a death in life, since, as the poet from Granada tells us, "*duende* won't come without the possibility of death, if one doesn't know that it is already prowling around one's house, if one isn't sure that it will shake those branches that we are all carrying and leave us inconsolable" (Lorca 2003). *Duende* is attracted to the very edge of the pit, in open war with the creator (Lorca 2003). It moves through a horizon of events, bearer of the mystery of a "naked feminine and re-creative singularity" that inhabits every soul. Spain, says García Lorca, is the only country in which "death has become a national spectacle, where death greets spring with long bugle calls, and its art is always ruled by a heightened *duende* that renders it distinctive and inventive" (Lorca 2003). Death comes to the lovesick, the wounded heart. St. John of the Cross says, "Love is like dying". "The wound and dying are two ways of suffering death" (Cruz 2002, Cc. 7: 3,4,5, pp. 764–65).

Appropriating an expression of Lorca's, and referring to Saint Thérèse of Ávila, it can be affirmed that the Sufi master Ibn 'Arabi of Murcia, the Hebrew Moses de León, and the Christian mystic St. John of the Cross, were all "*in-duended*", possessed by *duende*, enamored of an unsayable that summoned them through that indescribable impulse: the desire that awakes Eros in all of creation. There is an extract in the *Zohar* that speaks, like the *Song of Songs*, of a "Love strong as death, hard as the parting of the soul from the body" (Laitman 2015, p. 204), and that severing presupposes a lack, a longing, a discontent, a fatal attraction that impels the human being beyond reason and disjunctive logic. Love has wings, it raises its creatures above the everyday; it delights in risk, accepts suffering, leads to the invisible, brings upper and lower worlds into conjunction. And the "kiss of Love", in one paragraph of the *Zohar*, is said to "expand in the four directions and the four directions join together and unite" (Laitman 2015, p. 204). That death accompanying a spiritual

[11] Bernhard Riemann's (1826–1866) interest was not limited to flat, plane two-dimensional surfaces inhabited by the triangle and the circle, nor to the three-dimensional spaces inhabited by the cube and the sphere, nor even to four-dimensional mathematical spaces (difficult to visualize, but possible to define and manipulate mathematically). He was also engaged in characterizing "n-dimensional spaces," which makes the Riemannian methodology much more universal than previous concepts of geometry. Exactly like the imaginal world of the mystics, Riemann's conception concerns unbounded spaces capable of containing an infinite number of dimensions, in particular the dimension of the *Barzakh*, in which, as Ibn 'Arabi affirms, bodies are spiritualized and spirits materialized.

[12] Gematria is an ancestor of numerology. It originated among the Syrians, Babylonians, and Greeks but was pursued assiduously within Jewish mysticism, and particularly in kabbalistic studies. In Gematria, a certain value is assigned to each letter of the alphabet. The Hebrew Kabbalists said that "*Belzebiel*", the builder of the Tabernacle, was an initiate of Gematria, or the relationship between words and numbers and their effect on both heaven and earth. By adding up the value of the letters within a word, one obtains another number and a meaning, which is then compared with the total numerical value of other words.
For students of the Kabbalah, the divine creation is based on the power of Hebrew words and letters, as well as the numbers associated with them. There is also an Islamic Kabbalah, similarly associated with the Science of Letters, and practiced by Ibn 'Arabi. See Note 16 in this article.
The body of the invisible God materializes in the letters of the *Torah*, an icon of the invisible in which the numbers and the letters converge. God reveals himself to the Kabbalist, the illuminated exegete, as a secret, both in interior reality and in the esoteric sense.
See Gamliel Belk Akiva, *Gematria And Mysticism in Genesis* (Journey Through Genesis Book 1), B'Nai Toach Institute, LLC, Miussouri, 2012: http//www.bnti.us (accessed on 21 June 2020).

journey, in Ibn 'Arabi's (Chittick 2003, pp. 70–71) opinion, presupposes the abandonment of characteristically human limitations, so that the individual becomes "erased", or in the words of St. John of the Cross, "effaced", and within that "annihilation", nothing remains, according to Ibn 'Arabi, except the face of the *Wujud* (Chittick 2003, pp. 70–71), turned toward the Creation.

Lorca asks, And where is *Duende*? "Through the empty arc there enters a mental breeze that blows insistently over the heads of the dead, searching for new landscapes and unknown accents: a breeze that smells of a child's saliva, of bruised grass and the stinging medusa's veil, that announces the continual baptism of newly created things".

In sum Ibn 'Arabi, Moses de León and St. John of the Cross are captives of a *duende* that speaks of dying of love, of surrender and personal extinction. That is the only way to understand Ibn 'Arabi's nocturnal "Journey to the Lord of Power"; the *Zohar*'s "Opening of the Eyes", the "I die of not dying" of St. John of the Cross, and Santa Teresa's "I live without living in myself".

3. The Supremacy of the Mystical Nuptial: Antecedents

The Great Queen (NINGAL) of queens born for the rightful Me,
born of a fate-laden body,
you are even greater than your own mother,
full of wisdom, foresight, queen over all lands,
who allows existence to many,
I now strike up your fate-determining song!
All powerful divinity suitable for the ME,
That which you have said magnificently is the most powerful!
Of unfathomable heart, oh highly driven woman
of radiant heart, your ME, I will list for you now! (Vázquez 2016, vss. 60–65)

The texts of ibn 'Arabi's *Interpreter of Desires* and of the *Zohar* attributed to Moses de León contain ancient wedding images derived from the cult of the Feminine Power of Divinity.

The first written testimonies of the heavenly wedding and its projection in the earthly order are originated in Sumer. The mythico-theosophical system of Mesopotamian and Egyptian religions, which was inherited by the Mediterranean culture, consists of an internal structure based on a reconciliation of sexual polarities. Therefore, at the heart of the divinity there is a dynamism imbued with eroticism and seduction. The channels established between the higher worlds and the one inhabited by created beings are generated by the projection of that bi-directional desire. True, the disparity and diversity of the models, as well as cultural tensions and their conceptual complexities, display remarkable differences, but from the Hispanic perspective, the most striking phenomenon is the existence of several paradigms of mystical eroticism that share a common theme: the Union of the Holy (masculine), with its Divine Presence (feminine).

To establish this fact, it will be necessary to return to the primal origins of the feminization of an aspect of the Divinity related to desire, love, and the creation of the universe; and which is, in addition, the depository of the laws that govern it. Noah Kramer's works, *The Sacred Marriage Rite* (Wolkstein 1983, p. 62), *History begins in Summer* (Kramer 2010), and *Inanna Queen of Heaven and Hearth* (Kramer and Wolkstein 1983, p. 107), contain numerous poems about the courtship of the Sumerian goddess Inanna and her betrothal to her husband Dumuzi, as well as accounts of the deity's descent into the Underworld ("Inanna and Dumuzi", "Inanna queen of heaven and hearth"). While the first cuneiform texts describe the earthly erotic tensions of the nubile goddess, in those hymns that depict her journey to the underworld the divine Inanna, in her maturity, is obliged to pass through seven successive gates. To gain passage through each one, she must abandon something precious to her, until at the last she remains totally naked and silent, as the laws of the underworld are implacable. Nothing remains in her heart except an intense desire for rebirth. In this poem we find the three characteristic phases of the female power: the young woman, the mature woman, and the crone. These three figures could be correlated, in turn, with the three Marys of the Gnostic Gospel of Philip and with the three *Seefiroth* of the *Zohar*'s Tree of Life: the Mother, *Binah;* the wife, *Tiferet;* and the daughter, *Malkuth.*

In the case of St. John of the Cross, there are also three female representations: the Soul (*Spiritual Canticle*); the Church (*Romance 4;* pp. 150–55) (as the bride of Christ, similar to the Community of Israel, *Keneset Yi-ra-el*); and Wisdom herself (*Spiritual Canticle*), associated with a higher realm, the breath of the Holy Spirit. This indicates the fundamental importance of these ancient bridal images in the development of a Hispanic mysticism that incorporates the three currents that demarcated the relationships between God and human beings. The entire mystical journey is oriented toward that encounter, the union of opposites, which—despite the various forms it may assume—is never destructive. Instead, it consists of a harmonious entanglement on an equal plane (Cruz 2002, Cc.26,5–13;16–17, pp. 842–47) or even of a reversal of roles, as St. John of the Cross' commentary on the *Spiritual Canticle* demonstrates (Cruz 2002, Cc. 27,1, p. 848).

In this respect, it is appropriate to add that, contrary to current opinion, parity between the two celestial polarities was quite common in the religions of the near East, and markedly so in primordial erotic relations. At the same time, we note a strikingly explicit description in the *Sefiroth* of the Tree of Life in the *Zohar* (see Section 5 of this article), in that the male projections of the right branch of the Tree may at some point take on a female role or vice versa, so that the amatory functions are often interchangeable, which indicates that there is no radical separation of "genders" in this matter. Something similar occurs in the texts of St. John of the Cross[13] in relation to the figure of Wisdom, which sometimes resembles Christ. As for Ibn 'Arabi, he draws no distinction between the Lord and the Lady (Corbin 1958, p. 109).

In regard to the historical background, and in accordance with Moshe Idel and other researchers, there are grounds for questioning Gershom Sholem's nationalist thesis which asserts that the divine feminine and creative principle (the *Shekhinah* of the Kabbalists) first assumes her role at the birth of the Kabbalah. On the contrary, its suspect that it is the contact sustained between the Hebrews and the Sumero-Akkadian culture after the Babylonian exile that explains why the figure of the hieros gamos, in which the amorous female power assumes a leading role, is so deeply imbued in the biblical *Song of Songs.* That text, in turn, retained an indisputable authority for our three Hispanic mystics. Consequently, it is possible to affirm that the amatory progress derived from these pathways of Spiritual Knowledge clearly originates in the Sumero-Akkadian compositions that inspired the supremely erotic biblical outpouring of the *Song of Songs*, as well as the remarkable references to the mystical wedding in the Hispanic spiritual traditions linked to the three religions of the Book. As for the distinctive role of the feminine in both the 13th-century texts of the *Interpreter of Desires* and the *Zohar* and in the 16th-century *Spiritual Canticle*, from the Middle Ages onward there is a parallel preeminence of women in western mystical traditions that should not be overlooked.

Arthur Green (Yom et al. 2006, p. 350) defends the unmistakable feminization of the *Shekhinah*[14] in the Kabbalah of the thirteenth century, as a Jewish response to the heyday of devotion to Mary in the Western Church of the time. This inevitable transfer and mutual influence might explain, in part, three of the most striking creations of our Hispanic mystics: the powerful representation of the *Zohar's Matronita* (Patai 1947); the projection of the feminine divine in *Nizam* (Harmony) in the *Interpreter of Desires;* and

13 See: Cruz, S Juan *Obras Completas*, Ruano de la Iglesia Lucinio, Ed. Biblioteca Autores Cristianos, Madrid 2002, Introduction to "Espiritual Canticle"; Hodar Manuél, *La Sabiduría en S. Juan de la Cruz*, Monte Carmelo, CITeS, Ávila 2011.

14 In a verse from sura Fussilat, it is said, "We will make them see our signs on the horizon and in their souls" (41:53). The meaning of the statement for the Shaykh Al-Akbar is explained in a chapter where he deals with *sakina* (peace, "serenity," but also, like the Hebrew *Shekhinah*, the divine "Presence"): the *sakina* that God sent down for the children of Israel in the Ark of the Covenant [Qu'ran 2:248] "was made to come down in the hearts of believers of Muhammad's community [Qu'ran 3:110] ... " (Chodkiewicz 1993, p. 96).

the central role of Sophia, which penetrates both creation and the soul in love in the texts of St. John of the Cross. In the prologue to *the Interpreter of Desires, Tarŷumān al-Ašwāq*, the Murcian Sufi master relates an encounter with the divine feminine (Arabî 2002, pp. 16–19) during one of his circumambulations in Mecca. This apparition gives rise to a cryptic conversation that inspires the first qasida of that mystical-erotic poem. As Carlos Varona (Arabî 2002, pp. 16–19) has emphasized, the composition is organized around the axis of the female protagonist. In fact, the beloved *Nizām* (Harmony), protagonist of the epithalamion, manifests as an epiphany, a tangible reflection of the unnamable and unsayable Here we have an example of a process in which divinity is revealed by the "emanation" of "celestial archetypes" (Arabî 2002, p. 16). By means of this revelation, indescribable spiritual forces take form and become embodied in creatures. From then on, this process enables Love to attain to the condition of a bearer of "knowledge" (Arabî 2002, p. 101), and "the lover/beloved becomes the paradigm of a beauty beyond comprehension" (Arabî 2002, p. 19).

Just as in Sumer, in the *Zohar* the "face-to-face" sexual relationship is exalted. Luria (Yom et al. 2006, p. 77) suspects that this is the basis of a cosmogonic principle of continuity that is accessed through Wisdom, itself a transformative genesis. Ibn 'Arabi sustains this same principle, as he avows that "creation is continuous and occurs at every instant, renewing itself" (Chittick 2003, pp. 52–53). In the case of the Sufi master, this convergence between the masculine and the feminine is the cause of a "constant metamorphosis", an alchemical transmutation of the self; therefore, it is also responsible for the continuous and "endless fluctuations" suffered by the heart (*qalb*) and experienced by "perfect human beings" who are subjects of a "self-revelation that is never repeated" (Chittick 2003, p. 52).

The same can be said of the prodigal compliments and laments uttered by the bride, (the soul) in the *Spiritual Canticle*, where she takes the initiative, just as in the *Song of Songs*. In his commentary to songs 4 and 5 of the *Spiritual Canticle*, St. John of the Cross exalts the divine creative act, telling us, in his exegesis of these poems, about the path to Knowledge, which springs from "a concern for creatures" (Cruz 2002, Cc. 5,1, p. 760). This concern arises once the human being has descended/ascended "to the infernal regions", and has acquired self-discernment, a process of rebirth/re-creation described in the *Ascent to Mount Carmel*. The saint places it above all things, saying "this concern comes first of all, on that spiritual path of a progression toward the knowledge of God" because of "his greatness and excellence" (Cruz 2002, Cc. 5,1, p. 760).

In the *Zohar* as well, both the descent of the divine and the human ascent driven by the desire for union involve passing through several steps, *Sefiroth*, which act as screens or veils of the Light of the Creator. Ascending to know oneself, through those ten steps, during which the obstacles that prevent true *sight* are removed, also implies "being born at every moment of the process". This "birth at every moment", both of God in human creatures and of the creatures in God, is a consequence of the desire for transformation implanted by the Creator in order that the intensive point sheltered in the heart of the desiring self should be unfurled until it attains its goal: the perfect human being, free of restrictions and prejudices, free of self. That is the object of the spiritual path. For all three mystics, it marks the end of all "correction" and re-creation. In the texts under consideration, the "Creation", or rather the "re-creation", *autopoiesis*, is the result of a divine, feminine, and mysterious exhalation through which a willing, speaking, and living God is revealed through theophanies. Accepting this premise leads us to a new "re-enchantment of the world", to a sacrificial dimension that implies the abandonment of self, by the hand of *Sophia*, in order to "die" into the primordial Self. This knowledge implies the perception of a state of equilibrium between the two functions of divinity, which in turn permits all creatures to respond to the soul with sympathy, "overflowing with thanks", and "bearing witness", as St. Augustine says, "to the greatness and excellence of God" (Cruz 2002, Cc. 5, 1, p. 760).

The celestial marriage, the hieros gamos, descended to Earth and denoted the union between the upper and the lower worlds. In this way, two potencies combined to guarantee the order of the city and of the universe in a bi-directional movement of ascent (the human being) and descent (the divine). The driving force of this metaphysical dynamic was desire, brother of death, borne by the goddess who let her heart lead her along a path from which no traveler returns. The poet who speaks to us in the *Descent of Inanna to the Underworld* describes the seven levels of knowledge, which correspond to a host of other dwellings, stations and interior wine-cellars (Cruz 2002, Cc. 26, p. 842) celebrated by our mystics. The first line of the poem *The Descent of Inanna* says: "From the Great Up she directed her mind to the Great Down", and to the question that the poet Dianne Wolkstein asked N. Kramer: "What exactly does 'mind' mean?", he replied, "Ear", adding that in sum, the words "heard" and "wisdom" mean the same thing. "From the Great Up the goddess, 'opened her ear,' the recipient of wisdom, to the Great Down" (Wolkstein 1983, p. 5). Inner hearing and vision, assume the function of senses that contribute knowledge, giving way later on to prophecy and to visionary states. In Sumerian, then, the "ear" is equivalent to Wisdom and this Wisdom is transmitted precisely through "auditions" and "visions", fundamental theosophical devices in the works of Ibn 'Arabi, Moses de León, and St. John of the Cross. The *Song of Songs* recapitulates many of the metaphysical figures and symbols of the Sumerian accounts, full of mystery and seduction, such as the gallantries between the spouses, kisses sweet as honey, references to the orchard, to the bridal bed, and particularly to the initiative assumed by the wife. Today, the influence of this biblical text on the mystical vision of Ibn 'Arabi, Moses de León, and St. John of the Cross is widely acknowledged.[15]

A reaffirmation of the deity's power and her entanglement with the male aspect, topics that will recur in later bridal mysticism, is reflected in the hymn of Inanna and the God of Wisdom, topics that will recur in the later bridal mysticism. Thus, we see that the divine Inanna assumes the role of Queen of Heaven "in equality" with Enki, God of Wisdom, who, having "drunk beer" with her, imparts to the Queen (who is both his wife and his daughter) "the secrets": the seven Mé that generate the order of the creation (Kramer and Wolkstein 1983, pp. 101–3) The "journey" of ascent to Heaven also implies a descent, to the *Abzu*, the lower regions. "Water", as a *source of life*, receives prayer. "Hearing" (Cruz 2002, Cc. 14–15, pp. 793–97) is linked to knowledge. The *"festival"* derived from the "intoxicating union" implies, in turn, the surrender of her crown, the giving up of all her powers: The poem says (Wolkstein 1983, p. 10),

> I, the Queen of Heaven, must visit the God of Wisdom(...)/I must honor Enki, the God of Wisdom, in Eridu./I must say a prayer in the deep fresh waters (Idel 2009, p. 248).» "He, whose ears are widely open, He, who knows the Me, the sacred laws of heaven and earth, Enki, the God of Wisdom, who knows all things (. . .)

As we can see, the goddess assumes a primary role in everything pertaining to Knowledge. The feminine presupposes creative action, dynamism, movement, and the deployment of Wisdom, gifts ceded by the god Enki, husband and father. Similarly, in *Proverbs 8*, 22–30, there is an explicit reference to the timelessness of the feminine aspect of divinity.

> 22 Aren't you calling WISDOM? (. . .)
>
> From the beginning, the Lord possessed me;
>
> from before his works began.

[15] See: Luce López Baralt (2009): Simbología mística musulmana en San Juan de la Cruz y Santa Teresa de Jesús/Luce López Baralt | Biblioteca Virtual Miguel de Cervantes (cervantesvirtual.com);
Simbología mística musulmana en S. Juan de la Cruz, Biblioteca Virtual Miguel de Cervantes, Alicante 2009, p. 3: Referring to San Juan, López Baralt asserts: "In the case of the reformer, there are traces of Castilian poetry, of the Italianizing trend, of popular speech and song, of divine lyric, of the *Song of Songs;* this was demonstrated some time ago by Dámaso Alonso, Father Crisógono, María Rosa Lida, Colin Thompson."
(López Baralt 2020). *Asedios a lo Indecible, S. Juan de la Cruz canta al éxtasis transformante.*
Lucinio Ruano de la Iglesia, Introducción *Obras completas S. Juan de la Cruz*, Biblioteca Autores cristianos, Madrid 2002.
Juli Peradejordi, *El Cantar de los cantares comentado por el Zohar*, Ed. Obelisco S. L., 2015.
The "Shir haShirim," or *Song of Songs*, is a supremely important referent for the Spanish Kabbalah, as indicated in the following quotations from the *Zohar*: "It is a song that contains all of the Torah," and [It is["Kodesh Kodashim," or the Holy of Holies.

30 I was next to you, ordering everything,
dancing joyfully every day,
always enjoying its Presence.

The "idea" of beauty linked to Wisdom exercises a theophanic function, and it is this intuition that gives rise to the Feminine Creator, where the spiritual and the sensitive converge. As it is already well known, these Platonic texts were adopted by Neoplatonism, a philosophical current that permeates the *Tūryuman*, the *Zohar* and also the *Spiritual Canticle*, insofar as it regards the role of divine emanations, which in these texts are linked to the knowledge that Wisdom bestows on the traveler, the seeker.

4. Ibn Al 'Arabî: The Interpreter of Desires: *Tarŷumān al-Ašwāq*

> My heart can assume all forms: a pasture for gazelles, and for the monk a monastery/It is a temple for idols, a Ka'aba for the pilgrim, the Tablets of the Torah and the book of the Qur'an/I follow only the religion of Love, and wherever its camels go I take that way, for Love is my only faith and my only religion. (Arabî 2002, Cas. 11: vss. 13-14-15, p. 125)

Although Ibn al 'Arabi, also known as the "the teacher" and dubbed the "Red Sulphur" [*aš-Šayj al-Akbar wal-l-Kibrit al-Ahmar*], is the author of more than 350 works, one of his many poetical productions deserves separate mention. We refer to *The Interpreter of Desires* [*Tarŷumān al-Ašwāq*], whose intimate congruity with the *Zohar* and the *Spiritual Canticle* lies precisely in the preeminence of a "mystical wedding" in which the feminine aspect of the divinity plays a leading role. Reynold A. Nicholson translated this work into English in 1911. Later, in 1931, the Arabist Miguel Asín Palacios published his *Christianized Islam: a study of Sufism through the works of Abenarabi of Murcia*, which marks the beginning of a consideration of the convergence between *The Interpreter of Desires* and Christian mysticism.

Professor Luce López-Baralt also certainly associates the *Canticle* with the *Interpreter of Desires*.

> Writers like St. John of the Cross and St. Thérèse of Ávila—to mention only the most exalted figures—reveal an astonishing consanguinity: they share many of their symbols and their most important esoteric mystical language with their Middle-Eastern counterparts. From a literary point of view, this is extremely significant, as it implies that numerous references in the vocabulary of these two Christian saints must be sought among the Sufis. We are faced with the phenomenon of a European literature with numerous literary keys that originate in Arab, and even Persian sources. (López Baralt 2009, p. 1)

The *Tarŷumān al-Ašwāq* consists of 61 poems, qasidas [kasîdes], which the author himself deciphered in an interpretative key to a commentary on it [*al-djhajā ir*], where he clarifies some of the symbols and names. Like St. John of the Cross, Ibn 'Arabi performed an exegesis of his own work, which makes it somewhat easier to comprehend. The text, preeminent within the amatory poetry of Muslim spirituality, belongs to the category of epithalamia or love lyrics, part of the cult of beauty. As in the *Song of Songs*—similarly preeminent within Judeo-Christian wedding mysticism—the poem's plot revolves around a journey through complex geographies. It too is full of amorous nostalgia, intense sensations, laments for the flight of the beloved (the beautiful *Nizām*), states of intense dejection caused by separation, fleeting encounters, an urgent and longing search that follows a trail full of premonitory signs: birds, winds, mountains, trees, aromas, shops, fountains, roads, caravans, dunes, mirages, flowers, gazelles, fire, gardens, rain, lightning ... It is within this atmosphere that the celestial archetypes manifest, leading on to Knowledge through Love and desire and impelling the seeker to continue the search. She, the Beautiful One, subjugates the enamored traveler, seducing him through fleeting glimpses. Her elusive Presence is located between light and darkness, leading Her pursuer through a world of contrasts and indistinct appearances:

Even at night, we are next to Her in the light of day/and through the locks of her hair the night turns to day. (Arabî 2002, Cas. XXIX, vol. 8, p. 202)

This poem is as sinuous and difficult as the *Zohar* and *The Spiritual Canticle*, and like them, it reflects the painful soliloquy of an amorous soul, the soul of the pilgrim exiled from himself who wishes to apprehend an indefinable Reality: the Beloved. The text proceeds through a plethora of devices: symbolic codes, gnomic and knowing allusions that are suggestive rather than definitive [*Isāra*]; intimate events expressed through external and literal references [*Ibāra*]; obscure locutions (Arabî 2002, Cas. XXIX-8, p. 22) in which the hidden sense of words [*bâtin*] begets the existence of other, unfamiliar scenarios where words full of subtle nuances [*latīfa*] preside, scenarios that allude to realities [*haqīqa*] even more substantial than those operative in everyday life. All of these factors complicate the poem's interpretation, even for philologists who are experts in the Arabic Science of Letters[16]. It seems appropriate, therefore, to apply a spiritual hermeneutics [*tawîl*] in accordance with the revelations inspired by it. This is how the word acquires an existence beyond language, and the poet, an interpreter of himself, is inevitably overcome by what he intends to say.

The experience of an unsatisfied longing leads the author to describe an intense amatory process, a journey through "a shoreless sea" (Arabî 2002, p. 67), during which a restrained force seems to overflow the words themselves. Still, it is through words that he attempts to reveal the discovery [*kašf*] bestowed by an inner vision [*basira*] conferred by Wisdom [*hikma*], which becomes Presence [*hadra*], thanks to which it is possible to taste [*dawq*] a deep essential spirituality [*rūbāniyya*]. It is this deep spirituality that gives access to the secrets of an enigmatic Reality represented by the figure of a young Persian woman, [*Nizām*], an immaculate virgin. She is a true theophany of the Absolute, a prodigious and beautiful maiden suspended between reality and the interstitial world of creative imagination to which Ibn 'Arabi as the Sage, [*Šayj*] devotes verses of the highest mystical and symbolic tone. It can be deduced from these words that the true spiritual lover loves with both body and spirit, rejoicing in "the love of the seer, not of the blind" (Arabî 2002, p. 34), It follows that this love "with the entire being" embraces "the entire being of the Beloved". In his *Dhakari-al-a'laq* (*Treasure of the Enamored*), an interpretative introduction to the *Interpreter of Desires* sponsored by his followers, Ibn 'Arabi sets out to counteract the implicit danger of a literal, erotic-amorous interpretation of his verses, insisting explicitly (Arabî 2002, p. 99), that the events described were actually experienced. That is, he speaks about an intimate experience replete with such senses as taste [*dawq*], inner vision [*tabassur*], and hearing, associated with the creative word [*kun*].

It is precisely this inner vision that allows the opening [*futuhat*] of gateways to another world, an imaginal universe [*alam al mithāl*] linked to an intermediate plane of reality, the Isthmus, [*barzaj*], a dimension in which there is a convergence of the corporeal senses with that which transcends them. This is also the site of theophanies and revelations, not unlike the *Strange Isles* described by St. Jhon of the Cross. The qasidas radiate an elevated symbolic and gnostic content in which it is necessary to highlight the poet's interpretation of the unique transcendent, personal, and non-transferable experience deriving from a theory of "seeing" Knowledge inspired by theosophical forms [*maz-hir*]. As Henry Corbin maintains (Corbin 1958, p. 13), the poem emphasizes the transmutation of the visible world

[16] Ibn 'Arabi, the supreme master (*al-Shayj al-akbar*) of Sufism, who was influenced by Sahl al-Tustari (818–896), was one of the most the most prominent practitioners of the Science of Letters and dedicated some of his most important texts to it. The Science of Letters or *Ilm al-huruf*, is a traditional Islamic practice centered on meditations on the symbolism of the letters of the Arabic alphabet, considered according to their mystical significance as hermeneutic keys to every manifestation. The Islamic equivalent to the language in the Jewish Kabbalah, it is based on the correspondence between the process of revelation of the holy book and the creation of the cosmos, with the cosmos viewed as a book. The cosmos is revealed and created through the letters of His names, which are the manifestations of His essence, the archetype of all things and the point of return both for the spiritual path and one's individual path. Therefore, the Science of Letters is closely related to metaphysics, cosmology, and hermeneutics, as well as with the hierarchy (*walaya*), spiritual ascensions and magic. Ibn 'Arabi (560/1165-634-1240) dedicated some of his most important texts to the Science of Letters.

See: Mora Fernando, *Ibn Arabî, Vida y enseñanzas del gran místico andalusí*, Ed. Kairós, Barcelona 2011 and Garrido Pilar Clemente, *El Inicio de la Ciencia de las letras en el Islam*, Ed. Alquitara, Sevilla 2010

into symbols. The intuition of "something" occurs in an image that corresponds neither to a universal logic, nor to an entity endowed only with earthly senses. Therefore, the symbolic plane refers us to a differentiated dimension of consciousness; it reflects the cipher to a mystery, the only means of saying what cannot be apprehended or explained. The mystery can only be deciphered, and consequently postulates an esoteric knowledge wholly foreign to ordinary evidence. In these qasidas full of antinomies (Arabî 2002, Cas. XXIX, vs. 7, p. 238), paradoxes, contradictory and oscillating sentiments, Ibn 'Arabi describes a spiritual journey along a path that introduces the traveler step by step to a sacred vision of the cosmos that unifies the multiple dimensions of being in a search for the Eternal Feminine: Wisdom. The tracks, the traces, the signs that the Beloved leaves along the way [*išārat*], are clues that lead the traveler on towards the goal. These will inevitably differ according to the nature and disposition of the pilgrim, who searches for shortcuts and follows the signs without lingering overlong in the successive states [*hal*] and stations [*maqamat*].

The axis of the poem lies in the encounters and dissonances provoked by the fluctuations, absences, and Presences of the Beloved who cannot be seen by human eyes, but beheld only by a supra-sensible apprehension: "Human sight does not reach Him (...)" "It is subtle (...)" (*Koran* 6,103). The mystic is valiantly seeking the encounter with Her, propelled by the desire for Union. When the traveler reaches the corresponding stage along the way, the veils that cover the face of the Beautiful One are lifted to reveal the dawn, warning of the danger of unmediated vision:

> Cast her a glance through that veil/for (the vision) of such splendid beauty is too much to bear (Arabî 2002, Cas. XXIX, vs. 7, p. 238)
>
> When she raises her veil and uncovers her face she dims the brightest lights of morning. (Arabî 2002, Cas. XXVIII, 9, p. 222)
>
> Ah, Moon, who show yourself to us behind a veil/a timid blush on your cheeks!/You veiled yourself, for it would have been torture to behold your face unveiled. (Arabî 2002, Cas. XXV, vss. 5–6, p. 166)

The veils represent the insuperable distances between the diverse planes of Being. Whoever would to travel to "The Dwelling of No-Dwelling", the seventh station, the meeting point, must suffer a symbolic death that presupposes the annihilation of the self [*fanā*] as a step toward knowledge. The Gnostic [*sāalik*] attains to the Lady's subtle Presence through a tenuous mental perception in which the sacred—ever-present to the spirit—appears far away until it irrupts in the moment of theophany [*tayallī ŷalālī*]. Only through the development of the spiritual heart [*qālb*] can the spiritual and material worlds communicate with each other, and the approach toward spiritual knowledge and Wisdom, marked by successive states [*hal*] and stations [*maqām*], takes place within this heart.

Just as in the *Zohar* and the *Spiritual Canticle*, Solomon is a fundamental reference in the *Interpreter of Desires* [*Tarŷumān*], while Wisdom, in the person of *Nizām*, the beloved object of Ibn 'Arabi"s amorous laments, is a replica of Inanna, the Sumerian goddess praised by Akkadian poet Enheduanna in his '*Nin-Me-Sar-Ra*", who is clearly related to Light (Sab. 7, 29–30).

> For she is fairer than the sun, outshines every constellation of the stars/Compared to light, she is first of all/though night supplants light, wickedness does not prevail over Wisdom (Salomón 1992, vss. 29–30).
>
> At the end of the day, the radiant Star, the Great Light that fills the sky, the Lady of the Sunset appears in the heavens (Wolkstein and Kramer 1983, p. 101). (Inanna).
>
> You can see her, walking on crystal floors like a sun that mounts the firmament as far as the abode of Hermes (Arabî 2002, Cas. II, vs. 3, p. 107). (*Nizām*)

The profile of the Lady Wisdom, the object of the search in all three Hispanic spiritual traditions, is remarkably similar in each. She is remote; she inhabits the Mansion of Solitude, drawing near and then withdrawing, seducing the pilgrim by this game. She is elusive and majestic, demanding that her suitor await her patiently. Full of mystery, She is light and yet loves darkness and is sometimes deadly, as we will see in the next qasida where Ibn 'Arabi frames the description in this way: "You strike like Apollo's lightning." Her gaze is

> The murderous look that kills. The sun of the highest sphere in the bosom of Hermes. It is the Torah, the splendor of Beauty. It is savage, and in its Presence there is no rest. It maddened the sages of our religion, just as it did the singers of the Psalms of David, the Christian monks, and the followers of the Torah. (Arabî 2002, Cas. II, 4,5,7,8, p. 107)

As Henry Corbin says, *Nizām* is invested with the theophanic function of Beauty, and from this apprehension there arises the idea of the Feminine-Creator not only as an object, but as an exemplary image of the sympathetic *devotio* of the "faithful lover" [fideli d'Amore] (Corbin 1958, p. 24). She truly possesses the secret of divinity [*sirr al rububiya*], and she awaits the greeting of a lover maddened by nostalgia. She is the slender maiden whose smile radiates the *splendor of the sun* ((Arabî 2002, Cas. III-6, p. 111). Ibn Arabî speaks to us of an "orphan and exiled love", "besieged by desires and pursued by swift arrows" (Arabî 2002, Cas. IV-4, p. 113). There is no place for harmony in the time of rupture, and due to mis-encounters, the mountain and the riverbank diverge in a separation that only exacerbates the search for Union [*tawhid*].

In a verse from sura Fussilat, it is said, "We will make them see Our signs on the horizon and in their souls" (41:53). The meaning of the statement for the Sheikh Al-Akbar is explained in a chapter where he deals with *sakina* (peace, "serenity", but also, like the Hebrew *Shekhinah*, the divine "Presence"): "The *sakina* that God sent down for the children of Israel in the Ark of the Covenant [Qu'ran 2:248]" "was made to come down in the hearts of believers of Muhammad's community [Qu'ran 3:110] ... " (Chodkiewicz 1993, p. 96).

The poetic text tells us of a heart woun*ded* by the grief of absence that wishes to die in the gaze of its Beloved. There are symbolic references common to both Ibn 'Arabi's *Interpreter of Desires* [*Tarŷumān*] and Moses de León's *Zohar*, as well as St. John of the Cross' *Spiritual Canticle* and *Living Flame of Love*. One of them is the significant Presence of water, as in "To the Water" (Arabî 2002, Cas. XIII, 2, p. 130), where Ibn 'Arabi refers specifically to the sacred well of *Zamzam*, near the Ka'aba in Mecca, the meandering streams, the rain, the Living Source (Arabî 2002, Cas. XIII-2, p. 130) the "fire".

> Whoever ignites this fire, beware!/This fire of passion belongs to you/take some of its flames too. (Arabî 2002, Cas. VIII, p. 120)

He also tells us of the thickets, the breeze that spreads the fragrant odors, of "Love" among the flowers of the Garden (Arabî 2002, Cas. VIII, p. 121) of the "Ringdove" (Arabî 2002, Cas. XI: vs.1, p. 124), of the thirst for Love, of the "veiled gazelles" whose eyes send signals/who pasture in the breast" (Arabî 2002, Cas. XI, vs. 1, p. 124), of the *arāka* and *bān* trees (Arabî 2002, Cas. XI, vs. 12, p. 125). It is worth noting, as well, an erotic reference to hair (Arabî 2002, Cas. VII, 9, p. 118):

> When they feel fear they let fall their hair/and with their tresses steal away into the darkness.

In the *Zohar* and in the *Spiritual Canticle*, just as in the Sumerian texts, mystical terms refer to the erotic qualities of human life. The instinctive engine that drives the journey is a relentless desire. The mystical itinerary speaks to us of an incarnation of divinity that becomes palpable in all things. The image of the Beloved cannot be captured; it can be seen only in an epiphany, yet it is reflected in nature and in the soul as in a mirror. Language, the mediator between the two planes, is situated on that borderline where the body ends and the spirit reveals itself. The Beloved is the paradigm of intangible beauty and wisdom. The poet must travel far along the path that leads to Knowledge, a trajectory that must be read in terms of Presence/absence (Arabî 2002, p. 19) [*hadra/gayba*]. The absence of love is fatal, and once the encounter takes place, the heaviest burden of love becomes easy to bear. As in the *Spiritual Canticle*, the poet in love "dies of love and melts like snow" (Arabî 2002, Cas. XV, vs. 2, p. 135).

For Ibn 'Arabi, all Creation is essentially a theophany [*tajallî*], and as such is an act of the divine power of imagination. Accordingly, the organ of active human imagination is identical to the organ of the absolute theophanic imagination itself. The creative act proceeds from the primordial sadness and loneliness of an isolated divine being that impel Him to make Himself manifest in human beings who, in turn, reveal Him to Himself. Divine compassion leads Him to reveal Himself to the creatures through which His Names and attributes will operate and testify to his creative ability and supreme Wisdom. This epiphany is evident in the passage from the state of concealment, of implicit potency, to the luminous, externalized and revealed state (Corbin 1958, p. 139). In short, life results from the dynamism linked to a recurring creation, renewed at every moment and resulting from an incessant theophanic imagination projected through a succession of theophanies [*tajalliyât*], thanks to which beings are in continuous ascent. The names hidden in the Supreme (the "beautiful noble ones" referred to in sura 59:24), longed to manifest themselves. In a *hadith*, Allah declares: "I was a hidden treasure and I loved to be known; so I created the creatures ... "

Ibn 'Arabi holds that divine compassion embraces the God created within the creeds (Corbin 1958, p. 91). The Gnostic, however, is faithful to his own Lord, to the divine name with which he is invested. Therefore, each form or path implies a link to a certain phenomenology of prayer [*dhikr*] that refers to the emergence of the invisible being [*bâtin*] and its eternal individuality, as manifested through the compassionate and merciful breath in the Worlds of Divinity and Humanity: *lâhût* and *nâsût*. Each being is an epiphanic form [*mazhar*] of the divine being, his Lord, who manifests himself in a creature under one or more divine names (Corbin 1958, p. 93). "The Supreme Secret" has two aspects: The Name manifests thanks to the servant, who is the fulfillment of that *pathos*. There is, therefore, necessarily a correspondence between the Divine and His creature. This covenant of sympathy exists from pre-eternity, so that once beings come into existence, God praises Himself in all creatures; they are His theophanies. Consequently, the life of the mystic tends to realize this union of *simpatheia* by which God establishes a dialogue with himself in terms of Love. Ibn 'Arabi describes this beautifully when he says, "All creatures are wedding beds where God manifests Himself". Furthermore, the Andalusian master does not shrink from asserting something that may sound heretical to the ears of the jurist: God is present only insofar as he is recognized (an assertion difficult to reconcile with absolute divine sufficiency). That is the magic of creation. It is precisely such participation that potentiates the perfect human being, *Insam al Kamil*, who bears all the Names of God and realizes them harmoniously. This concept is also found in Jewish tradition, as well as in Fray Luis de León.

5. The Light of the Zohar: The Book of Splendor

> With the help of the Zohar we will emerge from exile (89–90): "The wise will shine like the splendor of the firmament" (Laitman 2015, p. 235)

The *Sefer ha-Zohar* is one of the most representative texts of the Jewish Kabbalah. Like the much older *Sefer ha-Yetzirah* (the Book of Creation, or alternatively, of Formation), the text applies a metaphysical-poetic hermeneutics of the first moments of creation to the accounts found in the Torah, where the erotic echoes of the *Song of Songs* resound:

> The flowers are blooming, the season of the singing birds has come, and the lullaby of turtledoves fills the air. The fig trees begin to form their fruit, and the fragrant vines are in bloom.
>
> Rise up, my beloved! Come with me, my beautiful woman! (*Song of Songs* 2:8–13)

The *Sefer Ha-Zohar*[17], hereinafter the *Zohar*, the greatest recopilation of the Hispanic Kabbalah, is transmitted in a theosophical-theurgic treatise that advocates a mystical path of knowledge and action whose objective is to describe the various manifestations of the God of Israel as implied in the revealed wisdom of the Torah. Not surprisingly, the *Zohar*, as a kabbalistic text, was long regarded as a cryptic and unapproachable collection of secret knowledge, lacking a necessary exegesis and reserved solely for a minority of rabbinical scholars. It is unanimously acknowledged as "the deepest, darkest, most mysterious, and the principal work of all the books of the Kabbalah" (Laitman 2011, p. 16).

There has been some controversy over the origins of the work, but Gershom Scholem (Wolfson 2001, p. 171), Elliot R. Wolfson, and Moshe Idel, after a rigorous and methodical examination of the historical and philological elements, adhere to Isaac of Acre's theory, which ascribes the work to Moshe ben'em Tob of León. Known to us as Moses de León, he was an outstanding figure within the world of Hispanic Sephardism. Most probably born in León, he lived in Guadalajara and died in Ávila, where some researchers believe that part of the *Zohar* was written. In Abraham Zacutor's *Sefer Yuhasim*, Isaac of Acre (Wolfson 2001, p. 171) attests that the work was written by Moses de León. For his part, Yehuda Liebes (Wolfson 2001, p 173) promotes a widely accepted theory that attributes the work not to a single author, but to a mystical group or fraternity operating in Castile under the direction of Moses de León. This group of rabbis and leading Kabbalists would include Joseph Gikatilla (Yosef Chiquitillia) and Abraham Abulafia himself.

Whether the product of a single author or a group of mystics, most researchers agree that the book was written in Castile, and some, like Yitzahak Baer (Wolfson 2001, p. 174), further assert that it is a reflection of the Castilian society of the time. The *Zohar*, in this view, would be a record of the actual experiences of Jews living in that complex, multi-cultural Spain. Gershom Scholem takes a similar and perhaps more extreme position, maintaining that the authors of the *Zohar* describe, under a kabbalistic interpretation, a compendium of the habits and customs of the Jewish community in 13th-century Castile (Wolfson 2001, p. 174).

The work purports to consist of the teachings given by R. Simeon ben Yohai in the second century C.E. to a group of his followers hiding in a cave in Galilee during the Roman occupation. Its structure consists of sections and imaginative stories that interpret the Torah. As previously noted, researchers like Yitzahak Baer and Gershom Scholem regard these stories as reflections of the Jewish life of the time, with Scholem claiming that Moses de Leon dressed his interpretation of Judaism in "archaic robes" to make it more acceptable, while Isaac of Acre credits him with having, in a sense, turned the story into a novel to assure that it would be more widely read.

It is worth noting that Michel Laitman is a principal dissenter who prefers to ignore the prodigious and unique Castilian contribution and attributes the *Zohar* to the author identified as Simeon ben Yohai, declaring that the text remained concealed until "at the right time" (somewhere between 1930 and 1940), when the world was prepared to receive the teachings and his own teacher, Rabbi Yehuda Aslagh, composed the exegetical commentary *Ha Sulam* ("The Staircase").

In order to penetrate the meaning of the stories related in the *Zohar*, one must remember that in it the Torah is perceived as a game of enigmatic puzzles through which the

[17] Ariel Bension, may be described as the last of the great Jewish Sufis. That is, he was the last of the Kabbalists who was also thoroughly learned in and sympathetic to *tasawwuf*, and who wrote with deep perception on outstanding Muslim Sufis, as well as on the Kabbalists influenced by Sufism. The most important work of Rabbi Bension is *The Zohar in Moslem and Christian Spain* (First edition in English: London, G. Routledge, 1932; second edition, New York, Sepher-Hermon Press, 1974. Translations apparently exist only in Spanish and Portuguese.)—written in English with its title referring to the preeminent classic of *Kabbalah*. *Zohar*, meaning "*Splendor*," is known in Arabic as *Kitab Al-Zawhar* or *Kitab Al-Zuhar*.
The distinguished 20th C Spanish philosopher Miguel de Unamuno, in his prologue to the 1934 Madrid edition of Bension's work, compared the *Zohar* with the Castilian classic, *Don Quijote*, thus underscoring its fundamental literary character. (Miguel de Unamuno, "Prólogo", in Bension, Dr. Ariel, *El Zohar en la España Musulmana y Cristiana*, Ediciones Nuestra Raza, Madrid, 1934, p. 13).
The brilliance of Rabbi Bension's commentary on the *Zohar* and its relationship to *tasawwuf* as well as to Christian spiritual traditions resulted in his election to the Royal Academy of History in Spain, and his book is cited in the bibliography of the most significant work of Jewish metaphysical historiography, *Major Trends in Jewish Mysticism* by Gershom Scholem. Before publishing his book on the *Zohar*, Rabbi Bension issued a work in Hebrew in Germany in 1925, deploring the decline of Kabbalah as he perceived it.

Creator speaks to His people by means of an esoteric numerology of letters, Gematria, in which the secrets of life and of the Universe are first concealed and then gradually revealed by the ascending knowledge of the emanations of divine Wisdom: the *Sefiroth*, which shape de Tree of Life[18] (see Figure 1). Any analysis of the work requires the reader to implement a poetic and symbolic hermeneutics and apply them to the elucidations of the Torah performed by the Kabbalists around Moshe de León. The truth is that the "Book of Splendor" speaks to us of a resplendent reality linked to the mystery of divine Wisdom (*Sod Hokmah 'Ĕlohit*) and to an enlightening gnosis that aims to decipher the most sacred of the divine names: YHWH, the Unity of God. The mystic who "receives" enlightenment, *Maśkil*, understands with "the eye of the intellect" (*'En Ha-Ŝekel*) thanks to a noetic, intuitive, and non-discursive knowledge. This sense of mystical "vision" is as fundamental to Hebrew mysticism as to the corresponding Sufi or Christian version, a fact emphasized by Fabio Samuél Esquenazi (Esquenazi 2020, p. 417 and ss.). Specifically, in the *Zohar* the ecstatic experience is associated with divine powers that are visualized as refulgent letters inscribed in a book written by God, the Torah, identified with the Tetragram.

Wolfson (2001, p. 165), considers that within the minority group in which the *Zohar* was born, this mystical interpretation of letters bears an unmistakable imprint of Joseph Gikatilla.[19].

Concerning the wisdom of the Kabbalah, Wolfson suggests that it may be a theosophical application of what was originally a philosophical concept (Wolfson 2001, p. 169). As Wolfson himself indicates, although Moses de León began as a follower of Maimonides, and particularly his *Guide for the Perplexed* (an antithesis to the *Zohar*[20]), he subsequently evolved through a process of ecstatic contemplation. This process took the form of the emanating images that shape The Tree of Life, a diagrammatic ordering in which the feminine and masculine components complement each other, constituting a complex geometry through which the sacred manifests in the ten *Sefiroth*, or planes of being. These *Sefiroth* are polar powers derived from the sacred tetragram YHWH, and their interplay represents the dynamic and bi-directional flow of the original principle whose objective is to make the Primordial Man, the *Adam Qadmon*, sprout from the depths of each soul. In this way, the *Zohar* provides a guide to the spiritual path, a map condensed in the Tree of Life, in which Reality is structured in terms of descending spiritual worlds.

The *Tree of Life* also corresponds to an anthropomorphic image of the relationship between divine energies. Not in vain "God created man in His image and likeness", and

18 Rabbi Simeon taught many forms of meditation on the spheres. One example of the multiple possibilities he expounded was to imagine the attributes as a series of dancing lights against the branches of the tree. Closing his eyes, the Kabbalist visualized.
Epstein, Perle, *KABBALAH: The Way of the Jewish Mystic*, Shambhala, Boston & London, 1988. pp. 57–58.

19 Joseph Gikatilla (Yosef Chiquitilla) was originally a disciple of Abulafia; together, the two practiced linguistic Kabbalah, employing gematria, notorikon and *temurah*, that is, the Hebrew meditation known as *Tseruf*. Since the letters of the Hebrew alphabet—like those of the Greek—are assigned different numerations that also represent quantities with their consequent inter-relationships, they open a wide gamut of possibilities to the practitioner. This type of speculation also characterizes much of Ramon Llull's study of permutations in *Ars Magna*, also known as *Ars Combinatoria*, although that does not necessarily imply the direct influence of Gikatilla, or even of Abulafia's methods.
Gikatilla and Moses de León, in contrast, placed less emphasis on combinations and calculation; instead they applied the linguistic Kabbalah to the theory of emanations, or *Sefiroth*. Thus, they were able to enter the field of theosophy and immerse themselves in cosmogony and the emanations, or Divine Names, according to the school of Provence and Gerona.
In the third quarter of the XIII century, Gikatilla wrote an extraordinary treatise entitled *Gates of Light* (*Sha'are' Orah*) in which he ranks the *Sefiroth* in inverse order to the previously prevailing arrangement. Almost all of the many earlier texts described them as emanating from the Supreme Principle in several equally sacred stages culminating in the *Queen-Bride*, that is, *Malkuth*, the universal receptacle. In opting for an ascending pathway from *Malkut* to *Keter*, Gikatilla inverted this order.
See R. Yosef Chiquitilla, *El Secreto de la Unión de David y Betsabé*, Introduction, translation, notes and the the Hebrew test by Charles Mopsik, Riopiedras Eds. Barcelona 1996.
Rabbi Yoseph Gikatilla, *Gates of Light (Sha ́are Orah)* translation by Avi Weinstein, Harper Collins Publishers, New York 1994

20 Maimonides' *Guide for the Perplexed* is based on medieval aristotelianism, tinged with Neoplatonism and with Seneca's Stoic philosophy. It supports creationism and searches for logical Unity, maintaining that reason and revelation coincide. A philosophical and moral guide, it accepts Aristotle's metaphysics, though not his method. The *Zohar*, in contrast, is an esoteric kabbalistic, and symbolic text that employs a hermeneutical method to interpret the *Torah*, and is therefore alien to aristotelian logic. Neither systematic nor philosophical, it relies on a poetic, eccentric, mystical, and intuitive rationality unrelated to the methods of medieval aristotelian rational philosophy. In the Kabbalah, biblical texts are analyzed and interpreted with a view to extracting their occult meaning.

the Hebrew letters form a model equivalent to the structure of a human body, which is identified as the actual celestial metaphysical order, seen through its reflection in a mirror. Thus, the human representation can be understood only in the light of the divine, composed of the letters of the Hebrew alphabet summarized in the *Tetragram*, a theonym, or proper name of God, the root of the mystical language and essence of the Torah.

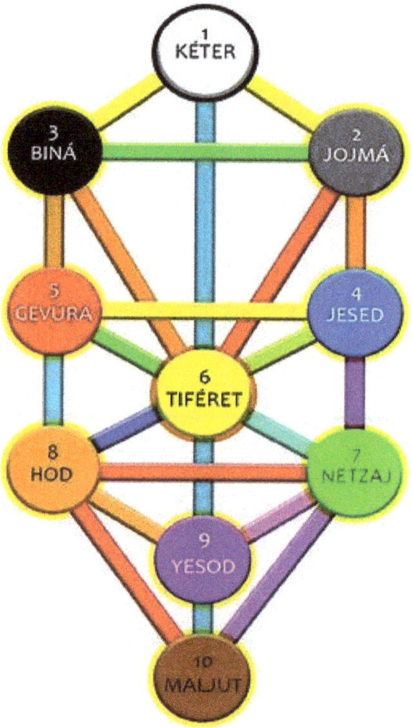

Figure 1. The tree of life.

Aritmosophy is essential to Zoharic rhetoric, just as it is to Sufism. In this sense, we refer to the connotation of the very term *Sefiroth*, ($s_e$far), "numbers", referring to the separate intellects and linked to each of the celestial spheres (Wolfson 2001, p. 167). The word can also represent *sappir* (luminosity, sapphire) and *sefer* (speech, book). Wolfson explains (Wolfson 2001, p. 176) that "no word in English or Spanish can account for all the semantic connotations associated with the term *Sefiroth*".

The group of planes or *Sefirah* that Laitman denominates as ZA[21], *Zeir Anpin*, (Laitman 2011, p. 31), are located between the upper spheres *Keter*, *Hokmah* and *Binah;* while the lower position, *Malkuth*, is united around the Balanced Beauty of *Tiferet* to *Yesod* (ego sphere), *Hod* (mental sphere), *Netzah* (emotional sphere), *Yesod* (compassion) and *Gevurah* (severity). As the heart, *Tiferet* occupies a central place in the Tree of Life, replica of the human being. It is in *Tiferet* that the spiritual correction of the process occurs, and thus it is the culmination of the lower *Sefiroth*.

Therefore, taking ZA into account, the *Sefiroth* can be reduced to five. *Keter* (the Crown, first *Sefirah*) represents the pure potential of any manifestation. Prior to all mental processes, it cannot be captured by thought, and it occupies a central position in the Tree. *Hokmah* (Unlimited *Wisdom*) is the second *Sefirah* and the first intermediate step between *Keter*, the Crown, and the other *Sefiroth*. It represents the primordial idea. *Binah* (black as the womb

21 *Malkuth* can realize a *zivug* (coitus, sexual union) with ZA and thus receive the Light of *Hokmah*.

where life thrives) is the rational process and the limitation of *Hokmah*. ZA represents the meeting of the five *Sefiroth* previously mentioned in *Tiferet* with *Malkuth*. The last of the 10 *Sefiroth*, *Malkuth* symbolizes the desire to receive for oneself. It is the lowest layer of manifestation and corresponds to the human being). All the *Sefiroth* are interconnected, and each of the 10 *Sefirah* includes all the others within itself. Each, in turn, consists of ten sub-*Sefiroth*.

The process of knowledge is as follows: ZA requests the Light for *Malkuth* from *Binah*, who receives it from *Hokmah*, the depository of the Creator's Light; and subsequently "the opening of the eyes" occurs in *Malkuth*. During this entire process, it is imperative to "listen" to action, and then "the eyes" will rise from *Malkuth* to *Hokmah*, in a return to the Creator. The "One who opens the eyes" is MI, and above Him no questions remain. In this process *Hokmah*, the seed of creation, acts through *Binah* which is placed under *Hokmah* and transfers the Light to *Malkuth*, the feminine recipient. *Malkuth* must rise up to *Binah*, who receives the Light of *Hokmah*. Once *Malkuth* (thanks to the fact that *Keter* and *Hokmah* descend toward her) "becomes equal to" her husband with "the opening of the eyes", she rises beside Him to AVI (*Hokmah* and *Binah*), where the spouses are enveloped in its Light.

As we have previously indicated and as we will later see, the erotic echo of the *Song of Songs* continues to resonate through the text attributed to the circle of Moses de León. Thus, we can affirm that the axis of the *Zohar* is articulated around a polar structure of the single divine principle in the form of a couple. This polarity implies a theosophical sensuality in which the feminine power acquires supreme relevance, as a luminous beacon and complementary creator of the unreachable and unspeakable essence. It is this that allows the human soul redeemed from exile and vivified by the Light of Wisdom, after a relentless search, to fully adhere to the incomprehensible primal Force,

Under the auspices of the two polarities, male and female, and full of encounters and dis-encounters, laments, compliments, and seductions, the search is marked by the tireless and desiring vivacity of the spiritual Sapience, the Divine Presence or Glory of God: the *Shekhinah*. She loves secrecy, enigma, and concealment, as can be seen in the discourse that Simeon ben Yohai develops from the parable of the princesses (Ex. 21–24), the principal passage of the Zoharic story of the *Princess* who is identified with the *Torah*, pure love.

> The *Torah* can be compared to a beautiful and majestic maiden who is held in an isolated bedroom of the palace, and has a lover of whose existence only she is aware. For her love, he continually passes by the bars and looks around in all directions to discover her.
>
> She knows very well that he is forever hanging around the palace and what does she do about it? She opens wide a small door in her secret bedroom, for an instant reveals her face to the lover, and then quickly retreats. Only he, no one else, is aware of this; but he knows that it is out of love that she has revealed herself to him, and his heart, soul and everything within him are directed towards her. (Sholem n.d., p. 30)

A striking characteristic of this parable is the way in which it addresses themes identical to those of Ibn 'Arabi and Saint John of the Cross, two other Hispanic mystics, as well as the frequency of symbols common to all three: signs, messengers, locutions behind the veil, and glimpses. In all of them, there is a traveler who gradually progresses along a journey motivated by longing, and as he progresses, a "maiden" appears, hinting at some of the secrets behind the veil. The Zoharic story speaks of a beautiful princess (*reto behimatá* = beloved) who, concealed in her palace, cries out for the love of her lover (*rehimáh*) and in her lament brandishes her seductive weapons. Her goal: a union in marriage. This account, replete with a powerful and archaic symbolic density, emphasizes the game of concealment and incitement that constitutes the axis of the "journey" in both Ibn 'Arabi's *The Interpreter of Desires* and the *Spiritual Canticle* of St. John of the Cross.

In this sense, Fabio Samuel Esquenazi[22] conducts an extremely refined hermeneutics of the *Zohar*. In lingering over a particularly enigmatic *Zoharic* story, the "Maiden without Eyes" (the Torah in this example) he refers us to Is. 43,4. The exegete is the sage "full of eyes" (Esquenazi 2020, p. 432) who knows how to decipher the riddles of a serpent that flies through the air with an ant lodged in its mouth; of an eagle that inhabits a tree that never existed, from whose nest were stolen eaglets as yet unborn, and which remain, still uncreated, somewhere; and the enigma of the maiden without eyes, whose body is both hidden and revealed, who emerges in the morning, is in hiding during the day, and is adorned with non-existent ornaments. The ancient rabbis considered the Torah to be the *Shekhinah* herself, the manifestation of Yahweh's Presence in the world, not only as the wife, bride, or daughter sent into exile by the King in the context of the Jewish symbology of exile, but as a maiden who "has no eyes", or has cried her eyes out on account of the loss of her homeland (Esquenazi 2020, p. 432). This celestial sexual polarity in which the metaphysical senses of sight and hearing become more acute, is in all these terms consubstantial with the human being. As a reflection of the supernal entity, and like Her, the human being conjugates within itself both the masculine and feminine.

Consequently, our world as a whole is a mirror, a reflection of what occurs in metaphysical realms, in boundless spaces in which the entire existing universe is combined in pairs, as implied by the Pythagoreans with their list of opposites. That said, the whole of the *Zohar* can be seen as a wager on the Union of those opposites without synthesis or annulment of the extremes; it envisions the coupling of different elements in the One. To know oneself involves discovering the complexity and concurrence of opposites within an identity that is assumed as one's own, and to this end, metaphysical vision and hearing are activated.

The *Zohar* says:

> Man is the inclusion of masculine and feminine, for he in whom masculine and feminine are joined is called, "Adam" and then worships God. Moreover, there is humility in him. And even more, there is mercy in him. (Laitman 2015, p. 221)

Upon the arrival of the Messiah (the Union in Light, or the reception of supreme Knowledge), humanity, the Eyeless Maiden, will abandon spiritual exile, which is linked to the self and to the destruction of the Temple. Access to transcendent knowledge requires irradiation and luminous reception in each of the souls that act as "vessels" brimming with the desire to receive: that is what the term *Kabbalah* means.

Thus, the content of the work deals with the loving relationship of the human being with his Creator, through the mediation of his wife, who assumes various identities: Wisdom; *Shekhinah* (Presence of the Glory of God); *Binah* (understanding and delimitation of the Light); *Keneset Yiśra'el* (The Community of Israel); and *Malkuth* (the human being). The feminine, Wisdom, is the Creator of life both in heaven and on earth and in the *Zohar* the earthly is a reflection of the heavenly. The heavenly copulation (*Zivug*) means the union of the supernal spiritual male and female, from Light to Light. (*Zohar*, 201): "The union of the sexes in this world will be from body to body, and the righteous who follow the right path "are rewarded with the pleasures of that world" (Laitman 2015, p. 219). (*Zeir Anpin*, the masculine, joins *Nukva*, the feminine = *Malkuth*, once it has descended to the last step.)

Ishah, woman, means *Esh*, fire of the Creator, *Alef-Shin*, and this fire is connected with the letter *Hey*, which is *Nukva* (Laitman 2015, p. 216), the feminine, plenitude of the left enlightenment due to the irradiation of the male *Sefiroth Hokmah*. Thus, the feminine receives the "luminosity of *Hokmah*" (*Or Hokmah Hasidim*), and from that "reception" (*Lekabel=Kabalha*) arises a similarity (Laitman 2011, p. 107) to the Creator, derived from the enjoyment of pleasure in the Union. "And the Light of the Creator will turn to fire" (Laitman 2015, p. 216).

22 Esquenazi, Fabio Samuel "Ya bien puedes mirarme después que me miraste". "Isomorfismo de la visualización divina en la parábola de la hermosa doncella sin ojos de *Zohar*" II, 94b–95a and in stanza 24 of St. John's *Spiritual Canticle*" http://www.2010.cil.filo.uba.ar/sites/2010.cil.filo.uba.ar/files/186.Esquenazi.pdf (accessed on 2 September 2020). III World Congress *Cántico Espiritual*, CITeS, Ávila, Sept. 2019.

Zohar 146 states that when the *Sefiroth Sefira Binah*, located on the left, which represents the understanding that comes out of Eden, and *Hokmah*, the Light of Wisdom on the right, come together under *Keter*, the Crown of the Tree of Life, they adhere to the world of *Malkuth* in Love. Therefore, the Congregation of Israel, the *Nukva*,[23] the wife, appears in the *Zohar* (Laitman 2015, p. 207) as the bride of the *Song of Songs*. "Thus, the Congregation of Israel said, set me as a seal upon your heart (*Zohar* 731) because "Love is strong as death" (*Zohar* 732) (Laitman 2015, p. 207).

Another line from the *Song of Songs* (1, 2) collected in the *Zohar* (371), says: "He will kiss me with the kisses of his mouth" (Laitman 2015, p. 204). Laitman/Aslach (Laitman 2015, p. 204) interprets this verse[24] as King Solomon's words expressing the love between the higher world, *Zeir Anpin*, and the lower world (ZON) composed of all the *Sefiroths* linked to *Tiferet* and the last *Sefiroth* of the Tree of Life, *Malkuth*. One spirit adheres to another with a "kiss" and "the kiss on the mouth is the entrance and exit of the Spirit". When the kiss occurs, the Union is sealed.

The wife is the essence of the house, (*Zohar*, 231) and when "the house is corrected", the masculine and feminine are also corrected, and then the masculine comes to the *Nukva*, and they join together (Laitman 2015, p. 216). Laitman/Aslagh (Laitman 2015, p. 205) refers to another fragment of the *Zohar* (70): "See life with the woman you love", in which "the woman you love" denotes the Congregation of Israel, the *Nukva*, because love is written about her; and it is linked to Mercy, the *sefiroth Yesed*, on the right side of the Tree of Life. The Assembly of Israel assumes the role of wife of the divinity, just as the Church does when it becomes the bride of Christ. However, when the community disobeys the Creator's precept, the *Shekhinah* retires, and the exile and rigor of the masculine prevail over Mercy.

In this respect, there is a curious paragraph (327) in the *Zohar* referring to women. When Israel deviates from its path, the prophet says,

> Indolent women, how can they be calm? Why do they remain seated and do not wake up the world? Stand up and rule over men. (Laitman 2015, p. 218)

A further mention of the feminine souls in the *Zohar* (195–202) describes six Palaces of feminine souls and four others that are hidden (*Zohar* 201): the ten *Sefiroth*. Every midnight all the Palaces, both heavenly and terrestrial, are gathered together (*Zohar*, 201) in the heavenly and terrestrial coupling (*Zivug*):

> I was shown six palaces with various pleasures and delights in the place where the curtain is displayed in the garden, since no man may enter beyond that curtain (...) There are four hidden palaces of the holy mothers that were not taken into exile. Every day they are alone. (Laitman 2015, pp. 218–19)

The *Zohar* (145–147) also reproduces the Palace of Love (Laitman 2015, p. 205): the Temple, Paradise, the Garden, the *PaRDéS*. Everything is founded on amorous coupling. When the higher, *Zeir Anpin*, and the lower, *Malkuth*, adhere to each other, they do so in love as "the groom and the bride" (Laitman 2015, p. 205). Thus, it is within that coupling that the connective game of the Hebrew letters composing the theonym YHWH (*Yod, Hey*, and *Vav*) is created through Love (*Zohar*, 147).

As we can see, the awakening of desire and passion permeates many of the stories that compose the *Zohar*, which continuously refers to the secrets of the Wisdom of the Hidden, inevitably surrounded by a halo of mystery in which Love is the force that drives the progress toward Knowledge and the consequent annulment of the individual. Since

23 "When the *Zohar* speaks of the *Shekhinah* as feminine—it quite frequently uses the term *'alma-de-nukva*, "the world of the female" in this connection—this is more than a mere circumlocution for the passive and receptive elements among the divine attributes".
" ... for him [the author of the Zohar], as for other Kabbalists, the *Shekhinah* is regarded as the "celestial Donna;" (*ha-isha ha-'elyonah;* cf.II, 54b) or the "Woman of Light"(*iteta da-nehora)*, in whose mystery are rooted all the females in the earthly world. In brief, she is the eternal feminine".
"But the feminine quality of the *Shekhinah* is understood, first and foremost, and emphatically, in her role as female partner in the sexual union, *zivuga kaddisha*, whereby the unity of the divine potencies is realized through the union of male and female".
See (Sholem 1991, chp. 4).

24 Ibid., p. 204.

Love signifies death to mundane reality, it is rebirth to another life; it "opens the eyes" while at the same time blinding the eyes of reason.

The *Zohar* also speaks of rupture, of heartbreak (344), of the exile of the *Shekhinah*, Wisdom, the Glory of God, on the day the Temple (Paradise) was destroyed and Israel went into exile. The proverb of the King and the Queen (Laitman 2015, p. 229) (*Zohar* 78–81) deals with the King's anger at his queen, whom He expelled from his palace for having distanced herself from Him. Nevertheless, in His desire for union, the King sought her out, lifted her up, and took her back to His palace, swearing that He would never part with her again. Thus, the Congregation of Israel awaits the wife, so that He will come and take her by the hand, lifting her up and reconciling with her.

In clear correspondence with the Christian gnosticism of the first centuries it says

> "The divinity was expelled from the King's house". "When she descended she saw that her abode had been destroyed (...) and the sacred Palace and house burned by fire" (Zohar 345). "And then she raised her voice in lament (. . .) because the holy King ascended and is not there" (Laitman 2015, pp. 224–25). All the passion of the wife, Israel, is concentrated in a desire to keep the *Shekhinah* from leaving, wishing that instead she run like the "deer or a little antelope", turning her head back towards the place where she was before: in the heart of His servant.

The *Shehkinah* is also the place where the mystic experiences the Creator. She is associated with the Rose, in which Elohim is present in the 13 petals (13 attributes, 13 words) that encircle and protect the assembly of Israel, and in five sepals (equivalent to so many other "doors" and to the five fingers that raise the cup of blessing: The Rose). In another pre-Zohar text, *'Orzaru'a*, "Light Sown" referred to by Wolfson (2001, p. 168), Moses de León highlights the word *Ehad*, which refers precisely, though cryptically, to those thirteen attributes:

> The secret of the Kabbalah is explained by another secret, that of the matter of the thirteen attributes, thirteen letters (6,4) and one left over, *'Alef*, which is the number One, which corresponds to the One, *'Ehad'*. (Wolfson 2001, p. 169)

He, the bridegroom, is the Secret (Laitman 2011, p. 41), but who is He? The one who "opens the eyes". MI = Who[25] has created this and is found along a hidden path and does not reveal himself and receives the name MI, the extreme end of heaven. The human being in the last degree, *Malkuth*, asks: MA = What are the heavens and the created earth, that is the testimony that lies in the wife, Jerusalem, the quintessence of beauty. So MI is the upper end of the sky, as MA is the lower, to which the Light (*Or*) of *Hokmah*, which illuminates *Binah*, descends, in order that Wisdom may penetrate *Malkuth*.

> A deep structure arose from a thought (a point that ascended, when the occult wished to reveal itself) and is called MI, principle of the erected structure, reserved in the most unfathomable of the Names, the Tetragram. He wished to reveal himself: ÉLEH: ELOHIM, that is the name that ascended. (Laitman 2011, p. 42)

The Rose (*Malkuth*), which is the assembly of Israel, rests upon and is sustained by the union of these letters: ÉLEH and MI.

> I am a rose of Sharon, a lily of the valleys" [*Song* 2:1] (...) the Community of Israel is called the rose of Sharon because its desire is to be watered by the deep spring that is the source of all spiritual rivers, it is called the lily of the valleys. And also because it is in the deepest place it is called lily of the valleys. At first, she is a rose with yellowish petals, and then a lily of two colors, white and red, a six-petalled lily that changes from one hue to another. She is called "pink" when she is about to meet the King, and after she has joined him in their kisses, she is called "lily". (Sholem n.d., p. 41)

[25] As for the interpretation of the Zoharic text, specifically of the *Tree of Life*, we continue Laitman's exegeses, supported by Yehuda Aslagh and reviewed in his work.

The Creator is the Source of Light and of Pleasure; and the Kabbalah, which dismantles the self into parts and comes from the verb *Lekabel*, to receive, is a way to absorb that Light from the Creator by providing an explanation of the way, an indicative map of the world (*Olam*) (Sholem n.d., p. 49), In which the concealment (*Ha Alamá*) of the Light occurs. The desire for love invites one to receive the invisible in the heart through a continuous transformation and transmutation, a gradual process of approaching from MAN to MI, where the Light resides. The human being has been created to voluntarily acquire that desire for elevation.

It is, finally, this map of the process of occultation, along with the corrective process of "receiving the invisible in the heart", that is described in the *Zohar* through the means of the Tree of Life and its *Sefiroth*, as well as through its illustrative and often mysterious parables. Emerging from the land of Castile and from Moses de León's mystical visions, the book creates a metaphysical realm that requires elucidation by the techniques of aritmosophy, hermeneutics, and the tradition of rabbinical interpretation. Following these clues, however, one finds that it depicts a spiritual pathway to mystical union with ancient roots, one that is quintessentially present in the *Song of Songs*, from which it entered the Christian tradition, but is also found in Ibn 'Arabi and other Sufi mystics. Ultimately, the soul is similar to a vessel, a womb, that wants to be filled with Light and the coming of the Light is determined by desire and interrogation. Thus, the *Zohar* offers, in addition to its many challenges, a vision of light and a source of inspiration.

6. Conclusions

The force behind the metamorphosis that leads to the Union of differences is both *Himma*, the creative energy referenced by our Sufi master Ibn 'Arabi and also the *Shekhinah*, the Presence and Glory of the Lord in the Kabbalah. This feminine energy, directly linked to the creation of beings, assumes several names and genders. She is the hidden secret. She is the *Duende*. Only those who are capable of "glimpsing" Her own incommensurable dimension may contemplate the Lady of the Numbers who dwells in the sacred geometry of the abyss, concentrated in a single point within the depths of the human heart. The hermetic, allusive, clandestine, and poetic Word, pregnant with gnostic resonances and musical murmurs, sweeps the "pilgrim" along with it on paths that cross and bifurcate at mysterious points before leading to an ancient, luminous knowledge. Once arrived, the seeker is introduced into the "germinal vessel", the "inner sanctum", the *Initium*, the *Matria*. She is *Nizam* [Harmony], and also *Hokmah* [the ninth Sepher, in the Tree of Life, full of light and *Malkuth*]; She is Wisdom in Proverbs 8, the *Mater de filius sapientae*, who through an alchemical transmutation becomes a song to Sophia, both absent and present throughout the Creation.

Sophia is diffused throughout the cosmos, in all living things. She trembles with the impetus of a desire freed from ego which demands access to this redemptive Force, pregnant as it is with a knowledge capable of raising the human being exiled from its true self into unsuspected dimensions. The Lady is dynamic and manifests in a "creation that renews itself each instant, in an incessant succession of theophanies" (Chittick 2003, p. 138). She remains veiled to the sight of the multitude, and yet She longs for transparency, even as She retreats. She potentiates a rebirth at every instant. She is the fruit of Divine Compassion (Corbin 2015, pp. 106, 109, 110), opposed to the masculine rigor that rules the normative religions. She manifests in the infinite diversity of the creatures whose function is to adore the Creator. She lives within us, yet is elusive. She is the Lover and the Beloved. She may assume all forms, being both Lord and, simultaneously, Lady. She promotes hidden encounters in pleasant gardens, accompanied by silent music and sonorous solitudes that continuously vibrate in the Temple, the *Ka'aba* of the heart.

Funding: This research received no external funding.

Conflicts of Interest: The author declares no conflict of interest.

References

Arabî, Ibn al. 2002. *El Intérprete de los deseos, Tarŷumān Al-Ašwāq*. Murcia: Editora Regional de Murcia.

Bachelard, Gastón. 2000. *Poética del Espacio*. Argentina: Fondo de Cultura Económica.

Borges, Esteves Paulo. 2008. *Da Saudade como via de libertação*. Lisboa: Quidnovi.

Chittick, William C. 2003. *Mundos Imaginales, Ibn Arabî y la diversidad de las creencias*, 1st ed. Madrid: Mandala.

Chodkiewicz, Michel. 1993. *An Ocean without Shore: Ibn 'Arabi, The Book, and the Law*. Translated by David Streight. Albany: State University of New York Press.

Corbin, Henry. 1958. *L'Imagination créatrice dans le soufisme d 'Ibn Arabî*. París: Flammarion.

Corbin, Henry. 2015. *Acerca de Jung, el budismo y la Sophia*. Madrid: Siruela.

Cruz, San Juan de la. 2002. *Obras Completas*. Madrid: Biblioteca de Autores Cristianos.

Esquenazi, Fabio Samuel. 2020. Cántico Espiritual. In *Las Parábolas del Zohar y sus Resonancias en el Cántico Espiritual*. Edited by Grupo Editorial Fonte-Monte Carmelo-Universidad de la Mística-CITeS. Ávila: CITeS.

Farré, Miguel Morey. 1998. Available online: https://dialnet.unirioja.es/servlet/articulo?codigo=5757574 (accessed on 13 June 2020).

Idel, Moshe. 2009. *Cábala y Eros*. Madrid: Siruela.

Kingsley, Peter. 2016. *En Los Oscuros Lugares del Saber*. Vilahur: Atalanta Edic.

Kramer, Noa. 2010. *La Historia empieza en Summer*. Madrid: Alianza Edit.

Kramer, Noah. 2001. *El matrimonio sagrado*. Sabadell: Ed. Ausa.

Kramer, Noah, and Diane Wolkstein. 1983. *Inanna Queen of Heaven and Hearth*. New York: Harper & Row Publishers.

Laitman, Michael. 2011. *El Zohar*. Toronto: Laitman Kabbalah Publishers.

Laitman, Michel. 2015. *Abriendo el Zohar*. Toronto: Laitman Kabbalah Publishers.

López Baralt, Luce. 2009. *Simbología mística musulmana en S. Juan de la Cruz y Sta. Teresa de Jesús*. Available online: http://www.cervantesvirtual.com/obra-visor/simbologa-mstica-musulmana-en-san-juan-de-la-cruz-y-santa-teresa-de-jess-0/html/021e4a2a-82b2-11df-acc7-002185ce6064_11.html#I_0_ (accessed on 20 July 2020).

López Baralt, Luce. 2020. *Asedios a lo Indecible, S. Juan de la Cruz canta al éxtasis transformante*. Madrid: Trotta.

Lorca, Federico García. 2003. Juego y Teoría del Duende. Available online: https://biblioteca.org.ar/libros/1888.pdf (accessed on 22 July 2020).

Mora, Fernando. 2019. *El Perfume de la existencia*. Córdoba: Almuzara.

Patai, Raphael. 1947. *Man and Temple in Ancient Jewish Myth and Ritual*. New York: Nelson & Soons.

Pessoa, Fernando. 2001. *Alberto Caeiro, Poesia*. O Pastor de Rebanhos. Lisboa: Assírio&Alvim.

Ryan, Kerry. 2008. HIEROS GAMOS: Sacred Sexuality Ancient and Modern. How Is Sacred Sexuality Manifesting in a Current Western Milieu and What Is the Emerging Role of the Modern Day Sacred Prostitute? Available online: https://www.academia.edu/10272231/HIEROS_GAMOS_Sacred_Sexuality_Ancient_and_Modern_How_is_sacred_sexuality_manifesting_in_a_current_western_milieu_and_what_is_the_emerging_role_of_the_modern_day_Sacred_Prostitute (accessed on 20 August 2020).

Salomón. 1992. Sabiduría 7. Available online: https://www.bible.com/es-ES/bible/442/WIS.7.dhhed (accessed on 3 August 2020).

Sholem, Gershom. n.d. *El Zohar*; el Libro del Esplendor. Available online: http://www.72.jaimegalo.tv/Zohar.pdf (accessed on 20 July 2020).

Sholem, Gershom. 1991. *On the Mystical Shape of the Godhead: Basic Concepts in the Kabbalah*. New York: Schocken Books.

Vázquez, Ana. 2007. *Historia del Mundo Antiguo Vol. I (Próximo Oriente)*. Madrid: Sanz y Torres.

Vázquez, Ana. 2016. Available online: http://www.uned.es/geo-1-historia-antigua-universal/ACADIOS/ESHEDUANNA_1.htm (accessed on 20 May 2020).

Wolfson, Elliot R. 2001. *Moisés de León y el Zohar*. Cuenca: Pensamiento y Mística Hispanojudío y Sefardí.

Wolkstein, Diane. 1983. *Inanna*. New York: Harper Collins.

Wolkstein, Diane, and Samuel Noah Kramer. 1983. Cantos e Himnos de Sumeria. Available online: https://litsdelaant.files.wordpress.com/2013/03/e2809ccantos-e-himnos-de-sumeriae2809d.pdf (accessed on 21 August 2019).

Yom, Tov Assis, Idel Moshé, and Senkman Leonardo. 2006. *Ensayos sobre Cábala y Misticismo Judío*. Buenos Aires: Edic. Lilmod.

religions

MDPI

Article

Giqatilla's Philosophical Poems on the Hebrew Vowels: Poetry, Philosophy, and Theology in Giqatilla's *Ginnat Egoz* and *Sefer ha-Niqqud*

Federico Dal Bo [1,2]

[1] SFB 933 "Material Text Cultures", Heidelberg University, 69117 Heidelberg, Germany; fdalbo@gmail.com
[2] Hochschule für Jüdische Studien, 69117 Heidelberg, Germany

Abstract: In the present paper, I will examine Yosef ben Abraham Giqatilla's philosophical poems on the Hebrew vowels that are included in his three early works on "punctuation:" the third section from the larger *Ginnat Egoz* ("The Nut Garden"), the longer version of *Sefer ha-Niqqud* ("The Book of Punctuation"), and a short version of the latter. Scholarship on the chronology of these three texts has been inconclusive. I will argue that a textual comparison of Giqatilla's philosophical poems and an analysis of their paratextual function allow for a solution, and therefore a possible chronology of their composition.

Keywords: Giqatilla; Hebrew punctuation; Jewish poetry; Spanish Kabbalah

check for updates

Citation: Dal Bo, Federico. 2021. Giqatilla's Philosophical Poems on the Hebrew Vowels: Poetry, Philosophy, and Theology in Giqatilla's *Ginnat Egoz* and *Sefer ha-Niqqud*. *Religions* 12: 554. https://doi.org/10.3390/rel12070554

Academic Editor: Cristobal Serran-Pagan Y Fuentes

Received: 10 June 2021
Accepted: 12 July 2021
Published: 20 July 2021

Publisher's Note: MDPI stays neutral with regard to jurisdictional claims in published maps and institutional affiliations.

1. Introduction

Yoseph ben Abraham Giqatilla (1248–c. 1325) was a prominent figure of the 13th-century Spanish Kabbalah. He devoted himself almost entirely to the investigation of the mysteries of the divine names and to the development of an onomatology—a "science of the divine names."[1] He presupposed that each name of God—from the Tetragrammaton to the several appellatives used in Scripture—pointed to a specific aspect of divinity. Therefore, the existence of God could only be understood by examining the divine names. In particular, Giqatilla distinguished between the "proper names" of God (*shemot*) and His "surnames" (*kinnuyim*), the former being related to upper and lower cosmological elements, as stated in his early works, or to the 10 upper and lower metaphysical entities in the system of emanation, called *sefirot* ("spheres"), as stated in his later works.[2] In any case, Giqatilla always maintained that there is a specific hierarchy in the several "names" of God and therefore that they had to be examined accordingly.

This short paper of mine follows on from my longer work on Giqatilla that was recently published to offer a comprehensive overview of his thought (Dal Bo 2019). My longer work aimed to discuss, in particular, the historiographical assumption that there would be a clear distinction between an "early Giqatilla" and a "later Giqatilla." On that occasion, I was able to argue that the lexical and conceptual differences between the two phases of Giqatilla's thought are quite obviously important and yet should only be considered a part of an overall speculative evolution. In other words, it was my contention to integrate the two perspectives into a comprehensive insight into his thought. As a result, Giqatilla's thought appeared to pass from a "top-down description" of the divine world according to the principles of a "philosophical system" (the so-called "early Giqatilla") to a "bottom-up description" according to the principles of a "theosophical system" (the so-called "late Giqatilla", Dal Bo 2019, pp. 1–96). In this sense, lexical and conceptual differences shall not be considered mutually exclusive, as if the "late Giqatilla" were to be read apart from the "early Giqatilla" (and vice versa). On the contrary, I argued that it was necessary to read Giqatilla's thought as a whole and therefore to rather distinguish between a first and a second phase of the same speculative attitude, despite the obvious differences in terminology and notions. In this sense, these differences would be functional to allow an evolution from the description of an almost static, if not "alchemical," system of the names of God to a "dynamic" one that presupposes the participation of the believer in the system of emanation. As I mentioned there, I was inspired mainly by Lachter and implicitly by Maier, as they both insisted on the political potentiality of Giqatilla's thought (Lachter 2008; Maier 1987).

On that occasion, I also dealt with two treatises that Giqatilla wrote on Hebrew "punctuation" (*niqqud*), the diacritical system commonly used for marking vowels in Hebrew,[3] and I included them as two *excursa* between the examination of the "early Giqatilla" and the "late Giqatilla". Yet, I was mainly concerned with showing the important specific differences between these texts, leaving out the continuity between them, especially between those written at the time of the composition of *Ginnat Egoz*, as discussed below. I would like to take the opportunity of this publication precisely to investigate more deeply the relationship not so much between the various treatises on Hebrew "punctuation" written by Giqatilla but rather on another neglected aspect of his production: the nature and function of some poems contained in the first two treatises on Hebrew "punctuation."

This short paper aims to offer for the first time a philologically reliable comparison between these three compositions but also to pay attention to the rhetorical variations between them. In doing so, I hope to be able to clarify a specific aspect of the chronology of Giqatilla's works. I am positive that this specific analysis may allow the reader to discern how mystical poetry had progressively become prominent in Giqatilla's long intellectual journey. In this respect, the question of chronology should not be mistaken for a simple erudite question on the written production of a 13th-century Kabbalist but rather be appreciated as the effort to localize Giqatilla's poems within his larger mystical production and therefore, by implication, as the effort to understand the role of poetry in Jewish mysticism.

My starting assumption is that these poems had an important rhetorical-argumentative function. Giqatilla used poetry to complement his speculative thought and in doing so he diverged from the typical indications of Greek Neo-Platonism that typically tended not to appreciate poetry in force of Plato's disapproval of it in the *Republic*. On the contrary, Giqatilla is shown to be influenced by Arabic thought and especially by the figure of Ibn Gabirol. He especially shares with him the assumption to use poetry as a complementary means of exposition. It is, however, a means of expression that Giqatilla will ostensibly leave behind, especially with the publication of the major later works: *Sha'arey Tzedeq* and *Sha'arey Orah*. In the present paper, I will examine Yosef ben Abraham Giqatilla's philosophical poems on the Hebrew vowels—the so-called "punctuation"—that are included in his three early works: the third section from the larger *Ginnat Egoz* ("The Nut Garden"), the longer version of *Sefer ha-Niqqud* ("The Book of Punctuation"), and a short version of the latter. Scholarship on the conceptual, textual, and chronological relationships between these three texts has been inconclusive, mostly due to the strong affinity between these texts, and yet their apparent diversity in the way of treating the nature of the Hebrew vowels.

I will attempt to offer a solution to the question about the possible chronology of these three texts, especially by avoiding a broader, almost exhausting philological confrontation between them, but rather by examining the paratextual material that included in each of them: namely, philosophical poems that have the function to introduce the reader to the philosophical treatment of the Hebrew vowels. These poems are not simply encapsulated into the main text but rather have the main function of introducing the reader to it, and therefore operates "more than a boundary or a sealed border, the paratext is, rather, a threshold" (Genette 1997, pp. 1–2). Giqatilla's poems are discrete literary entities but also play such a paratextual function, exactly because they prepare the reader to learn the quite technical and difficult philosophical content.

By publishing synoptically the philosophical poems, I will try to prove that Giqatilla first authored the third section of larger *Ginnat Egoz*, then made the first abridgement in the longer version of *Sefer ha-Niqqud*, and finally refined the latter text in the conclusive, shorter version of the latter text.

2. The 13th-Century Context and the Interest in Hebrew Vowels

Here, I cannot delve too deeply into describing the context in which Giqatilla wrote his treatises on Hebrew "punctuation" and I will mostly rely on the excellent recent scholarship on the matter: Rachel Orna Wiener's doctoral thesis on the mysteries of vocalization in the Kabbalah of Castilla (Wiener 2008) and the more complex work of Tzahi Weiss that follows a complex interdisciplinary path, combining Talmud, history, philosophy, Kabbalah, and Lacanian psychoanalysis (Weiss 2015). Both these scholars have emphasized that Jewish scholars especially began carefully investigating the secret of Hebrew vowels, "punctuation," and letter from a very specific time in Jewish history: after the development of a proto-modern study of Comparative Semiotics in Castilla following the Islamic domination. It is well known that Islamic society had a great interest in the literary quality of the Quran and was therefore encouraged to study Arabic together with its cognates languages: Hebrew, Aramaic, and Syriac. In particular, Wiener has investigated the activity of the *'Iyyun* circle: a 13th-century Provencal circle that meditated on combinations of letters and permutations of the divine name, mostly linked to the *Sefer ha-'Iyyun* ("The Book of Contemplation") that was disseminated in many versions.[4] He has emphasized that the theoretical contribution of the *'Iyyun* circle was precisely to show that each vocalization of the divine names reflects a particular cosmic force and thus an aspect of the divine essence, which, however, is also to be associated with the attainment of a special state of meditation.[5] This sentiment still expressed in occasional and unsystematic terms is then transmitted to later thinkers as well. It is not necessary here to reiterate all the points addressed by Wiener. It will be sufficient to say that this speculative investigation of the Hebrew language—its grammar, spelling, and vocalization—was particularly rooted in Castilla and clearly followed on from the development of Comparative Semitic in the Islamic milieu.[6] Therefore, it is hardly surprising that Wiener was able to examine and compare many Castilian authors: the aforementioned *'Iyyun Circle*, Rabbi Ya'qov ben Ya'qov ha-Kohen, Rabbi Ya'aqov ha-Kohen, Rabbi Itzhaq ha-Kohen, the Zohar, Giqatilla himself, Moshe de Leon, and Yosef ha-Bah mi-Sudhan ha-Birah, aka Yosef of Hamadan. Concerning this large constellation of authors and "schools," it clearly emerges that Giqatilla was not unique in offering a philosophical and mystical examination of Hebrew "punctuation."

An examination of these different approaches to Hebrew "punctuation" largely escapes the purpose of the present paper. However, it will be worthwhile to briefly recall the matter of fact that Giqatilla operated within a circle of Kabbalists—whose best-known representative probably was Moshe de Leon, the principal editor of the Zohar but also an author, as recently shown by Avishai Bar-Asher, of several treatises on Hebrew "punctuation" (Bar-Asher 2020). This circumstance has made it possible to argue that Giqatilla's intellectual collaboration with Moshe de Leon was more complex than expected and was stretched to the point that both of them wrote some treatises on the mysteries of the Hebrew alphabet and "punctuation." As in Giqatilla, also in the case of Moshe de Leon, one can appreciate a speculative evolution from the system where the *sefirot* are hardly mentioned or, more correctly, from a system in which the *sefirot* are considered a sort of cosmic powers to a system in which the *sefirot* are parts of a complex system of emanation. Bar-Asher has correctly said that the reason for this evolution is still "a mystery" (Bar-Asher 2020, p. 357). However, this does not mean that it is not possible to try to explain this evolution if not philologically or historically (in the absence of documentation or evidence) at least speculatively, as I have tried to do in my text in the case of Giqatilla. What is particularly important is to note that Moshe de Leon also wrote several treatises on Hebrew "punctuation" before turning to writing his Hebrew-Aramaic theosophic works and editing the Zohar. Bar-Asher has primarily drawn attention to a series of texts that deal with several themes similar to those in Giqatilla himself: speculation on the forms of letters, vowel points, divine names, etc. according to Aristotelian philosophical principles. This clearly is a common ground to Giqatilla's early works on Hebrew "punctuation" as well.

3. Yosef Ben Abraham Giqatilla and His Three Works on Hebrew "Punctuation"

Rabbi Joseph ben Abraham Giqatilla (1248–c. 1325) is considered the most representative figure of a stream of Jewish mysticism devoted in particular to the investigation of the mysteries of the divine names. Despite his fame, information about the life of Giqatilla is sparse. He was born in the Castilian city of Medinaceli. There, he might have been educated in the mystical speculation of Abraham Abulafia (1240–c. 1292). In Medinaceli, he seems to have written most of his works.[7]

The most characteristic of Giqatilla's statements is the absolute individuality of divine names: he believes that any appellative which Scripture attributes to God does not simply represent a linguistic reality among other profane objects but, rather, the very matrix of the universe. Thus, his mystical speculation essentially consists in formulating a 'science of the divine names' and in deciphering the hidden secrets of each—a divine onomatology. Giqatilla never contradicts this fundamental assumption in his speculative itinerary but modulates it according to his different means of investigation: numerology, acrostics and permutations. Both hermeneutical methodology and speculation converge in his belief—reality consists in the articulation of the divine names. Joseph Giqatilla was a prolific writer—particularly active in the 1280s and 1290s. Giqatilla wrote several texts, exclusively in Hebrew: liturgical poems, philosophical speculations on the names of God, a short Talmudic encyclopedia, commentaries on Hebrew "punctuation," theosophical speculations on the names of God, commentaries on Scripture, commentaries on Jewish liturgy, commentaries on some specific commandments, and some other collateral topics. The influence of other sources, such as German Pietism, the philosophical work of Rabbi Jacob ben Sheshet Gerondi (in particular his *Meshiv Devarim Nekohim*) and the "School of Gerona" have to be added to these mainframes. For instance, the notion of "inner point" as a designation for the focus of Scripture, which is inaccessible to the Nations that spin around it—just a circumference that rotates around an "inner point," the Tetragrammaton—is derived by Jacob ben Sheshet Gerondi and populates the theology of Giqatilla's *Ginnat Egoz.*[8] Both Rabbi Jacob ben Sheshet Gerondi and the 'School of Gerona' respectively influenced Giqatilla in interpreting mystically the Hebrew "punctuation," in developing the doctrine of emanation from a metaphysical primordial point and in using the term *hamshakah* ("emanation") to describe the ontological proceedings of the Tetragrammaton. Emanation would then emerge as the act of pouring of the divine effluence pouring—by means of a sort of "divine water"—from the upper world into the lower world.[9] Giqatilla applies this notion of an internal point to both Hebrew "punctuation" and the cosmic-ontological distinction between three worlds. Accordingly, he assumes that Hebrew vowels—especially the ones designated with a single dot—localize the divine essence and its activity with respect to the three worlds: the upper, middle, and inferior world. Giqatilla constantly insists on the symbolic power of these single-dotted vowels, which are a modulation of the same divine reality, frequently described as "point," "simple point," and "internal point." This is a phraseology that Giqatilla derived, again, from Jacob ben Sheshet and the School of Gerona and preludes to his own later notion of "mental point" (*nequdah mahshavit*), which occurs both in *Sha'rey Tzedeq* and *Sha'arey Orah,* with some slight differences (Dal Bo 2019, pp. 87–88). Giqatilla authored at least three main works on the Hebrew vowels in different phases of his life: the third section from *Ginnat Egoz,* the longer version of *Sefer ha-Niqqud,* the shorter version of *Sefer ha-Niqqud,* and a later text called *Perush ha-Niqqud* ("The Commentary on Punctuation"), see also (Wiener 2016). The similarity in title and expression often produced some confusion between these different texts, as reflected in manuscripts, prints, and catalogues.[10] I will discuss them separately for clarity's sake.

3.1. Giqatilla's Ginnat Egoz

Ginnat Egoz is a long work that was unquestionably written in 1273–1274, as reported in several transmitted manuscripts. The text provides a very detailed description of the divine structure of "emanation" (*hamshakhah*) that descends from the upper into the lower world in an increasingly complex chain of divine names. Accordingly, Giqatilla provides also a long, detailed exposition of the names of God in Scripture on account of a specific hermeneutical methodology that involves three fundamental ways of interpretation: numerology, combination of letters, and acrostics. The text is divided into three parts: a philosophical description of the nature of God based on Maimonides' rationalism (Book I), a cosmological description of emanation flowing into the physical universe (Book II), and a philosophical-theosophical description of the Hebrew vowels (Book III). The third part of *Ginnat Egoz* is known under the title *Sha'ar ha-Niqqud* and is connected to the independent text *Sefer ha-Niqqud*.

3.2. *Giqatilla's Sefer ha-Niqqud*

Giqatilla was a prolific writer, and the dating of his works is quite challenging. The only certain dates in the history of his works are the following ones: *Ginnat Egoz*, written in 1273–1274; *Perush ha Merkavah*, written in 1286; and *Sharey Orah*, written in 1291. All other dating is only conjectural (Dal Bo 2019, p. 338). In particular, the dating of *Sefer ha-Niqqud* is difficult. Gottlieb argued that this text was a manuscript version before the third Book of *Ginnat Egoz* and is therefore datable between 1270 and 1273, but Blickstein maintained that this reconstruction would be conjectural and insisted on dating this work after the composition of *Ginnat Egoz*—between 1274 and 1275 (Gottlieb 1976, pp. 101–3). Annett Martini was inconclusive on this point and simply assumed that Giqatilla wrote it "at a very early stage of his creative life" (Martini 2011a, p. 57) but also suggests that "Giqatilla most probably composed [*Sefer ha-niqqud*] before the *opus magnum* of his early period, the *Ginnat Egoz*" (Martini 2011b, p. 208). On the contrary, I assume that this text was possibly written either at the time of finishing the third Book of *Ginnat Egoz* or immediately after it and, therefore, the text might have been written between 1273 and 1275. The text is found in a long and a short recension that I treat here as a longer and shorter *Sefer ha-Niqqud*. My understanding of the textual similarities between these texts has persuaded me that the poem published in *Ginnat Egoz* was probably the inspirator of the one published in the longer version of *Perush ha-Niqqud* that was further elaborated into the shorter version. I will discuss this hypothesis below.

3.3. *Giqatilla's Perush ha-Niqqud*

The dating of *Perush ha-Niqqud* falls between the composition of *Sha'arei Tzedeq* (that it explicitly mentions) and *Sha'arei Orah* (that it does not mention). Therefore, this text has presumably been written between 1286 and 1291, most probably in early 1290.

The dating of *Perush ha-Niqqud* mainly depends on the temporal location of the so-called *sodot*—a series of short treatises that represent a complex of texts, that were presumably written all together with *Sha'arey Tzedeq* or immediately after it or perhaps even written together with *Sha'arey Tzedeq* but eventually discarded for some unknown reasons. The spectrum of composition of these texts is quite broad and lies after the 1280s. If one accepts this view, *Perush haNiqqud* must necessarily have been written after these various works.

Giqatilla wrote this second treatise on Hebrew "punctuation" in the later stage of his speculation. Accordingly, he employs several references to the system of *sefirot* that are pictographically depicted by the placement of Hebrew vowels above, within, or under the line of writing.

4. Giqatilla's Philosophical Poems

Before examining Giqatilla's philosophical poems included in his three works on Hebrew "punctation," it might be useful to briefly consider the role of Jewish philosophical poetry in the Iberian Peninsula in the Middle Ages.

4.1. *Hebrew Poetry in Medieval Spain*

There are few doubts that Hebrew literature—even during its golden age between the 10th and 12th centuries—was deeply influenced by Arabic literature, as it is still evident from the rhymed-prose that were delivered quite later, during the 12th to the 15th centuries. In general, contemporary scholarship is usually conclusive about the nature of Jewish poetry in Spain during and after Muslim domination. It is apparent that medieval Jewish poets were particularly dedicated to writing in Hebrew and also to develop the notion of "poetry" as an independent topic of its own—not too dissimilar from what it can be encountered in modern poetry. Several Jewish poets from Spain—Shemuel ha-Nagid,[11] Solomon ibn Gabirol,[12] Moshe ibn Ezra,[13] and Judah ha-Levi[14]—were quite determined to write poetry only in Hebrew. Jewish intellectuals in Medieval Spain were involved in an ongoing rivalry with the Muslim socio-religious environment, its rising importance in the study of an incipient Comparative Linguistics,[15] and its assimilation of Aristotelian philosophy.[16] In this respect, Jewish poets were no exception. They were eager to relaunch the merits of the Hebrew against Arabic. They also wanted to prove "on-field" that Hebrew was as good as Arabic in producing both poetry and speculation.

Religious images, as well as Jewish liturgy, were deeply connected to the art of poetry. Particular attention to this topic has recently been paid by the prominent US-born Jewish poet Peter Cole in his auspicious collection *The Poetry of Kabbalah* (Cole 2012). This important anthology of mystical verses and liturgical hymns allowed us to appreciate how poetry had progressively emerged since early Jewish mysticism of the divine "Palaces"[17] and was progressively disseminated in Al-Andalus, Spain, and Ashkenaz—stretching back to the Galilean Kabbalah. It is indeed poetry that allowed stellar figures—for instance, the prominent Salomon ibn Gabirol—to infuse philosophical notions with religious fervor. Poetry was a perfect way for improving Aristotelian mentality with poetic images. One of the most prominent figures in Jewish religious and philosophical poetry certainly was the prominent Jewish thinker Solomon Ibn Gabirol, whose works, language, and notions deeply inspired Giqatilla as well as many other Jewish and Christian scholars. In particular, Solomon Ibn Gabirol—undoubtedly the first and most influential Jewish philosopher and poet in Spain—intended to both preserve traditional rhetorical genres and innovate Jewish poetry by experimenting in form and style. Ibn Gabirol was courageous enough to combine tradition with innovation in his poetic writings. In particular, he especially relied on paronomasia—the art of juxtaposing two correlated terms—that also influenced the young Giqatilla and his philosophical poems. In truth, Ibn Gabirol was instrumental to setting the poetic and speculative tone for the entire school of Andalusian writers who simultaneously were poets and philosophers and intended to combine Neo-Platonism with rabbinic Judaism.

The use of poetry as a "complement" to "metaphysical" thought required negotiating in a complex way with the principles of Plato's disapproval of poetry and the systematization of poetry according to Aristotelian *Poetics* that was complexly receipted in the Islamic world. On a conceptual level, the influence of Ibn Gabirol on Giqatilla was not particularly strong. I have argued elsewhere that the influence of Ibn Gabirol on Jewish mysticism was mostly indirect and consisted in assimilating his pivotal notion of *ratzon* ("will") as the apex of the Godhead into a lower localization within the system of emanation, see (Dal Bo 2021a). One can easily say that Ibn Gabirol's influence was therefore oblique. In the case of Giqatilla, it probably consisted of instilling the idea that one could search for a "poetical thought"—by combining poetry and speculation. It is particularly significant that, in his later works, especially in *Sha'arey Tzedeq* and *Sha'arey Orah*, Giqatilla makes no longer use of poems to intersperse speculative argumentation. However, this does not mean that he had abandoned the idea of writing in flowery Hebrew, but rather that he no longer felt it necessary to make poetry "have a direct dialogue" with philosophy.

One must be careful not to project his own philosophical expectations—and thus "Western" philosophical ones—onto the Jewish–Arab milieu in Spain. In his survey of the history of poetry in the Arab world, Cantarino has emphasized the role of orality and thus poetry as a fundamental criterion for defining the identity of the Arab people—a characteristic that was not directly acquired also by their "co-resident" Jews but passed on to the Jewish world as a deep cultural appreciation of poetry, without considering it "an alternative" or a "contradiction" to "philosophy." In particular, Cantarino was able to show that Islamic culture excelled in identifying so clearly and self-consciously with "literature." He assumes that Arab writers pursued excellence in the art of poetry as a form of imitation of the Quran—the most perfect literary product as the full realization of the word of God in writing. As such, the Quran represented the perfection of the Arabic literary art but also introduced the subtle paradox that the Quran would not be a "poetical book" but rather the most perfect Revelation of God. In this sense, the Quran could not be "poetical" by definition. Nevertheless, as Cantarino insists, this implicit praise of poetry resulted in the elaboration of the theory of *takhyil* ("the imaginary"): this was a concept that denoted the creative flowering of Islamic literary theory. In other words, this notion presupposed that the poet had one primary goal: not imitating reality in literal terms, but rather encouraging virtuous action exactly by eulogizing goodness (Cantarino 1975).

In her careful investigation of the so-called "poetic syllogism," Tannyss Ludescher has been able to show that this particular device allowed the point of junction between poetry—that was "de-intellectualized" by virtue of *takhyil*—and its application to philosophical speculation (Ludescher 1996). In other terms, "poetic syllogism" was believed to effect on the reader's rational faculties and allow for the voluntary exercise of the human will. In his intriguing study of the development of "poetic syllogism," Tanyss Ludescher suggested that *takhyil* allowed for developing a poetics that was anti-realist and yet resulted into a paradoxical effect—allowing for the abstraction of speculative concepts: "as we have seen, the concept of *takhyil* crystallized a tendency in Arab thought which sought to divorce poetry from objective Truth. I would like to suggest that other factors contributed to this general tendency to sever poetry from Nature and treat it as an artefact that can be manipulated in syllogistic terms" (Ludescher 1996, p. 97). It is exactly within this context that one shall read Giqatilla's poems: not simply as a theoretical divertissement but rather as the effort to elaborate a properly speculative thought that is not alienated from poetry.[18]

4.2. Giqatilla as Both a Kabbalist and a Poet

In this respect, it should not be surprising that Giqatilla wrote philosophical poems. On the contrary, contemporary scholarship usually agrees with the assumption that he presumably began his literary production exactly by writing two poems—possibly redacted before 1273—that precede *Ginnat Egoz*: they are two pieces of poetry consisting of a sixty-nine verse mystical poem titled *Baqqashah* and the mystical poem titled *'Iqerey Emunah* that is very similar in content to the previous one and "saturated with ideas and themes culled from *Ginnat Egoz*" (Blickstein 1983, pp. 35, 151; Cf. Gruenwald 1966).

It appears that the young Giqatilla produced several pieces of poetry and emphasized the opposition between short and long vowels, probably under the influence of Arabic phonology. Scholars in Comparative Semitic Philology have frequently remarked the equivalence in rhyme and meter between Arabic long vowels and Sephardic full vowels in the Hebrew Poetry of Medieval Spain. Such metrical equivalence evidences, as *terminus ante quem*, the loss of the earlier long–short opposition within the full vowels of Hebrew. Giqatilla encapsulated his poems into his first major philosophical work: *Ginnat Egoz*.

Giqatilla's *Ginnat Egoz* is a large and difficult text: the supernal world is described with a very technical philosophical vocabulary that is mostly drawn from Maimonides' seminal *Guide to the Perplexed*. This already complex topic is further complicated by the use of *Sefer Yetzirah* as a book of cosmology, the use of a complex philosophy of language, a doctrine of the divine names (an onomatology), and, in the present case, also a doctrine of the Hebrew vowels. As result, *Ginnat Egoz* is a quite demanding text that requires a large number of competencies in theology, philosophy, cosmology, and linguistics. In this respect, the several poems that are to be found in this text play a specific role: easing the study of this difficult text by introducing and summarizing the content in a more fashionable, possibly more agreeable way. *Ginnat Egoz* is organized in several books, chapters, and sections. Giqatilla typically introduces each chapter and several sections with short poems, mostly a few verses that are organized in hemistichs. These texts often attempt to be alliterative and use some rhyme to better connect their philosophical content. Albeit of a literary nature, these poems have the paratextual function to educate the reader and, therefore, are phenomenologically similar to the mnemotechnical texts that are often to be found in rabbinic texts.

In the following table, I have synoptically edited these poems as they emerge from the three works on Hebrew "punctuation:" namely, the third book from *Ginnat Egoz*, the longer version of *Sefer ha-Niqqud*, and the shorter version of the letter text. This edition is a fabrication. I cannot emphasize enough that I myself have edited these texts together and that they are scattered in the text, in specific locations. Therefore, these texts should not be considered, strictly speaking, as a whole poetic composition but rather as a series of poetic introductions to each portion from the larger *Ginnat Egoz* and the shorter versions of *Sefer ha-Niqqud*. Given this philological precaution, I have edited these poetic texts but have still signalized the division among them with a blank row. Despite some philological difficulties, I maintain that these several poems can offer an interesting point of view on the relationship between these three tractates on Hebrew "punctuation" and possibly offer a solution to the historiographical enigma of their origin.

4.3. *Giqatilla's Philosophical Poems: Synopsis and Translation*

For clarity's sake, I have divided each verse into two separate hemistichs, consequently numbered each verse into an "a-verse" and a "b-verse" hemistich, evidenced in italics the most important lexical differences between the three versions, in bold the most important lexical similarities, and with the sign ø the absence of textual material in the respective versions. Lexical correspondences between the *Ginnat Egoz* and the longer version of *Sefer ha-Niqqud* are marked in italics, lexical correspondences between the longer and shorter version of *Sefer ha-Niqqud* are marked in bold, and the lexical correspondences between *Ginnat Egoz* and the shorter version of *Sefer ha-Niqqud* are marked in underlined text (see Table 1):

Table 1. A Synoptic Edition of Gikatilla's Poems on Hebrew "Punctuation".

Line	*Ginnat Egoz*	*Sefer ha-Niqqud* Short Version	*Sefer ha-Niqqud* Long Version
1a		ø	ø
1b		ø	ø
2a		ø	ø
2b		ø	ø
3a		ø	ø
3b		ø	ø
4a		ø	ø
4b		ø	ø
5a		ø	ø
5b		ø	ø
6a		ø	ø
6b		ø	ø
7a		ø	ø
7b		ø	ø
8a		ø	ø
8b		ø	ø
9a			
9b			

Table 1. *Cont.*

Line	*Ginnat Egoz*	*Sefer ha-Niqqud* Short Version	*Sefer ha-Niqqud* Long Version
10a			
10b			
11a			
11b			
12a			
12b			
13a			
13b			
14a			
14b			
15a		ø	ø
15b		ø	ø
16a		ø	ø
16b		ø	ø
17a			
17b			
18a			
18b			
19a			
19b			
20a			
20b			
21a			
21b			
22a			
22b			
23a			
23b			
24a			
24b			
25a			
25b			
26a			
26b			
27a			
27b			
28a			
28b			
29a			
29b			

Table 1. *Cont.*

Line	*Ginnat Egoz*	*Sefer ha-Niqqud* **Short Version**	*Sefer ha-Niqqud* **Long Version**
30a	ø		
30b	ø		
31a			
31b			
32a			
32b			
33a		ø	ø
33b		ø	ø
34a		ø	ø
34b		ø	ø
35a		ø	ø
35b		ø	ø
36a		ø	
36b		ø	
37a		ø	
37b		ø	
38a		ø	
38b		ø	
39a		ø	
39b		ø	
40a		ø	ø
40b		ø	ø
41a		ø	ø
41b		ø	ø
44a		ø	ø
44b		ø	ø
45a		ø	ø
45b		ø	ø
1a	My dear soul, know wisdom and intelligence	ø	ø
1b	Thus you'll bring to light all the mysteries.	ø	ø
2a	Investigate to find the foundation of all foundations	ø	ø
2b	And to find the mystery of the primordial Being	ø	ø
3a	You'll see, foundation is conception and matter	ø	ø
3b	You'll feel, [they] are married with nothingness	ø	ø
4a	He makes the sons of intellect to lead	ø	ø

Table 1. *Cont.*

Line	*Ginnat Egoz*	*Sefer ha-Niqqud* Short Version	*Sefer ha-Niqqud* Long Version
4b	The sons of matter in perfect leadership.	ø	ø
5a	Narrow to them the image of dot and letter	ø	ø
5b	To lead action and signifying.	ø	ø
6a	If intellect and dot are souls	ø	ø
6b	A single world soul to soul	ø	ø
7a	In the moment, the foundation is pleasant words	ø	ø
7b	It is attached to three dots	ø	ø
8a	They alone are the fundaments of meaning.	ø	ø
8b	He alone the foundation of all mysteries	ø	ø
9a	If you intend, my son, to ascend to the level of intellect	If you, my son, intending to ascend to the level of intellect	My son, if you intend to ascend to the level of intellect
9b	to study the secret levels of wisdom, be vigilant.	to study the secret levels of wisdom, be vigilant.	to study the secret levels of wisdom, be vigilant.
10a	The altar with fire, wood and knife prepare	The altar with fire, wood and knife prepare	the altar with fire, wood and knife prepare
10b	Sand upon it with bound foot and hand.	Stand upon it with bound hand and foot.	Stand upon it with bound foot and hand.
11a	Contemplate the verse and its conception	Contemplate the verse and its conception	Contemplate the verse and its conception
11b	By the letters, value and secret of punctuation.	By the letters, value and secret of punctuation.	By the letters, value and secret of punctuation.
12a	Explore cholam and *explore* qamatz and tzere	Explore cholam and *explore* qamatz and tzere	Explore cholam and also qamatz and tzere
12b	Explore chiriq and shuruq, the miracles.	Explore chiriq and shuruq, the miracles.	Explore chiriq and shuruq, the miracles.
13a	Explore alef and *explore* taf, beginning and completion,	Explore alef and *explore* taf, beginning and completion,	Explore alef and taf, beginning and completion,
13b	Divide into five the *host of each letter*.	Divide into five the *host of each letter*.	Divide into five all hosts.
14a	Set five opposite five,	Set five opposite five,	Set five opposite five,
14b	Understand, you will find in them *miracles*	Understand, you will find in them *miracles*	Understand, you will find in them fullness
15a	Raise to the way to chariot and prosperity	ø	ø
15b	To understand the fundament, a	ø	ø
16a	The name to understand all those with intellect	ø	ø
16b	The world stands on an eternal covenant	ø	ø

Table 1. *Cont.*

Line	*Ginnat Egoz*	*Sefer ha-Niqqud* **Short Version**	*Sefer ha-Niqqud* **Long Version**
17a	You will stand forever and being	You will stand forever and being	You will stand forever and being
17b	Placed towards life and resurrection,	Placed towards life and resurrection,	Placed towards life and resurrection,
18a	The time you'll be placed in cholam,	The time you'll be placed in cholam,	The time you'll be placed in cholam,
18b	Who is, who was and who will be.	**Who was, who is and will be.**	**Who was, who is and who will be.**
19a	Consider and look at the secret of qamatz	**Consider and look** at the secret of qamatz	**Consider and look at** the secret of qamatz
19b	Perhaps you will be worthy to gaze	Perhaps you will be worthy to gaze	Perhaps you will be worthy to gaze
20a	See through the nice secret of its motion	See through the nice secret of its motion	See through the nice secret of its motion
20b	Every letter; and it is round and surrounds.	Every letter; and it is round and surrounds.	Every letter; and it is round and surrounds.
21a	Look at shuruq an image of a bridge,	Look at shuruq an image of **tightened knot**	Look at shuruq an image of **tightened knot**
21b	for the middle is like conciliation.	**whereupon the foundation is set to support.**	**whereupon the foundation is set to support.**
22a	It descends to the sheol below	**It's the secret of waw in form of disposition,**	**It's the secret of waw in form of disposition,**
22b	and it flies high like an eagle.	**for, indeed, its levels are like six.**	**for, indeed, its levels are like six.**
23a	Indeed, it knots together the dispositions.	**Look at chiriq like ice** and the electrum	**Look at chiriq like ice** and upon it are
23b	wherefore it is called by the name 'knot.'	**the image of intellect and the secret of splendors.**	**the image of intellect and the secret of splendors.**
24a	Explore chereq and look before you		
24b	Explore the supreme and highest miracles		
25a	Compose and compound of spices is the image of powder		
25b	Very far away from the levels of splendor		
26a	Ice is ice and pure is.		
26b	But not as spheres and as God		
27a	Intelligence the foundation of edifice and every matter		
27b	And see foundation of morality at the beginning of matter		
28a	Know that the being that has acquired		
28b	The intellect of the spheres at the beginning of years		

Table 1. *Cont.*

Line	*Ginnat Egoz*	*Sefer ha-Niqqud* Short Version	*Sefer ha-Niqqud* Long Version
29a	All the beings without him together		
29b	They all are renewed and are an edifice		
30a	ø	The moment you consider its image	The moment you consider its image
30b	ø	you will consider the image of the chariot of God.	you will consider the image of the chariot of God.
31a	Understand and expose the *foundation* of tzere,	Understand and expose the *foundation* of tzere	Understand and expose the secret of tzere
31b	and you will find in it the image of the edifice of splendors.	**Thus you will ascend to the level of the highest.**	**Thus you will ascend to the level of the highest.**
32a	For if you understand its *foundations* and its secrets,	For if you understand its *foundations* and its secrets,	For if you understand its levels and its secrets,
32b	the spirit of God will prosper within you.	the spirit of God will prosper within you.	the spirit of God will prosper within you.
33a	Understand, my dear soul, the secret of punctuation		ø
33b	Teaching the foundation of sphere and it is round	ø	ø
34a	In *segol*, its punctuation is intended	ø	ø
34b	Therefore, know that it is the foundation of sphere	ø	ø
35a	And understand the intellects and the line of features therein	ø	ø
35b	Know that there are arrangements between the word of sheva	ø	ø
36a	Know wisdom and intelligence.	Know wisdom and intelligence.	ø
36b	Arranged in the secret of patach	Arranged in the secret of patach	ø
37a	And it will be your defense	And it will be your defense	ø
37b	Planted on water.	Planted on water.	ø
38a	And understand the intelligences	And understand the intelligences	ø
38b	In the series of receptacles.	In the series of receptacles.	ø
39a	And see [their] visions	And see [their] visions	ø
39b	Hosts of the skies.	Hosts of the skies.	ø
40a	Know, my friend, that our God	ø	ø
40b	Stretches out the celestial tends	ø	ø

Table 1. *Cont.*

Line	*Ginnat Egoz*	*Sefer ha-Niqqud* Short Version	*Sefer ha-Niqqud* Long Version
41a	Operates the intelligences at the level of the world	ø	ø
41b	So the movement on the abode of patach	ø	ø
44a	A fearful individual and dressed in fear.	ø	ø
44b	In front of a governor's closet without a hitch.	ø	ø
45a	Know that our Lord is the foundation of everything.	ø	ø
45b	And out of all is found and is one.	ø	ø

5. Analysis and Commentary

Before proceeding with a comprehensive analysis of these three texts, I would like to shortly analyze each of them separately.

The third book from *Ginnat Egoz* expectedly offers the more rich and complex textual material concerning the longer and shorter versions of *Sefer ha-Niqqud*. It is possible to divide this material into different sections: (1) a proem to the reader (vv. 1a–6b), (2) a shorter thematic introduction (vv. 7a–8b), (3) a series of minor introductions to each single Hebrew vowel (vv. 9a–39b), (4) a shorter "mid-introduction" to the reader that is inserted between these introductions (vv. 15a–16b), and (5) a coda that recalls the proem in form and style (vv. 40a–45b). In other words, each poetic introduction to each element from Hebrew "punctuation" is encapsulated into a proem (vv. 1a–6b) and a coda (vv. 40a–45b) that play a specific, but unique rhetorical role.

It is not surprising that the proem (vv. 1a–6b), the shorter introduction (vv. 7a–8b), the mid-introduction (vv. 15a–16b), and the coda (vv. 40a–45b) are not extant either in the longer or int the shorter versions of *Sefer ha-Niqqud*. No reason is given, but textual material was probably abridged only for brevity's sake. Interestingly, both Giqatilla's *Sha'arey Tzedeq* and *Sha'arey Orah* are introduced by a dedication to the reader, whom Giqatilla addresses with a stereotypical formula: "my beloved of my soul ... " Indeed, Giqatilla frequently uses this formula to introduce the reader both to works. As Charles Mopsik maintains, this formula is probably modelled on a Biblical expression: "I have given the dearly beloved of my soul" (*natati 'et yedidut naphshi*) (Jer 12:7). Probably due to Giqatilla's scholarly fame and moral authority, this formula might have been used both in Talmudic and liturgical context. On the one hand, this very formula is also employed by David ben Zimra in some Rabbinic responsa.[19] On the other hand, a similar formula—*yedidi nefesh* (literally: "my dear soul")—is also used in the homonymous liturgical poem, commonly attributed to Eleazar ben Moshe Azikri.[20] It is difficult to determine whether Giqatilla was referring to either an actual or fictitious reader. Nevertheless, it is important to note that Giqatilla assumes a particular tone towards this individual who asked for some clarifications about the secrets of God's names. Giqatilla familiarly addresses him in a familiar way but he also assures him that he is going to transmit to him all his knowledge "in black and white," with no hesitation. Such willingness to transmit mystical knowledge is of great importance—as clearly suggested by the imperative verb form: "know that ... " The longer version of *Sefer ha-Niqqud* mostly consists of a series of minor introductions to each single Hebrew vowel (9a–39b). Concerning the material from *Ginnat Egos* and the shorter version of *Sefer ha-Niqqud*, the longer version of *Sefer ha-Niqqud* exhibits these notable differences: (1) the removal of the proem, a shorter thematic introduction, the "mid-introduction," and the coda, as already anticipated; (2) the abridgement of a larger body of verses from *Ginnat Egoz* (vv. 21a–29b) into shorter, more condensed verses (vv. 21a–23b); and (3) a final summary that is included only in *Ginnat Egoz* but absent in the shorter version of *Sefer ha-Niqqud* (vv. 36a–39b). The shorter version of *Sefer ha-Niqqud* exhibits the same characteristics as the longer version except for the exclusion of a final segment (vv. 36a–39b).

This preliminary analysis of these philosophical poems is important to determine what are the crucial philological features for establishing the relationships between these texts. First of all, the removal of the proem, a shorter thematic introduction, the "mid-introduction", and the coda (vv. 1a–6b, 7a–7b, 15a–16b, and 40a–45b) is a dramatic change that suggests that the larger *Ginnat Egoz* was the archetype for both the longer and shorter version of *Sefer ha-Niqqud.* Yet this major change is still not sufficient to offer detail on the actual relationships between these texts. On the other hand, the removal of this textual material allows us to focus on the more relevant textual changes that take place in the remaining texts: the so-called single introductions to each Hebrew vowel (broadly speaking the material included between vv. 9a–39b). Given the philological proportions of this synoptic edition, I can pass now on a philological and stylistic analysis of these texts. There are three types of differences between the third book from *Ginnat Egoz,* the longer and shorter version of the *Sefer ha-Niqqud*: phraseological, phraseological-rhetorical, and textual. I will discuss them separately for clarity's sake.

5.1. Phraseological Differences

These differences mostly consist of small variations in word order—see for instance "if you, my son, intend to ascent" () in *Ginnat Egoz* and the longer *Sefer ha-Niqqud* and "my son if you intend to ascent" () in the shorter *Sefer ha-Niqqud* (v. 9a); the hemistich "The altar with fire, wood and knife prepare" () in *Ginnat Egoz* and the longer *Sefer ha-Niqqud* and "the altar with fire, wood and knife" () in the shorter *Sefer ha-Niqqud* (v. 10a). Each of these differences is negligible from a strict semantic point of view but undisputedly proves textual proximity between the third book from *Ginnat Egoz* and the longer version of *Sefer ha-Niqqud,* on the one hand, and the relative independence of the shorter version of *Sefer ha-Niqqud.* Some of these phraseological differences also involve changes in the rhetoric of the text, and therefore shall be discussed separately below.

5.2. Phraseological-Rhetorical Differences

These differences are very similar to the former ones, with the exception that they consist of variations in wording that have a sensible change in the rhetorical pattern of the three texts—see for instance the anaphoric use of "explore cholam and explore qamatz and tzere" () in *Ginnat Egoz* and the longer version of *Sefer ha-Niqqud* and their resolution into the unimpressive "explore cholam and also qamatz and tzere" () in the shorter *Sefer ha-Niqqud* (v. 12a); the simplification in wording from "explore alef and explore taf, beginning and completion," ( ) in *Ginnat Egoz* and the longer version of *Sefer ha-Niqqud* to "explore alef and taf, beginning and completion," () in the shorter *Sefer ha-Niqqud* (v. 13a).

5.3. Textual Differences

These differences are particularly relevant and shall be classified in two subtypes: (1) minor differences in wording and (2) larger textual differences. For clarity's sake, I will treat them separately below.

5.3.1. Minor Textual Differences

These differences typically pertain only a single word, and therefore can also depend on some scribal error—see for instance the reading "divide into five the host of each letter" () in *Ginnat Egoz* and the longer version of *Sefer ha-Niqqud* in with respect of the reading "divide into five all hosts" () in the shorter *Sefer ha-Niqqud* (v. 13b); the reading "understand, you will find in them miracles" () in *Ginnat Egoz* and the longer version of *Sefer ha-Niqqud* in with respect of the reading "understand, you will find in them fullness" () in the shorter *Sefer ha-Niqqud* (v. 14b); the reading "understand and expose the foundation of tzere" () in *Ginnat Egoz* and the longer version of *Sefer ha-Niqqud* with respect of the reading "understand and expose the secret of tzere" () in the shorter *Sefer ha-Niqqud* (v. 31a); the reading "for if you understand its foundations and its secrets" () in *Ginnat Egoz* and the longer version of *Sefer ha-Niqqud* with respect of the reading "for if you understand its levels and its secrets" () in the shorter *Sefer ha-Niqqud* (v. 32a).

Each of these differences usually pertains to small orthographical details and are fully compatible with scribal errors. Nevertheless, it is philologically relevant that each of these differences gathers the three texts into two groups: *Ginnat Egoz* and the longer version of *Sefer ha-Niqqud,* on the one hand, and the shorter *Sefer ha-Niqqud,* on the other hand. There are only two exceptions to this frame: first, the reading "look at shuruq, an image of a bridge" () in *Ginnat Egoz* alone with respect of the reading "look at shuruq, an image of a tightened knot" () in both longer and shorter version of *Sefer ha-Niqqud* (v. 21a). As anticipated, almost each of these differences in wording could be originated by a scribal error: words like ("host of letter") vs. ("hosts") (v. 13b), ("miracles") vs. ("fulness") (v. 14b), ("bridge") vs. ("knot") (v. 21a), and ("fundament") vs. ("secret") (v. 31a).

The distribution of these differences in wording is telling. With only one exception, they all suggest that *Ginnat Egoz* and the longer version of *Sefer ha-Niqqud* are closer to each other rather than to the third text (the shorter version of *Sefer ha-Niqqud*).

5.3.2. Larger Textual Differences

The larger textual differences pertain to the central segment of the poem that covers many verses in Ginnat Egoz (vv. 21a–29b) and only a few in both the longer and shorter version from *Sefer ha-Niqqud* (vv. 21a–23b). The longer version in *Ginnat Egoz* (vv. 21a–29b) is different from the notably shorter one in both longer and shorter *Sefer ha-Niqqud* (vv. 21a–23b): the lexical material extant in *Ginnat Egoz* is more or less reduced to one third in both longer and shorter *Sefer ha-Niqqud.*

A closer look suggests that Giqatilla has not simply shortened his previous version from *Ginnat Egoz* but rather condensed it into a shorter text that still uses most of the key-words that had previously employed. Notably, a portion from the longer text in *Ginnat Egoz* (vv. 24a–26b) was built on the alliteration of the root ("chereq") (v. 24a) that is permuted in all four possible ways: ("explore") (v. 24b), ("compose") (v. 25a), ("far away") (v. 25b), and ("ice") (v. 26a). The longer and shorter text from *Sefer ha-Niqqud* do not exhibit these wordplays but still refer to the term "ice" () (v. 23a) by oversimplifying the allusions concerning the original text. The term "fundament" () in *Ginnat Egoz* (v. 27a) is somehow alluded to in the longer and shorter version of *Sefer ha-Niqqud* with the almost homographic term "secret" () (v. 22a), while the expression "the levels of splendor" () in *Ginnat Egoz* (v. 25b) appear in the shorter and longer version from *Sefer ha-Niqqud* to be broken up into the terms "its levels" () (v. 22b) and "splendor" () (v. 23b). On the contrary, terms like "image" (), "knot" (), and "disposition" () occur in all three texts and therefore manifest interesting textual solidarity.

5.4. A Philological Comparison

Note the following textual congruencies between the three text versions:

5.4.1. Lexical Congruences between *Ginnat Egoz* and the Longer *Sefer ha-Niqqud* vs. the Shorter *Sefer ha-Niqqud* (Marked in Italics)

These are the most common ones—see the terms ... (v. 10b), (v. 12a) that is also reinforced by the expression and then (v. 13a), (v. 14b), and (v. 32a). Occasionally, lexical congruences between *Ginnat Egoz* and the longer *Sefer ha-Niqqud* vs. the shorter *Sefer ha-Niqqud* are quite thin and yet involve notable semantic changes. See, for instance, the expression (v. 13b) that is "hypercorrected" only in the shorter *Sefer ha-Niqqud* as or the term (v 31a) "hypercorrected" only in the shorter *Sefer ha-Niqqud* as In some other cases, there the shorter *Sefer ha-Niqqud* lacks material that is present in the other two texts (vv. 36a–39b).

5.4.2. Lexical Congruences between the Longer and Shorter *Sefer ha-Niqqud* vs. *Ginnat Egoz* (Marked in Bold)

These are not very frequent and mostly involve minor changes in word order. Note, for instance, the sequence (v. 18b) that occurs as , possibly due to a copyist's error, the correction of (v. 19a) with due to the—euphonic?—passage from the letter to the letter but also from to , the correction of (v. 21a) with , which is simultaneously a correction of with (cf. 23b) and the addition of and collectively all verses vv. 21a–23b, which show how both the version of *Sefer ha-Niqqud* diverge from *Ginnat Egoz.*

5.4.3. Lexical Congruences between *Ginnat Egoz* and the Shorter *Sefer ha-Niqqud* vs. the Longer *Sefer ha-Niqqud* (in Underlined Text)

This congruence is very rare and only occurs with the expression (v. 10b) that otherwise occurs as and plausibly is a copyist's error

5.4.4. Lexical Departures of *Sefer ha-Niqqud* from *Ginnat Egoz*

These are quite obvious and consist in the removal of lexical material (vv. 1a–8b, 15a–16b, 24a–29b, 33a–35b, 40a–45b).

5.4.5. Summary

The frequent lexical congruences between *Ginnat Egoz* and the longer *Sefer ha-Niqqud* vs. the shorter *Sefer ha-Niqqud* (marked in italics) (Section 5.4.1) vis-à-vis the less frequent lexical congruences between *Ginnat Egoz* and the longer *Sefer ha-Niqqud* vs. the shorter *Sefer ha-Niqqud* (marked in italics) and the occasional lexical congruences between *Ginnat Egoz* and the shorter *Sefer ha-Niqqud* vs. the longer *Sefer ha-Niqqud* (in underlined text) suggest that there is no appreciable *direct* connection between *Ginnat Egoz* and the shorter *Sefer ha-Niqqud.* This philological analysis makes it plausible—and yet still not unequivocal—to assume that the shorter version of *Sefer ha-Niqqud* was elaborated from the longer version of *Sefer ha-Niqqud* that was originally extracted from *Ginnat Egoz*, as is clearly shown by the lexical departures from these texts (Section 5.4.4).

This analysis suggests that the passage from *Ginnat Egoz* to the longer and shorter versions from *Sefer ha-Niqqud* clearly shows an apparent change in the versification but also an elaboration on almost the same lexical material. This suggests that Giqatilla's reworking was less radical than one would expect at first and, on the contrary, that he intended to elaborate the longer and shorter versions from *Sefer ha-Niqqud* based on the longer lyrics in *Ginnat Egoz.* Again, the similarity of the lexical material suggests that the third book of *Ginnat Egoz* was the basis for the longer version of *Sefer ha-Niqqud* and that the latter was the basis for the shorter version of *Sefer ha-Niqqud* as well.

6. Conclusions

The phraseological, rhetorical-phraseological, and especially the textual differences point to true textual variants between the three texts. These are particularly notable between the version from *Ginnat Egoz*, on the one hand, and the two versions from the *Sefer ha-Niqqud*, on the other hand. These variants clearly evidence that *Ginnat Egoz* presumably was the original text from which the two versions from the *Sefer ha-Niqqud* have been elaborated, specifically the longer version of the *Sefer ha-Niqqud* from the *Ginnat Egoz* and the shorter version of the *Sefer ha-Niqqud* from the longer version of the same text. Martini has argued that these differences would only show notable textual differences between these texts and claimed for their textual independence. In my opinion, these textual variants prove exactly the opposite: the version from *Ginnat Egoz* was the source text for both the longer and shorter version of *Sefer ha-Niqqud.* More specifically, some verses from *Ginnat Egoz* appear to be summarized and rephrased in both the longer and shorter version of *Sefer ha-Niqqud.* The derivation of the latter from the former can be appreciated from the fact that Giqatilla extracted some specific keywords from *Ginnat Egoz* (emphasized in bold above), reworked them in a notable different way, and finally encapsulated in two sensible different variants.

Funding: This research received no external funding.

Institutional Review Board Statement: Not applicable.

Informed Consent Statement: Not applicable.

Data Availability Statement: Not applicable.

Conflicts of Interest: The author declares no conflict of interest.

Notes

1 On the notion of "onomatology," see (Dal Bo 2019, pp. lix–lxi, lxii–lxx, 9–80, 238–336). For a comprehensive treatment of the name of God and the development of this motif in Jewish thought, see (Miller 2016).

2 The proper meaning of the term *sefirah* (here tentatively rendered as "sphere") is still disputed. Gershom Scholem famously argued that the Hebrew term *sefirah* would be correlated to the Greek term *sfaira* (Scholem 1987, p. 26). However, this etymology is inconclusive (Dan 1998, p. 8). Besides, Jewish mysticism traditionally associates the nature and activity of *sefirot* with several other entities, like "number" (*mispar*), "book" (*sefer*), "sapphire" (*sapir*), etc. Therefore, even a definitive etymology of *sefirah* would not exhaust the complex semantics of this term.

3 The English term *punctuation* designates the use of symbols—full stops, periods, commas, etc.—to divide written words in sentences but is only used here to translate the Hebrew term *niqqud* (literally "dotting," "pointing"). *Niqqud* is a system of diacritical signs that are traditionally used to represent vowels in Hebrew, whose writing system is consonantal, as it also happens in other Semitic languages. On the experience of reading a consonantal language, see the recent (Shimron 2006) and the debated (De Kerckhove 1990).

4 On the so-called 'Iyyun Circle and its literature, see (Scholem 1987) and especially (Verman 1992).

5 (Wiener 2008, p. 50). On similar topics, see also (Dal Bo 2021b).

6 On the development of a primitive Comparative Semitic Philology in the Spanish milieu, see (Maman 2004).

7 For a comprehensive discussion on Giqatilla's life and works, see (Dal Bo 2019, pp. i–lxx). See also (Blickstein 1983) and the more recent (Morlok 2011).

8 See again (Dal Bo 2019, pp. 121–22).

9 For a discussion of Giqatilla's notion of emanation, see (Dal Bo 2011). In particular, the term *hamshakhah* refers to taking drawn water and pouring it into the *mikveh* so that it flows over the ground before entering the pool (Soloveitchik 1978).

10 For a further discussion, see (Dal Bo 2019, pp. 81–98 and 203–14).

11 Samuel ibn Naghrillah (993–1056), better known as Shemuel ha-Nagid, was a Medieval Spanish Talmudic scholar and an influential Hebrew poet. See (Cole 1996).

12 Solomon Ibn Gabirol (1021–1070), also known as Avicebron or Avencebrol in the Latin world, was a prominent Jewish philosopher and poet. See (Cole 2001).

13 Moshe Ibn Ezra (1055–1138) was a Jewish philosopher and poet. See (Brody 1934).

14 Jehudah ben Shemuel ha-Levi (1075–1141) was Jewish philosopher and poet. See (Brody 1974).

15 Arabic linguistics and its interest in cognate languages—Hebrew, Aramaic, and Syriac—played a main role in awakening the study of Hebrew among the Arabic-speaking Jewry from Spain. Notable figures of Jewish scholars in Hebrew are Menachem ben Saruq (910–?), Dunash ben Labrat (920–990), and Abu Zakariyya Yahya ibn Dawūd Hayyūj (Hayyug) (940–1000), known as the founder of Hebrew linguistics. On the impact of Arabic on Hebrew poetry, see, for instance, (Martínez Delgado 2013).

16 Scholarship on the transmission of Aristotelianism into the Semitic and Latin world is very large. See, for instance, (Marenbon 2011).

17 The so-called "*Heikhalot* literature" is a form of early Jewish mysticism that was mostly concerned with the notion of ascending into heavenly "Palaces" by a complex process of purification. This genre also overlaps with the mysticism of the Chariot (*Merkabah*) and fills the historical segment between post-Talmudic and pre-Kabbalistic literature. Cf. (Schäfer 1992).

18 Scholarship on Jewish philosophical poetry in Spain and Ibn Gabirol abounds. I will limit here to the important text (Pessin 2013).

19 David ibn Abi Zimra or David ben Zimra (1479–1589), also known as Radbaz, was a prominent Talmud scholar and one of the most respected figures in the 16th century. See (Morell 2004).

20 For Mopsik's commentary, see (Mopsik 1994, n. 1, p. 47). On the poetic flair of this expression and its probable connection with Andalusian love poetry, see (Tanenbaum 2002); cf. (Dal Bo 2019, pp. 100–1).

References

Bar-Asher, Avishai. 2020. From Alphabetical Mysticism to Theosophical Kabbalah: A Rare Witness to an Intermediate Stage of Moses de León's Thought. *Revue des Études Juives* 179: 351–84.

Blickstein, Shlomo. 1983. Between Philosophy and Mysticism: A Study of the Philosophical-Qabbalistic Writings of Yosef Gikatila. Ph.D. thesis, Jewish Theological Seminary, New York, NY, USA.

Brody, Heinrich. 1934. *Selected Poems of Moses ibn Ezra.* Philadelphia: The Jewish Publication Society of America.

Brody, Heinrich. 1974. *Selected Poems of Jehuda Halevi.* Philadelphia: The Jewish Publication Society of America.

Cantarino, Vincente. 1975. *Arabic Poetics in the Golden Age. Selection of Texts Accompanied by a Preliminary Study.* Leiden: Brill.

Cole, Peter. 1996. *Selected Poems of Shmuel HaNagid.* Princeton: Princeton University Press.

Cole, Peter. 2001. *Selected Poems of Solomon Ibn Gabirol.* Princeton: Princeton University Press.

Cole, Peter. 2012. *The Poetry of Kabbalah. Mystical Verse from the Jewish Tradition.* New Haven: Yale University Press.

Dal Bo, Federico. 2011. The Theory of 'Emanation' in Giqatilla's "Gates of Justice". *Journal of Jewish Studies* 62: 79–104.

Dal Bo, Federico. 2019. *Emanation and Philosophy of Language. An Introduction to Joseph ben Abraham Giqatilla.* Los Angles: Cherub Press.

Dal Bo, Federico. 2021a. Appendix. Ibn Gabirol between Philosophy and Kabbalah. A Comprehensive Insight into the Jewish Reception of Ibn Gabirol in Medieval and Modern Jewish Scholarship. In *Unravelling Ibn Gabirol's Metaphysics: Philosophical and Historical Studies.* Edited by Nicola Polloni, Marienza Benedetto and Federico Dal Bo. Turnout: Brepols.

Dal Bo, Federico. 2021b. *La via Ebraica Alla Meditazione.* Milan: Corriere della Sera.

Dan, Joseph. 1998. *Jewish Mysticism. The Middle Ages.* North Bergen: Jason Aronson.

De Kerckhove, Derrick. 1990. *La civilisation vidéo-chrétienne.* Paris: Retz.

Genette, Gerard. 1997. *Paratexts: Thresholds of Interpretation.* Cambridge: Cambridge University Press.

Gottlieb, Ephraim. 1976. *Studies in the Kabbala Literature.* Tel Aviv: Tel Aviv University Press. (In Hebrew)

Gruenwald, Ithamar. 1966. Two Cabbalistic Poems of Yosef Giqatila. *Tarbiz* 36: 76–84. (In Hebrew).

Lachter, Hartley. 2008. Kabbalah, Philosophy and the Jewish-Christian Debate. Reconsidering the Early Works of Joseph Gikatilla. *JJTP* 16: 1–58. [CrossRef]

Ludescher, Tanyss. 1996. The Islamic Roots of the Poetic Syllogism. *College Literature* 23: 93–99.

Maier, Johann. 1987. Politische Aspekte der Sefirotlehre des Josef ben Abraham Gikatilla. In *Aspetti della Storiografia Ebraica.* Rome: Piattelli, pp. 213–25.

Maman, Aharon. 2004. *Comparative Semitic Philology in the Middle Ages. From Sa'adiah Gaon to Ibn Barūn (10th–12th C.).* Brill: Leiden.

Marenbon, John. 2011. Aristotelianism in the Greek, Latin, Syriac, Arabic, and Hebrew Traditions. In *Encyclopedia of Medieval Philosophy.* Dordrecht: Springer. [CrossRef]

Martínez Delgado, José. 2013. On the phonology of Hebrew in Alandalus as reflected by the adaptation of Arabic grammar and poetry. In *Archaism and Innovation in the Semitic Languages.* Edited by Juan Pedro Monferrer-Sala and Wilfred G. E. Watson. Cordoba: CNERU, pp. 73–86.

Martini, Annett. 2011a. Introduction, in Joseph Giqatilla. In *The Book of Punctuation. Flavius Mithridates Latin Translation, the Hebrew Text, and an English Version.* Turin: Aragno, pp. 17–162.

Martini, Annett. 2011b. Seven Mystical Poems on the Hebrew Vowels as Interpreted by Yosef Giqatilla and Modekhay Dato. *EJJS* 5: 205–18. [CrossRef]

Miller, Mike T. 2016. *The Name of God in Jewish Thought. A Philosophical Analysis of Mystical Traditions from Apocalyptic to Kabbalah.* London: Routledge.

Mopsik, Charles. 1994. *Le Secret du Mariage de David et Bethsabée. Texte Hébreu Ètabli. Traduit e Présenté.* Paris: Editions de l'Éclat.

Morell, Samuel. 2004. *Studies in the Judicial Methodology of Rabbi David Ibn Abi Zimra.* Dallas: University Press of America.

Morlok, Elke. 2011. *Rabbi Joseph Gikatilla's Hermeneutics.* Tubingen: Mohr Siebeck.

Pessin, Sarah. 2013. *Ibn Gabiol's Theology of Desire.* Cambridge: Cambridge University Press.

Schäfer, Peter. 1992. *The Hidden and Manifest God. Some Major Themes in Early Jewish Mysticism.* New York: SUNY.

Scholem, Gershom. 1987. *Origins of the Kabbalah.* Princeton: Princeton University Press.

Shimron, Joseph. 2006. *Reading Hebrew: The Language and the Psychology of Reading It.* London: Lawrence Erlbaum Associates.

Soloveitchik, Chayyim. 1978. Can halakhic texts talk history? *AJS Review* 3: 153–96. [CrossRef]

Tanenbaum, Adena. 2002. *The Contemplative Soul. Hebrew Poetry and Philosophical Theory in Medieval Spain.* Leiden: Brill.

Verman, Mark. 1992. *The Books of Contemplation. Medieval Jewish Mystical Sources.* Albany: SUNY.

Weiss, Tzahi. 2015. *Letters by Which Heaven and Earth Were Created.* Jerusalem: Bialik Institute.

Wiener, Orna Rachel. 2008. The Mysteries of the Vocalization of the Spanish–Castillia and Kabbalah in the Thirteenth Century. Unpublished Ph.D. thesis, Bar Ian University, Tel Aviv, Israel.

Wiener, Orna Rachel. 2016. *Sod ha-Nikud.* Tel Aviv: Idra Publishing.

religions

MDPI

Article

"One Kind of Water Brings Another." Teresa de Jesús and Ibn 'Arabi

María M. Carrión

Departments of Religion and Comparative Literature, Emory University, Atlanta, GA 30322, USA; mcarrio@emory.edu

Received: 18 September 2020; Accepted: 13 October 2020; Published: 21 October 2020

Abstract: Mystical literature and spirituality from 16th-century Spain engage religious images from the three most prominent religions of al-Andalus—Christianity, Islam, and Judaism: among others, the dark night, the seven concentric castles, the gazelle, the bird, the *sefirot*'s encircled *iggulim* or towering *yosher*, the sacred fountain, ruins, and gardens. Until the 20th-century, however, scholarship read these works mostly as "Spanish" mysticism, alienated from its Andalusī roots. This comparative study deploys theological, historical, and textual analysis to dwell in one of these roots: the figure of the garden's vital element, water, as represented in the works of Teresa de Jesús and Ibn 'Arabi. The well-irrigated life written by these mystics underscores the significance of this element as a path to life, knowledge, and love of and by God. Bringing together scholarship on Christian and Sufi mysticism, and underscoring the centrality of movement, flow, and circulation, this article pieces together otherwise disparate readings of both the individual work of these two figures and their belonging in a canon of Andalusī/Spanish mysticism. The weaving of these threads will offer readers a different understanding of early modern religion, alongside traditional readings of Spain's mystical literature and its place in the global context.

Keywords: Spanish mystical literature; Teresa de Jesús; Ibn 'Arabi; Christian/'Spanish' and Sufi mysticism; comparative theology; gardens; life; knowledge; path; proximity to God

---

1. Introduction

> "For each limb or organ there is a particular kind of spiritual knowledge stemming from the one source, which is manifold in respect of the many limbs and organs, even as water, although a single reality, varies in taste according to its location, some being sweet and pleasant, some being salty and brackish. In spite of this it remains unalterably water in all conditions, with all the varieties of taste." (Ibn al-'Arabi 2015).

In the late 1400s AD four events aligned to change the religious and cultural landscape of what had been known by different people as Hispania, Iberia, and al-Andalus: the "discovery" of the Americas; the publication of the first Spanish grammar, "partner of empire;" the Expulsion of the Jews from Spanish soil, and the transfer of Granada—the last Muslim Kingdom of the Peninsula—from Sultan Abū 'Abdi-llāh Muḥammad ath-thānī 'ashar, the last ruler of the Nasrid house, to Isabel and Fernando, Catholic Queen and King. Anyone remaining within the geographic confines of the modern Church-State of Spain was forced to convert to Catholicism, the official religion, and the first half of the 16th century witnessed myriad conversions (Havrey 2005; Catlos 2014). By the late 1560s the *Recopilación de las Reyes destos Reynos*, first unified legal compilation of the Peninsula, inaugurated Spain's centralizing code of citizenship organized around the backbone law of the land, "De la Santa Fé Católica" (On the Holy Catholic Faith):

> "The Holy Mother Church teaches, and preaches, that every faithful Christian reformed by the holy Sacrament of Baptism firmly believes and simply confesses that there is one only true

God, eternal, immense, unchanging, omnipotent, ineffable, Father, Son, and Holy Spirit three Persons and one essence, substance, or nature: the Father unattainable, the Son engendered, by the Father alone and the Holy Spirit breathed from high simplicity, proceeding likewise from the Father, and the son in essence, equal in omnipotence: and one beginning principle of all visible and invisible things." (*Recopilación*, Lib. I Tit. 1 Ley 1; my translation)

Despite the powerful rhetorical charge of this "First Law" and subsequent Laws"Laws" regarding the clearly defined hierarchical theological and political directives for Spanish citizens to profess their exclusive alliance and devotion to Catholicism, the lives of crown subjects in the Peninsula unfolded in many different directions. It was as if Ibn 'Arabi's *Bezels of Wisdom* foreboded, from 13th-century Murcia, that the Spanish body would be one, and that its organs and limbs—like different bodies of water in relation to the main source—would stem from one source, but would have a "particular kind of spiritual knowledge." There is ample evidence that substantial numbers of crypto-Muslims and *conversos—anusim* publicly declared their adherence to the Holy Catholic Faith while at the same time they continued to practice their Islamic or Jewish faith in underground spaces and secret times (Havrey 2005, pp. 102–21).[1] Against all odds, and true to their faith as best as they could in a soil declared inhospitable for their religion, the perseverance of these individuals and their communities preserved religious pluralism inside the peninsula. Be that as it may, the importance of their lives and voices has remained covered for centuries under the aegis of one national Catholic unity. The movement known as Spanish mysticism is one more example of this kind of survival of religious pluralism in 16th-century Spain and partial reception thereafter.[2] In an era when moral virtue and religious rituals and iconography were heavily regulated to favor a Catholic gestalt (food, clothing, movements, inspirational figures, devotional imagery, and rituals, among others), mystics in Spain engaged in traditions from the three most prominent religions of al-Andalus—Christianity, Islam, and Judaism. With this hybrid approach to religious life, they shaped a rich spiritual legacy being fully revealed only in the last century by scholarly work like that found in this special issue of *Religions*. Among numerous other images and practices, 16th-century mystics in Spain imagined and wrote of the dark night, the seven concentric castles, the gazelle, the bird, the *sefirot*'s encircled *iggulim* or towering *yosher*, ruins and gardens, and the sacred fountain to inspire others to seek their own spiritual paths.[3]

The ties that bind Spanish and Sufi mystics are, no doubt, numerous; at once, they are quite complex and, because of that, they require rigorous comparative theological, historical, and textual

---

1 Richard Pym's important edited collection of essays on various aspects of morality establishes a radical critique of conventional readings of Spanish history as fully compliant with this powerful rhetorical mainframe of Catholic-only virtue and morals. According to Pym (Pym 2006), the imperative tone and register of official documents—such as the *Recopilaciones* were, as well as the series of Edicts, published and delivered by *pregoneros* (town criers) throughout the peninsula to make these laws known to all citizens—they did not "map unproblematically onto the complex topography of everyday life, or the immediate experience of Spaniards," in which voices of skepticism, subversion, irony, survival, and other modes of resistance abounded (p. ix).

2 I emphasize "Spanish mysticism" to question the perception that the literature and history of these mystics is exclusively Spanish, which may seem to mean that they were produced in, or that it may only be significant for, an exclusive religious or linguistic sense, nationalist spirit, or imperial tone and register. These texts were produced in Spain mostly by Catholic monastic figures, to be sure; however, as it has and continues to become apparent, the spirit and the letters of their legacy is far from monolingual, monocultural, or inspired by one dogma alone. No doubt, to call them Andalusī mystics would not be fully correct, given the political changes that the Capitulation of the Nasrid Kingdom of Granada brought about after 1492. To decolonize the expression, the present study intently deploys the expression "Spanish mystics" as figures that inscribe the historical memory of al-Andalus in their literary and theological works.

3 Cynthia Robinson (2006, 2013) and Conde Solanes (2020) are two amongst many other scholars who have analyzed the fertile Medieval soil upon which the survival of this religious and spiritual pluralism is founded. Robinson's work shows how Iberian spiritual traditions such as the one of the "Cristo de la Cepa" (devotional figure of Christ carved in vine wood) were forged in Al-Andalus by the influence of 9th century Sufi mystics such Abū 'l-Muġīth Al-Ḥusayn bin Manṣūr al-Ḥallāj and Sahl al-Tustarī from Persia, and 'Abd Allah b. Masarra, from Córdoba; Conde Solares also points out that sustained devotional practices like this one reveal the existence of legacies "guarded in medieval Spain by Benedictine and Dominican monks whose main mission was converting Jewish and Muslim souls to Christianity" (Conde Solanes 2020, p. 4). The devotional system structured by trees of love and knowledge, key signs of the spiritual and artistic Andalusī garden, and their influence on Castilian devotional practices is explored thoroughly by Robinson (2006).

analysis.[4] Such analysis is particularly difficult because the means of transmission of ideas and practices of spirituality produced by the figures who compose these two groups has only recently begun to be analyzed. Given the histories of conflict and violence in the name of religion that characterized the latter Middle Ages and the Renaissance in Spain, and the underground religious practices such as those noted above, these means of transmission have proven hard to find. It may seem that Sufi and Spanish mystical practices did not bridge the religious divide. With that premise in mind, this study analyzes how water, a key material and spiritual element represented in the mystical works of Ibn 'Arabi and Teresa de Jesús, makes the theological worlds of Islam and Christianity overlap and, when read together, grow substantially.[5] The method here, then, is not one designed to contribute to the question of how the ideas of a Sufi mystic were transmitted to those of a Spanish mystic. The comparative approach is deployed here not to prove how their shared ideas travelled from 13th-century Murcia to 16th-century Ávila, or from a Medieval to a Renaissance spiritual journey, but rather to focus on reading their ideas together to explore the ways in which their separate and corresponding versions of water expand, and not limit, spiritual growth. To be clear, the "mutual" influences noted here between the works of these two mystics do not point to a two-way cycle of synchronous literary or religious exchange between them, or to an imagined scenario of one, the other, or the two having read each other's work; given the span of almost four centuries that separate them, that is simply not a logical choice.[6] Furthermore, because my argument does not seek to tease out an exact replica of the represented element of water in their texts, or an authoritarian one-way influence of Ibn 'Arabi over Teresa de Jesús (hereafter identified by her last name, De Jesús), or a posteriori, vice versa, the present reading will explore the ways in which their texts mutually inform each other in their representing water and its loving meanings, while discerning the respective contexts of production of these works. The well-irrigated life written by these mystics underscores the element of water as a path to life, knowledge, and love of and from God, as it foregrounds the spiritual value of movement, flow, and circulation. With the sign of water, readers can piece together otherwise disparate readings of both the individual works of these two figures and their belonging in a revised continuum of Andalusī and Spanish mysticism.

2. Teresa de Jesús's Four-Way Water

> Pay attention to the Most High's saying: "It is watered with one water" (Q. 13:4). The earth is one, but the tastes, fragrances, and colors differ. (Ibn 'Arabi, *Al-Futūhāt al-Makkiyya/The Meccan Illuminations* (III, 231, Chapter 351))[7].

Teresa de Jesús, a leading Spanish mystic, publicly declared and lived according to her affiliation with the Catholic Church. In 1622, about forty years after her death, Pope Gregory XV canonized her and she came to be Santa Teresa de Ávila for some, De Jesús for others, or just plain Santa Teresa for others; a few years later she was named Patron Saint of Spain by King Philip IV, which placed her next to St. James, *Santiago Apóstol*. By virtue of her theological contributions and religious writings, Pope Paul VI granted her and Catherine of Siena the degree of Doctor of the Church in 1970; they were the first two women to receive such an honor (Slade 1970; Bilinkoff 1989). Alongside her devotion to the *Santa Fé*, she lived and wrote not fully adhering to the religious exceptionalism and isolation

---

4 Recent debates about the beginnings of Sufi mysticism in Al-Andalus make the landscape of crypto-religious practices in 16th-century Spain even harder to read. See Yousef Casewit's "The Rise of the Andalusī Mu'tabirūn" (Casewit 2017, pp. 57–90). Francis X. Clooney's model of deep learning and comparative theology (Clooney 2010) across borders has inspired much of the thinking behind the present comparative study; the errors, of course, are all mine.

5 For an analysis of water in Islamic and Christian heterodox cosmogonies from the Ottoman Period, see Stoyanov (2001).

6 Muḥyī al-Dīn Muḥammad ibn 'Alī ibn al-'Arabī (560/1165–638/1240) is named here by his more common name Ibn 'Arabi, and Teresa de Cepeda y Ahumada (1515–1582), more commonly known as Saint Teresa of Avila and named here by her nom de plume, Teresa de Jesús, and de Jesús thereafter. On the names of Teresa de Jesús and their critical importance, see Carrión (1994, pp. 43–67).

7 Quoted in Jaffray (2008, n.p.).

promulgated by the *Recopilación* in the mid-1500s. Instead, she developed a literary and religious ethic that, at odds with the expressed will of the Church State, conjugated signs and practices from different spiritual and religious traditions. The words from Chapter XIX of her *Book of Life* cited in the title illustrates this revelation: "un agua tray otra" (one kind of water brings another) (pp. 138, 109).[8] With this sentence De Jesús concludes an extensive meditation on the image of the garden, a metaphor she develops to refer to the heart, and the importance of irrigation, or prayer, for its well-being. Her texts repeatedly say that water equals prayer, and in wrapping up her theological articulation of the garden, a sort of paradise regained, that one kind of water-prayer brings about (*tray-trae*), carries, or attracts another kind. The sorrow and pain that brought tears to her life before she built her own garden are conjugated with the waters of the source, the well, the fountain, the river, and rain—a liquid universe that she, like any of her readers, can learn to manage if they know how water can transform their tears into life. This conjugation of water, garden, and prayer opens up a way for readers to meditate on the significance of this element as a path to life, knowledge, and love of and from God. This well-watered garden grew in the context of underground religious pluralism in which De Jesús lived, and in which Sufi mysticism was more present than history has led readers to believe.

De Jesús devoted a good part of her life to reform the Discalced branch of the Carmelite Order in Spain; to that effect, she founded seventeen convents organized to support the welfare and betterment of spiritual life for women. She wrote four major volumes: *Book of Life*, *Way of Perfection*, *Interior Castle*, and *Book of Foundations*. Because of her religious experiences and the programs she instituted in her convents, her male confessors were adamant that she should write them down; the fact that her first book spent decades in the hands of the Holy Office of the Inquisition granted her writings a judicial-confessional layer that has not gone unnoticed by scholars (Llamas Martínez 1972; Egido 1986; Slade 1995, pp. 9–29). With the confessors palpably present in her life and narrative works, she openly spoke to her sisters, the nuns who lived in her reformed convents with whom she mobilized a highly sophisticated communication network that Alison Weber termed a "rhetoric of femininity" (Weber 1990). Because of her Jewish lineage, her work has been associated with Semitic traditions (Álvarez 1995; Connor 1986, pp. 43–81; López-Baralt 1985, pp. 120–41, 156–60). Because of her being a subject of the Church State in 16th-century Spain and a key reformer of the Carmelite Order, her work has been associated with Catholicism and, more often than not until recent times, with the "Spanish Golden Age", an era traditionally characterized—as Carlos Conde Solares aptly notes—as "the political and imperial heights of Spain" (Conde Solanes 2020, p. 1).[9] Because of the imagery of her books, such as the darkness of the soul, the seven concentric castles, the little bird, spaces such as ruins and gardens, as well as the design of her convents and her approaches to spiritual life, her work has been also associated with Islamic traditions (Asín Palacios 1946; López-Baralt 2002, 1985, 1981; Carrión 2017, 2016b, 2016a, 2013, 2012, 2010, 2009). Additionally, because her writings and conventual reform charted such new territory for women to define prayer, humility, and virtue, her work still provides great inspiration for many communities of women the world over (see (Dorgan 2015; Pérez 2013), among others).

The *Book of Life* and the *Interior Castle* place water at the center of De Jesús's spiritual quest. As a result, communities who read her closely and organize their spiritual lives with those readings in mind focus on the importance of this sign in spiritual life. Carol Ann Chybowski, for instance, offers an economic synopsis of "St Teresa of Avila's Four Waters" which, she notes, are "based on her own experiences with mystical prayer. It is a path that we all must follow each in our own way as we

---

8 Hereafter cited as *Life* (De Jesús 1995). The first page number in parenthesis corresponds to the original version in Spanish, quoted from the edition by Father Silverio de Santa Teresa O.C.D. (De Jesús 1915); unless otherwise noted, the quotes in English are from the translation by Edgar A. Peers (De Jesús 1995).

9 For a small, yet representative sample of analysis of the literary and spiritual work of De Jesús favoring these heights, which constitute the vast majority of the reception of her work until a century ago, see (Ricard 1965; Dickens 1970; García de la Concha 1978). For a fuller bibliography of these strands of reception, see (Carrión 1994).

make our journey home to God" (Chybowski 2015). The cultivated path begins with dipping a bucket into a well in the first stage; thereafter, it moves on to adding a water wheel to alleviate the labor's hardship; in the third stage a river or a stream appear, further lifting the spirit of the gardener; and finally, the blessed rain, and the letting go of the work to allow oneself be wrapped by God's rain. Immersed in a liquid journey, the gardener works through labor, distractions, spiritual advancement, difficulty, recollection, and sheer will to leave behind the deserted soil and reach a lush landscape of spiritual plenitude. The Carmelite Sisters of the Most Sacred Heart of Los Angeles compare De Jesús's water world with scriptural passages (the Book of Genesis, John 4: 10, 13–14 and Mark 10: 17–31), and with the works of John of Ruysbroeck and George Bernard Shaw, to highlight how that life of irrigation and prayer marks a sound path to know the gift of God (Carmelite Sisters of the Most Sacred Heart of Los Angeles 2015).

This experiential knowledge brought about by water contrasts with the little attention paid by scholarship to this important unit of Spain's mystical legacy. Scholars have analyzed copiously the presence and meanings of gardens in Spanish cultural history, from the tradition of the *hortus conclusus* in Iberian Hebrew rhymed narratives and Berceo's *Milagros de Nuestra Señora* to the sexual landscape represented in the paternal *huerto* of *Celestina*, among others (Decter 2007; Alchalabi 2004; Bailo 2016; Snow 2000).[10] There is also substantial scholarship about the centrality of gardens in life and culture in the peninsula; thus, for instance, the sophisticated Medieval hydraulic and irrigation systems that informed Spain's entrance to engineering modernity (Glick 1996), the singular presence of Islamic gardens, especially in what came to be the modern region of Andalucía (Ruggles 1997, 2003), and the design and development of royal gardens in El Escorial and Aranjuez to foster the early modern industry of water distilleries (Rey Bueno 2004, 2009), among others.

Only a few scholars have underscored the critical importance of gardens in the works of Spanish mystics (Lottman 2010; Carrión 2012, 2013). The work in which De Jesús more copiously uses the word *agua* is her first one, the *Book of Life*, although the *Castle* also devotes important passages to the presence and meaning of water for the heart. For the sake of brevity, this study will focus on the representation of water in the *Life*, structured as a four-way space and medium.[11] Used over 70 times between Chapter 6 and Chapter 22, this term forges a semantic field with which the author spells a pathway from tears of sorrow to a universe of joy and plenitude organized in four levels: underground water from the well, water extracted by mechanical devices, water flowing in a river or stream, and water from rain. These four stages of water, moving from below-ground level to high-up in the atmosphere from whence it falls as rain, correspond with ways to irrigate the soul that, in turn, are equivalent to four stages of labor, knowledge, and love in the garden; encompassing them all, a corresponding frame to that of the four stages of water stands a structure of four degrees of prayer. The first one, characterized by heavy labor and a foundation of sorrow and pain, asks readers to imagine themselves a gardener who lowers a bucket into the depths of a well to obtain water. Citing water as an element that after her near-death experience in Chapter 5 she could not even tolerate, De Jesús's voice eventually turns to a desire to reach the Prayer of Quiet. To that end, in Chapter 11 she suggests that a book may help focus on prayer, but that for her, "to look at a field, or water, or flowers" serve to remind her of the Creator, to awaken her, and to recollect herself (*Life* p. 66). The book she is given, Augustine's *Confessions*, does not quench her spiritual thirst because saints fall once and recover, but she falls many times and does not seem to go anywhere in life.

Because she is a woman "writing simply what I am commanded," she would prefer not to engage comparisons, but spiritual language requires that she does so and, hence, she turns to the basic practice of prayer:

10 Liz Herbert McAvoy (2014a, 2014b) and Naoë Kukita Yoshikawa (2014) offer important feminist critiques of the tradition of the *hortus conclusus* in Northern European literature.

11 The representation of water in the *Castle* lends itself to the possibility of a comparative theological reading with the buildings and gardens of La Alhambra. See Carrión (2017).

"The beginner must think of himself as of one setting out to make a garden in which the Lord is to take His delight, yet in soil most unfruitful and full of weeds. His Majesty uproots the weeds and will set good plants in their stead. Let us suppose that this is already done—that a soul has resolved to practice prayer and has already begun to do so. We have now, by God's help, like good gardeners, to make these plants grow, and to water them carefully, so that they may not perish, but produce flowers which shall send forth great fragrance to give refreshment to this Lord of ours, so that He may often come into the garden to take His pleasure and have His delight among these virtues." (*Life*, p. 73).

In this first stage of prayer where the paradise lost of the garden first appears, De Jesús immediately underscores the importance of water. Water not as a possession, as a commodity to trade, or as an entity to contain and control—that is the work of men. Rather, she literally conjugates water as a verb, *regar*, to irrigate, as proactive work that flows with knowledge and will to grant life to the garden of the soul: "Let us consider how this garden can be watered, so that we may know what we have to do, what labor will cost us, if the gain will outweigh the labor and for how long this labor must be borne" (*Life*, p. 73).

Here, the garden adopts the classic form of the four directions that also harbor four ways to irrigate; both reader and gardener must remember that they are already inside the first stage of the garden, where pain and sorrow prevail, a rhetorical strategy to which she will return in the *Interior Castle*, where she tries to explain how one is to enter the space in which s/he already is, and how to reach the innermost chamber where the Lord sits, and whence the mystical union is bound to happen. The secret to that paradox, De Jesús will reveal in the *Castle*, is to know that this chamber has many other mansions above and under, everywhere, where the Lord is with, for the castle is the soul and anyone, especially the nuns reading her text, can roam free in that space. The map is clear, while the maze is thick: there are four stages of the garden, which are four ways to irrigate, and four ways to pray.

"It seems to me that the garden can be watered in four ways: by taking the water from a well, which costs us great labor; or by a water-wheel [*noria*] and buckets [*arcaduces*], when the water is drawn by a windlass [*torno*] (I have sometimes drawn it in this way: it is less laborious than the other and gives more water); or by a stream [*río*] or a brook [*arroyo*], which waters the ground much better, for it saturates it more thoroughly and there is less need to water it often, so that the gardener's labor is much less; or by heavy rain, when the Lord waters it with no labor of ours, a way incomparably better than any of those which have been described." (*Life*, p. 73).

Alas, the application of these ways of watering is not quite as easy or simple as the layout seems. In fact, the first part of the garden, with the well at its core, is a place of fatigue, labor, distractions, "aridity, dislike, distaste and so little desire to go and draw water" that may drive both gardener and prayer to give up (*Life*, p. 74). In this desert, where echoes of despair may be heard, loss and sorrow may jeopardize the quest; however, the text reminds gardening readers of His Majesty's capacity to keep "the flowers alive without water" and to make "the virtues grow" (*Life*, p. 74). Water, in this case, shall be tears "or, if there be none of these, tenderness and an interior feeling of devotion" (*Life*, p. 74). Devotion here is fueled by humility, and through a series of reiterations of gestures of patience echoing the death of Christ in the Cross, St Jerome in the desert, and other Christian devotional imagery, the text encourages readers to not give up, but rather to think of the "firm foundation" being laid for the garden (*Life*, p. 75). Only this way all gardeners, prisoners in this dry land, and more precisely "poor women like myself, who are weak and lack fortitude" shall be able endure the pain and sorrow that extends through this three-chapter stage, to be able to find themselves literally in greener pastures (*Life*, p. 75). The watering and the garden, as we shall see, harbor the connection between the Christian characters of the narrative with the Sufi mystical signs.

Chapter 14 opens with a memory of the earlier description of the second degree of prayer, in which the structure of the windlass and buckets saves on labor so the gardener may be "able to take some rest instead of being continually at work" (*Life*, p. 86). In this stage, "grace reveals itself to the soul more clearly" while faculties recollect, remember with grace, and they move to the inner space where the gardener can do her Prayer of Quiet (*Life*, p. 86). The windlass and buckets move water with significantly less work from the gardener, transforming the place into a space where she can focus on receiving the grace of God. Will is captive, like the water in the buckets and, in turn, the buckets in the windlass. At the same time, other faculties—memory and imagination—join in the collection of water and recollection of the mind and soul, and as this second degree of prayer unfolds, the tears of despair and sorrow turn into a joyful flow, a "water of great blessings and favors which the Lord gives in this state," which makes the little buds of virtue grow stronger (*Life*, p. 87). In this space the common best practices of gardening—pruning, weeding, rooting—move center stage and temperature moves in, with the cold of winter being a decisive factor, as it was in the lands where De Jesús moved in life. When winter is coming, the text says, the spark planted earlier by God begins the fire that will warm up the place. Chapter 16 moves on to the third degree of prayer, the "third water" that flows without mechanical structures, a section of gardening life where labor dwindles to open up to pleasure, sweetness, delight, ineffable joy, and the water of grace that "rises to the very neck of the soul" in the fruition of God (*Life*, p. 95).

In this area of the garden the Creator of the water irrigates the soul boundlessly, yielding a fusion of the soul and the Lord reminiscent of the Hebrew *Song of Songs*. Humility fuses with joy, and an inexplicable event takes place: "what the poor soul could not acquire, even if it labored and fatigued its understanding for as much as twenty years, this heavenly Gardener achieves in a moment; the fruit grows and ripens in such a way that, if the Lord wills, the soul can obtain sufficient nourishment from its own garden" (*Life*, p. 99). The death-in-life and the sorrow present in the first degree of prayer, that point of despair where the tale of Melibea ends, is transformed in Chapter 18 of the *Life* into the fourth and last water, where another death takes place: that of the soul to the world. In this space "there is no feeling, but only rejoicing, unaccompanied by any understanding of the thing in which the soul is rejoicing" because here grace rains onto the soil/soul (*Life*, p. 95). The great reward of this rain may come in abundance, granted by the Lord when the will has been relinquished in prayer to Him, oftentimes "when the gardener is least expecting it" (*Life*, p. 104). The benefits for the soul of this kind of water are the greatest one can imagine in prayer. The following chapters address various possible caveats one can find in the application of this watering system; in the end, however, De Jesús concludes this segment of her mystical journey stating that the water is a key element for spiritual life.

3. Ibn 'Arabi's One and Many Ways of Water

> "But here it is like rain falling from the heavens into a river or a spring; there is nothing but water there and it is impossible to divide or separate the water belonging to the river from that which fell from the heavens." St. Teresa of Avila, *Interior Castle* (De Jesús 1961, p. 132).

The chronological, theological, and gender gaps that separate Teresa de Jesús from Ibn 'Arabi are sizeable. He left Spain to make the pilgrimage to Mecca in 1193, a journey that Claude Addas calls "the voyage of no return" because that first incursion led him to multiple travels around the Maghreb, Egypt, Mesopotamia, Anatolia, Palestine, and Syria, where he met the saints of his time and read books that marked his path, and because these experiences in turn led him to an inner voyage guided by his learning Sufi doctrines, sainthood, disciples and tribes, servitude and the nocturnal voyage, as well as teachings and legacies of the prophets, the unicity of Being, and the Seal of the Saints, among others (Addas 2000). De Jesús, on the other hand, travelled merely inside the peninsula once she started the Discalced Carmelite reform at a point late in her life, but did not have the encompassing exposure to other areas of the world like Ibn 'Arabi, or experiences comparable to those he acquired in North Africa and the Middle East. At once, from the distant points in space and time in which they lived and wrote, a common practice of questioning borders emerges in their works; after all, both of their lives were marked by cultural difference and religious reform.[12] On the one hand, De Jesús negotiated her Jewish lineage with memories of Al-Andalus and a mandated Catholic profession and devotion; Ibn 'Arabi, on the other, negotiated his Sunni lineage with Shia Islamic circles where his writings became popular, as well as Jewish and Christian stories and characters that commonly appear in is writings. Readers inside and outside the convents De Jesús reformed read her books, while Ibn 'Arabi's "teachings quickly spread throughout the Islamic world, and they kept on spreading wherever Islam went, from Black Africa and the Balkans to Indonesia and China," spanning into languages other than Arabic, such as Urdu, Turkish, and Persian (Chittick 2007, pp. 2–3). At the very core of their lives, they shared a number of interests and vocations, amongst which mysticism, writing, and theology stand out. In their theological education they also shared ground, for despite their formal education and the differences that separate them, they both favored friendship and their own experience when it came to religious training and growth. Rather than deriving their own religious principles and practices solely from a formal school of theology, both De Jesús and Ibn 'Arabi developed sophisticated theological structures by wondering, engaging discussion with others, and especially by devoting themselves to the spiritual path. As Scott Kugle notes that is characteristic of Sufi mysticism, both of these authors aimed to love God and to be loved by God than to articulate theological arguments to be debated or turned into law (Kugle 2007, p. 1).

In this life of loving devotion, water plays as much a critical role for Ibn 'Arabi as it does for De Jesús, if the two mystics articulate this role in different narrative and spatial ways. Ibn 'Arabi does not structure levels to irrigate the garden of the soul in four ways, as De Jesús does; however, his articulation of water as a spiritual path enumerates as many areas of representation, which Angela Jaffray recognizes as "life, knowledge, *sharī'a*, and purification" (n.p.). However, to say that the structure of garden segments, types of irrigation, and degrees of prayer precisely crafted by De Jesús in the *Life* is the same as the four areas of meaning of water represented in various books by Ibn 'Arabi is, in a nutshell, comparing apples and oranges. At once, the structured ways in which Ibn 'Arabi inscribes water in his understanding of the spiritual path is as much of critical importance to his quest to love God and be loved by God as it would be for De Jesús. Springing from the verse from the Qu'rān "It is watered with one water" (13: 4), his *Meccan Illuminations* engage the many waters of divine intention, ranking in excellence, essential oneness, and perceptual diversity; divine intention "is like water," and the many different flavors in fruits and vegetables occur both despite and because the one water that irrigates them, yielding a variety of tastes, fragrance, and colors, as well as a natural paradox, all indebted to one water. No doubt this most voluminous work by Ibn 'Arabi grants significant meaning and weight to water, as other of his works do. However, as it happened with the *Life* in relation to all the other works by De Jesús where water surfaced, the book where Ibn 'Arabi articulates a more tangibly structured approach to water is in the *Fuṣūṣ al-Ḥikam* or *Bezels of Wisdom*, where the splendor of 27 prophets is generated by means of a different gemstone equal to a divine virtue (such as patience, oneness, heart, or being) and, in some cases, clearly defined by water.[13] As in the previous section, for the sake of brevity this section will concentrate on reading water in the *Bezels*.

According to Todd Lawson, each one of the 27 bezels that thread this book in as many chapters "is given to a particular community in the narrative and poetic person of a particular prophet. The bezel or prophetic reality is shaped to receive the particular divine virtue in the same way the mark of friendship on Ibn 'Arabī's back was shaped to receive the prophetic seal on the back of the Prophet Muḥammad" (Lawson 2016, p. 35). In other words, inside the structure of their individual chapters, these gemstones speak of relations between each prophet and their communities; as a whole, though, they establish a "rhyme of friendship" or "identity-blurring that may occur through the imitation of

---

[12] Manuela Ceballos argues that the issue of borders, and especially a clear understanding the porosity of such borders, is a critical one for the interpretation of the works of Sufi and Christian mystics from Al-Andalus and North Africa (Ceballos 2016).

[13] Hereafter cited as *Bezels*. Unless otherwise noted, the quotes in English are from the translation by Binyamin Abrahamov.

the Prophet," which here "is between the particular prophet and the Divinity" (Lawson 2016, p. 36). This structure is a sonorous rhyming space in Arabic that reveals a community of prophets related to each other, as Lawson points out, in alphabetical order and in a non-hierarchical manner: "As a special group of *awliyā'* (all messengers and prophets are *awliyā'* but not all *awliyā'* are messengers or prophets) they comprise the linguistic elements, the spiritual vocabulary, for the new/old language and revelation of Islam in which every community that has ever existed has had a prophet and in which each prophet speaks to their community in that community's language" (Lawson 2016, p. 37). The non-hierarchical circulation between them, signaled in the rhythmic pattern of poetic rhyme and alphabet strikes in the titles, harbors echoes of the garden and Mansions that De Jesús will lay out four centuries later, where readers could find themselves and at once lose themselves in the lull of poetic prayer:

> The wisdom of divine praise in Noah's teaching is:
> Ḥikma subūḤiyya fī kalima nūḤiyya
> The wisdom of divine mercy in Solomon's teaching is:
> Ḥikma raḤmāniyya fī kalima sulaymāniyya
> The wisdom of divine Being in David's teaching is:
> Ḥikma wujūdiyya fī kalima dā ʾūdiyya
> The wisdom of divine uniqueness in MuḤammad's teaching is:
> Ḥikma fardiyya fī kalima muḤammadiyya. (Lawson 2016, p. 36)

This is certainly not a random circulation; language, prophet, community, God, the *walī* and Muḥammad, as Lawson notes, constitute a tightly interlocked tapestry of correspondences and relations in the *Fuṣūṣ*, "truly, an ocean without shore, that attests to the reality and veracity of the vision" and the glue that weaves them into that one tapestry of *walāya* or loving friendship is *sarayān*, meaning "current, flux, circulation, emanation, and permeation" which causes rhythm, movement, and beauty (Lawson 2016, p. 37, 45).

The bezel of "the prophetic wisdom that exists in the essence of Jesus" conjugates aspects of water and, more particularly, its flow or circulation after leaving the source, with the spiritual path. Right after the bezel of "the wisdom of predetermination in the essence of Ezra", and right before the bezel of "the wisdom of mercy in the essence of Solomon", Ibn 'Arabi places the chapter on Jesus, the only one bearing prophetic wisdom because, as Chittick points out, his prophecy is eternal (Chittick 1984, p. 25). To characterize that timelessness of Jesus and his prophecy, Ibn 'Arabi uses a term related to liquidity: "This measure of life which permeates things is called divine nature (*lāhūt*), and human nature (*nāsūt*) is the substrate in which this spirit dwells. *Nāsūt* is called spirit, because of that which inheres in it" (*Bezels*, p. 105; my emphasis).[14] The intertwining of water in this narrative articulation of divinity is even more palpable in Yoshihiko Izutsu's translation: "The (universal) Life which flows through all things (*wa sarat al-ḥayāt fīhi*) is called "divine aspect" (*lāhūt*) of Being, while each individual locus in which that Spirit (i.e., Life) resides is called the "human aspect" (*nāsūt*). The "human aspect", too, may be called "spirit", but only in virtue of that which resides (*al-qā'im*) in it" (quoted by (Lawson 2016, p. 46); my emphasis). The eternal life of Jesus permeates and flows, like water, from the divine nature or aspect to exist essentially or permanently, to inhere, to reside in who receives him.

The incarnation of Jesus in the Bezels is the point at which water fully surfaces in this chapter: "the body of Jesus was created from the real water of Maryam and the imaginary water of Jibrīl, which pervaded the moisture of his breath, because of the breath of an animate being contains humidity, and element of water in it. The body of Jesus was composed of imagined and real water" (*Bezels*, p. 105). The birthing scene generates a sustained meditation on spiritual breath, revivification, luminosity, loftiness, and the attributes of what Ibn 'Arabi will call the Perfect Man mobilized by water as process, as element, and as divine essence. "The cosmos," says Ibn 'Arabi, "emerged in the form of its originator,

14 For a comparative philosophical analysis of water as life in Taoism and Sufism, see Izutsu (1984, pp. 141–51).

that is, the divine breath. When it is hot, it rises, and when it is cold and humid, it falls, for precipitation belongs to coldness and humidity, and when it is dry, it is stable without trembling. Precipitation drives from coldness and humidity" (*Bezels*, p. 108). This scientific articulation of the divine origin and subsequent flux of water seems to fuse *walāya* and *wilāya*, those terms of relation and command, of love and authority that preoccupied Ibn 'Arabi and other mystics.[15] On the one hand water freely circulates, like the friendship or relationality indicated by *walāya*, while on the other, the immutability of the originator who is the bearer of authority indicated by *wilāya*, both fusing in the act of creation.

Alongside Jesus, several prophets mobilize their gemstones by means of water, thus forming a discrete body of knowledge within the larger ocean without a shore. Thus, for instance, "The bezel of the wisdom of the unseen exists in the essence of Job" (*Bezels*, p. 132). Job's gemstone chapter begins with an encompassing view on water: "Know that the mystery of life permeates through water, since it is the root of the elements and foundation (*'anāṣir, arkān*). For this reason God makes "of water every living thing" (Qur'ān 21:30) [...] And the root of everything is water" (*Bezels*, p. 132). Indeed, for Ibn 'Arabi water is the material that sustains life. The Throne, like The Castle for De Jesús, is heart and center of all spiritual activity.[16] That space and material of divinity rests on water "because it was composed of water; it floats on the water, and the water preserves it from beneath" (*Bezels*, p. 132). Water sustains the world, and it signifies knowledge. The gemstone of knowledge exists in the essence of Moses, as he possesses "many kinds of wisdom" (*Bezels*, p. 156). Encoded in an interpretation of the story of Moses, Ibn 'Arabi offers the correspondence of water and knowledge:

> "As for the wisdom of putting him (Moses) in the ark (*tābūt*) and throwing him into the river, (its meaning is the following): The ark alludes to his humanity, while the river is the symbol of the knowledge he attained through his body. Only via the body, composed of the four elements, can the human soul be supplied with the faculties of reasoning, sensation, and imagination. [...] When he was cast into the river in order to attain different kinds of knowledge through these faculties, God taught him that even though the spirit which directs him is his ruler, the spirit directs him through these faculties." (*Bezels*, p. 157).

As De Jesús would see the gardener's death of the self to receive the rain of God, Ibn 'Arabi sees the immersion in the waters of knowledge experienced by Moses when he is put in the ark and cast in the river as "an external form of destruction," a moment in his life that, at once, saves him from getting killed later (*Bezels*, p. 158).

Moses, the man named after the Coptic *mū* (water) and *sā* (tree), brings life to Pharaoh and his wife, only to be woven into a thick tapestry of exaltation of wisdom where Adam, Pharaoh, and al-Khiḍr play roles with water, knowledge, life, and death. Noah's bezel of wisdom of exaltation engages transcendence, limitation, and restriction, and it how curtailing the flow of his call to his people—a flow that could have combined the call for a transcendent God and an immanent God—made them run and not listen to his call. By not combining both calls, Noah resorts to the Qu'ranic quote "He will send down abundant rain from the sky for you" but fails to reveal that the rain represents various kinds of intellectual knowledge and, because of that, rain and knowledge end up "far removed from the fruits of reflection" (*Bezels*, p. 39). Because of their heeding his call, Noah's folk ends up perishing in the very same waters of the sea of knowledge of God.

Knowledge and love must be part of the process of immersion into water, so the spiritual path can be cleared to move towards the union with Allah. For Angela Jaffray, water plays more than a mere supporting role to other theological narrative units (like character, place, action, and so on). In her teasing out Ibn 'Arabi's vision of the One and the Many she argues that water holds powerful clues, because it "performs a function in Ibn 'Arabi's writing that is analogous to his treatment of

[15] Vincent Cornell thoroughly examines the critical importance of these "semantic fraternal twins that coexist symbiotically, like yin and yang" to understand sainthood in Morocco (Cornell 1998, p. 1).
[16] On the heart of the faithful as The Throne of the All-Merciful, see Nasr (2002).

Islam, the Qu'ran, and the prophet Muhammad" (n.p.). As water is one of four elements, Islam is one of four monotheistic religions noted in the Qu'ran, alongside Judaism, Christianity, and Sabianism. For the faithful, water is as much an element tied to scientific discourse, pregnant with formulas, abstractions, or engineering material objects such as wells, watering holes, windlasses, or oases, as it is a metaphysical principle, because "it stirs the imagination as a polyvalent symbol expressing many things" (Jaffray 2008, n.p.).

4. Conclusions

Water, indeed, stirs the imagination and unfolds in a protean manner to mean many different things in the written works left for us by Teresa de Jesús and Ibn 'Arabi: life, path, fuel for prayer, knowledge, death, process, origin, and divine breath, amongst many others. In their many books devoted to writing about the spiritual path and their experiences of love for and from God, they inscribe various versions and forms of water that present their readers and spiritual interlocutors with ways to envision and grasp, intellectually and somatically, such divine love. Water, then, is not merely an element to manage or control or to know rationally; as Lawson points out, water is that sign and element by virtue of whose permeation and constant flow "all of the various oppositions are resolved, dissolved and even reversed. In short, the fearful symmetry of our lives is shown to be ephemeral while the substance of our lives, *walāya*, is shown to be timeless and permanent" (Lawson 2016, p. 48). This timeless and permanent meaning of water certainly shares symbolic and semantic ground with the theological and legal aspects of Baptism, the Christian Sacrament designed to bring divine life to a believer, and to secure a permanent presence of God in their soul. Ibn 'Arabi also mentions the ritual aspect of ablution as a means to cleanse and ready the body to receive God. These are ritualized and institutionalized approaches to water that no doubt occupy an important place in the beliefs of these two mystics.

However, water is also a source of life, knowledge, *poesis*, beauty, and a path to experience and express their love of God and the love of God for them. As such, water is nexus, structure, link, source, and channel of and for the spiritual path. Humility and piety are atoms that constitute this element, and so is relating to others such as the confessors, the nuns, and God in the case of De Jesús, or to the prophets, the members of their communities, and God in the case of Ibn 'Arabi. This relationality is marked by a desire to communicate the centrality of water, prayer, and knowledge in terms of beauty. Although their works are both written in narrative form, the previous pages show the benefits of reading them both as *poesis*, for beauty is an integral part of the happening of water as a spiritual path and in a believer's spiritual path. Rhyme, structure, patterns, numbers, and relations between signs and peoples compose the systematic aspect of these spiritual strands. By letting the life of the body and spirit negotiate divine and human, *walāya* and *wilāya*, humidity and dryness, one bezel and the other, a prophet with the other, or a segment of the garden with the other in constant motion, flowing like and with water, readers can capture a substantial part of their message.

Much work remains to be done to fully understand the impact and meaning of water in relation to gardens, prophecy, friendship, authority, sainthood, sacraments, rituals, and so on, both in Christian/Spanish and Sufi mysticism. Hopefully, the growth experienced in Andalusī studies in the past few decades will continue to reveal discoveries in the ways of transmission of Andalusī Sufi mysticism and its continuum in latter historical periods in the Iberian Peninsula. Until that time, the evidence provided by these mystical texts offers proof of a textually-based dialogue with imagery provided by their authors to fuel their spiritual quest and that of their readers. The presence and meaning of water in the works of these two and other mystics can be interpreted as an invitation to loosen up the reins of strict national, religious, historical or philological argumentation, a sign in their texts that, when read in dialogue with each other, can contribute to a better understanding of the great spiritual paths they built. Alongside the Cristo de la Cepa, the bird, the seven concentric castles or mansions, and myriad other mystical signs and symbols that link Christian/Spanish mysticism with Sufi mysticism, water shows another instance of how these great spiritual legacies inform each other.

With four ways, or one and many waters, every believer can find their own way to and with God. With Ibn 'Arabi, I conclude reciting from the *Fuṣūṣ*: "If the believer understood the meaning of the saying "*the color of the water is the color of the receptacle*", he would admit the validity of all beliefs and he would recognize God in every form and every object of faith."[17] Let us let them flow together.

Funding: This research received no external funding.

Conflicts of Interest: I declare no conflict of interest as author of this article.

References

Addas, Claude. 2000. *Ibn 'Arabī, the Voyage of No Return*. Cambridge: The Islamic Text Society.

Alchalabi, Frédéric. 2004. La fin de l'errance: Le jardín dans les *Milagros de Nuestra Señora* de Gonzalo de Berceo. *Le jardín: Formes et Representations*, 87–91. Available online: https://halshs.archives-ouvertes.fr/halshs-00608859/document (accessed on 15 July 2020).

Álvarez, Father Tomás. 1995. Santa Teresa de Ávila en el drama de los judeo-conversos castellanos. In *Judíos. Sefarditas. Conversos. La Expulsión de 1492 y sus Consecuencias*. Edited by Ángel Alcalá. Valladolid: Ámbito.

Asín Palacios, Miguel. 1946. El símil de los siete castillos del alma en la mística islámica y en Santa Teresa. *Al-Andalus* II: 267–78.

Bailo, Florencia E. 2016. Sobre la tradición del *hortus conclusus* en la "Introducción" de los *Milagros de Nuestra Señora*. *Letras* 73: 65–74.

Bilinkoff, Jodi. 1989. *The Avila of Saint Teresa. Religious Reform in a Sixteenth-Century City*. Ithaca: Cornell University Press.

Carmelite Sisters of the Most Sacred Heart of Los Angeles. 2015. If you Only Knew ... Saint Teresa and the Waters of Prayer. Available online: https://carmelitesistersocd.com/2015/st-teresa-and-prayer/ (accessed on 15 June 2020).

Carrión, María M. 1994. *Arquitectura y Cuerpo en la Figura Autorial de Teresa de Jesús*. Barcelona: Anthropos.

Carrión, María M. 2009. Scent of a Mystic Woman. Teresa de Jesús and the *Interior Castle*. *Medieval Encounters* 15: 130–56. [CrossRef]

Carrión, María M. 2010. Home, Sweet Home. Teresa de Jesús, *Mudéjar* Architecture, and the Place of Mysticism in Early Modern Spain. In *A New Companion to Hispanic Mysticism*. Edited by Hilaire Kallendorf. Leiden: Brill, pp. 343–65.

Carrión, María M. 2012. *Primero huerto*. Iconography, Anamorphism, and the *Idea* of the Garden in Sixteenth-Century Spanish Mysticism. *La Corónica* 41: 61–92. [CrossRef]

Carrión, María M. 2013. Amor a Dios, por amor al arte. Arquitectura e iconografía en la narrativa de Teresa de Jesús. In *Repensando la Experiencia Mística Desde las Ínsulas Extrañas*. Edited by Luce López-Baralt. Madrid: Editorial Trotta, pp. 277–308.

Carrión, María M. 2016a. After Thought. Devotion, Teresa de Jesús, and Miguel de Cervantes. *Romance Quarterly* 63: 40–50. [CrossRef]

Carrión, María M. 2016b. Alborada de las ruinas. Teresa de Jesús y la memoria histórica de al-Andalus. *eHumanista* 33: 178–90.

Carrión, María M. 2017. Jardines del alma. Hacia una teología comparada de La Alhambra y el *Castillo interior*. Special Issue: *Islam. Siwô'. Revista de Teología* 11: 81–105.

Casewit, Yousef. 2017. *The Mystics of al-Andalus. Ibn Barrajān and Islamic Thought in the Twelfth Century*. Cambridge: Cambridge University Press.

Catlos, Brian A. 2014. *Muslims of Medieval Latin Christendom, c.1050-1614*. Cambridge: Cambridge University Press.

Ceballos, Manuela. 2016. Communal Boundaries and Mystical Violence in the Western Mediterranean. Doctoral Dissertation, Laney Graduate School of Arts and Sciences, Emory University, Atlanta, GA, USA. Available online: https://etd.library.emory.edu/concern/etds/qn59q4563?locale=en (accessed on 15 May 2020).

Chittick, William C. 1984. The Chapter Headings of the *Fuūs*. *Journal of Muhyddin Ibn 'Arabi Society* II: 41–94.

[17] Quoted in the Introduction to Ibn 'Arabi in The Muhyddin Ibn Arabi Society's website. https://ibnarabisociety.org/introduction-muhyiddin-ibn-arabi/ (accessed on 11 October 2020).

Chittick, William C. 2007. *Ibn 'Arabi: Heir to the Prophets*. Oxford: Oneworld Publications.
Chybowski, Carol Ann. 2015. St Teresa of Avila's Four Waters. *New Evangelizers. Know Your Faith. Live your Faith*. Available online: https://newevangelizers.com/blog/2015/08/01/st-teresa-avilas-four-waters/ (accessed on 15 June 2020).
Clooney, Francis X. 2010. *Comparative Theology. Deep Learning Across Religious Borders*. Malden: Wiley-Blackwell.
Conde Solanes, Carlos. 2020. The Moral Dimensions of Sufism and the Iberian Mystical Canon. *Religions* 11: 15. [CrossRef]
Connor, (Swietlicki) Catherine. 1986. *Spanish Christian Cabala: The Works of Luis de León, Santa Teresa de Jesús, and San Juan de la Cruz*. Columbia: University of Missouri Press.
Cornell, Vincent. 1998. *Realm of the Saint. Power and Authority in Moroccan Sufism*. Austin: University of Texas Press.
De Jesús, Teresa. 1915. *Libro de la vida*. In *Obras de Santa Teresa de Jesús*. Edited by P. Silverio de Santa Teresa C. D. Burgos: El Monte Carmelo, vol. 1, Available online: https://archive.org/details/obrasdesantatere01tere/page/n7/mode/2up (accessed on 20 June 2020).
De Jesús, Teresa. 1961. *Interior Castle*. Edited by P. Silverio de Santa Teresa C. D. New York: Doubleday.
De Jesús, Teresa. 1995. *The Life of Teresa of Jesus. The Autobiography of Teresa of Ávila*. Translated and Edition by E. Allison Peers. From the critical Edition of P. Silverio de Santa Teresa C. D. Available online: http://www.carmelitemonks.org/Vocation/teresa_life.pdf (accessed on 20 June 2020).
Decter, Jonathan P. 2007. Landscape and Culture in the Medieval Hebrew Rhymed Prose Narrative. *Jewish Studies Quarterly* 14: 257–85. [CrossRef]
Dickens, Trueman. 1970. The Imagery of the Interior Castle and Its Implications. *Ephemerides Carmeliticae* XXI: 198–218.
Dorgan, Margaret, Sr. 2015. St. Teresa of Avila and Water. Carmelite Nuns of Eldridge Iowa. Available online: http://carmelitesofeldridge.org/St.TeresaofAvilaandWater.htm (accessed on 10 July 2020).
Egido, Teófanes. 1986. *El Linaje Judeoconverso de Santa Teresa (Pleito de Hidalguía de los Cepeda)*. Madrid: Editorial de la Espiritualidad.
García de la Concha, Víctor. 1978. *El Arte Literario de Santa Teresa*. Barcelona: Ariel.
Glick, Thomas. 1996. *Irrigation and Hydraulic Technology: Medieval Spain and Its Legacy*. Aldershot: Variorum.
Havrey, L. P. 2005. *Muslims in Spain, 1500 to 1614*. Chicago: University of Chicago Press.
Ibn al-'Arabi. 2015. *Ibn al-'Arabi's Fuūs al-Hikam. An Annotated Translation of "The Bezels of Wisdom"*. Translated by Binyamin Abrahamov. London: Routledge.
Izutsu, Toshihiko. 1984. *Sufism and Taoism: A Comparative Study of Key Philosophical Concepts*. Berkeley: University of California Press.
Jaffray, Angela. 2008. "Watered with One Water"—Ibn 'Arabi on the One and the Many. *Journal of Muhyddin Ibn 'Arabi Society* 43. Available online: https://ibnarabisociety.org/watered-with-one-water-by-angela-jaffray/ (accessed on 10 July 2020).
Kugle, Scott. 2007. *Sufis and Saints' Bodies: Mysticism, Corporeality, and Sacred Power in Islam*. Chapel Hill: The University of North Carolina Press.
Lawson, Todd. 2016. Friendship, Illumination, and the Water of Life. *Journal of the Muhyiddin Ibn 'Arabi Society* 59: 17–56.
Llamas Martínez, Enrique. 1972. *Santa Teresa de Jesús y la Inquisición Española*. Madrid: CSIC.
López-Baralt, Luce. 1981. La simbología mística musulmana en San Juan de la Cruz y Santa Teresa. *Nueva Revista de Filología Hispánica* XXX: 21–91.
López-Baralt, Luce. 1985. *San Juan de la Cruz y el Islam. Estudios Sobre Filiaciones Semíticas de su Literatura Mística*. México: El Colegio de México, Río Piedras: Universidad de Puerto Rico.
López-Baralt, Luce. 2002. La arquitectura mística secreta de santa Teresa, en deuda con el Islam. *Sigila. Revue Transdisciplinaire Franco-Portugaise sur le Secret* 10: 85–110.
Lottman, Maryrica Ortiz. 2010. The Real and Mystical Gardens of Teresa of Ávila. In *A New Companion to Hispanic Mysticism*. Edited by Hilaire Kallendorf. Leiden: Brill, pp. 323–42.
McAvoy, Liz Herbert. 2014a. The Medieval *Hortus conclusus*: Revisiting the Pleasure Garden. *Medieval Feminist Forum* 50: 5–10. [CrossRef]
McAvoy, Liz Herbert. 2014b. 'Fluorish Like a Garden:' Pain, Purgatory, and Salvation in the Writing of Medieval Religious Women. *Medieval Feminist Forum* 50: 33–60. [CrossRef]

Nasr, Seyyed Hossein. 2002. The Heart of the Faithful is the Throne of the All-Merciful. In *Paths to the Heart: Sufism and the Christian East*. Edited by James S. Cutsinger. Bloomington: World Wisdom, Inc., pp. 32–45.

Pérez, María José. 2013. De la Rueca a la Pluma. Available online: https://delaruecaalapluma.wordpress.com/ (accessed on 20 August 2020).

Pym, Richard, ed. 2006. *Rhetoric and Reality in Early Modern Spain*. London: Tamesis.

Rey Bueno, Mar. 2004. Juntas de herbolarios y tertulias espagíricas: El círculo cortesano de Diego de Cortavila (1597–1657). *Dynamis* 24: 243–67. [PubMed]

Rey Bueno, Mar. 2009. *La Mayson pur Distiller des Eaües* at El Escorial: Alchemy and Medicine at the Court of Philip II, 1556–98. *Medical History Supplement* 29: 26–39. [CrossRef]

Ricard, Robert. 1965. Le symbolisme du *Château intérieur* chez Sainte Thérèse. *Bulletin Hispanique* LXVII 9: 25–41. [CrossRef]

Robinson, Cynthia. 2006. Trees of Love, Trees of Knowledge: Toward the Definition of a Cross-Confessional Current in Late Medieval Iberian Spirituality. *Medieval Encounters* 12: 388–435. [CrossRef]

Robinson, Cynthia. 2013. *Imagining the Passion in a Multiconfessional Castile*. University Park: Penn State University Press.

Ruggles, D. Fairchild. 1997. The Eye of Sovereignty: Poetry and Vision in the Alhambra's Lindaraja Mirador. *Gesta* 36: 180–89. [CrossRef]

Ruggles, D. Fairchild. 2003. *Gardens, Landscape, and Vision in the Palaces of Islamic Spain*. University Park: Penn State University Press.

Slade, Carol. 1970. *Santa Teresa de Jesús, Doctora de la Iglesia. Documentos oficiales del Proceso Canónico*. Madrid: Editorial de la Espiritualidad.

Slade, Carol. 1995. *St. Teresa of Avila. Author of a Heroic Life*. Berkeley: The University of California Press.

Snow, Joseph. 2000. The Sexual Landscape of *Celestina*: Some Observations. *Calíope* 6: 149–66.

Stoyanov, Yuri. 2001. Islamic and Christian Heterodox Water Cosmogonies from the Ottoman Period: Parallels and Contrasts. *Bulletin of the School of Oriental and African Studies* 64: 19–33. [CrossRef]

Weber, Alison. 1990. *Teresa of Avila and the Rhetoric of Femininity*. Princeton: Princeton University Press.

Yoshikawa, Naoë Kukita. 2014. The Virgin in the *Hortus conclusus*: Healing the Body and Healing the Soul. *Medieval Feminist Forum* 50: 11–32. [CrossRef]

religions

MDPI

Article

St. John of the Cross and the Monopolar Concept of God in the Abrahamic Religions in Spain

Daniel Dombrowski

Philosophy Department, Seattle University, Seattle, WA 98122, USA; ddombrow@seattleu.edu

Received: 16 June 2020; Accepted: 16 July 2020; Published: 21 July 2020

Abstract: The aim of this article is to philosophically explore the tension between "the God of the philosophers" and "the God of religious experience." This exploration will focus on the mystical theology of the 16th c. Spanish mystic St. John of the Cross. It will be argued that a satisfactory resolution of the aforementioned tension cannot occur on the basis of the monopolar theism that has dominated the Abrahamic religions. That is, a better understanding of mystics in Judaism, Christianity, and Islam can occur via dipolar theism as articulated by contemporary process philosophers in the Abrahamic religions, especially the thought of Charles Hartshorne.

Keywords: mysticism; St. John of the Cross; Spain; monopolar theism; dipolar theism; process theism; Charles Hartshorne

1. Introduction

A largely neglected feature of the famous *convivencia* or coexistence of the three major Abrahamic religions in Spain is that at a very abstract level, thinkers in Judaism, Christianity, and Islam shared certain questionable assumptions regarding the concept of God. These shared assumptions were due to the way that Jewish, Christian, and Muslim thinkers interpreted their sacred scriptures in light of ancient Greek philosophers, especially Plato and Aristotle (the problem is not so much with Plato, in particular, but rather with the way he was interpreted). One key assumption will be examined here, which is the view of divine attributes as monopolar (to be described momentarily). Influential Jewish thinkers like Philo (1st c.—Philo 1929–1962) and Maimonides (12th c.—Maimonides 1885), Christian thinkers like Saints Augustine (4–5th c.—Augustine 1998), Anselm (11th c.—Anselm 1962), and Thomas Aquinas (13th c.—Thomas Aquinas 1972), and Muslim thinkers like Al-Ghazzali (11–12th c.—Al-Ghazzali 2000) and Averroes (12th c.—Averroes 1954) were all classical, monopolar theists. They shaped a concept of God that influenced mystics in all three Abrahamic religions in Spain (see Hartshorne 1953, pp. 76–164). Maimonides and Averroes are especially noteworthy because they lived in Spain (also see Solares 2020).

The problem with the classical, monopolar concept of God that I will analyze is that it is at odds with the God experienced by the great mystics in the Abrahamic religions, including the great mystics in 16th c. Spain, which is something of a peak era in mystical theology, much like Renaissance painting or 19th c. symphonic orchestration. In the present article, I will concentrate on St. John of the Cross (John of the Cross 1948, 1973), but other mystics in the Abrahamic religions (including kabbalistic and Sufi mystics) would serve just as well, given the pervasiveness of classical, monopolar assumptions regarding the concept of God. It is familiar to hear of "the God of the philosophers" at odds with "the God of religious experience." My thesis in the present article is that this opposition is actually between "the God of classical, monopolar theism" and "the God of religious experience," in that an improved neoclassical, dipolar concept of God can be developed that lessens the tension between the *concept* of God and the *experience* of God. Charles Hartshorne is the neoclassical, dipolar theist on whom I will rely most heavily in the development of my thesis in philosophy of religion.

The hope is that I will be able to advance the evolving legacy of Spanish mysticism by providing a concept of God more in line with what the great Spanish mystics, especially John of the Cross, say about the God they experienced. That is, I do not see the present article as a piece of antiquarian lore, but as an exercise in ongoing engagement with the concept of, and experience of, God. Further, in addition to Hartshorne, there are other neoclassical, dipolar theists who have come out of Christianity on whom I could rely, such as Alfred North Whitehead (Whitehead 1967, 1996), John Cobb (Cobb 2007), David Ray Griffin (Griffin 2001), Teilhard de Chardin (Teilhard de Chardin 1971), and Carol Christ, as well as contemporary Jewish neoclassical, dipolar theists like Abraham Heschel and Bradley Artson, and contemporary Muslim thinkers with similar views like Mohammed Iqbal and Mustafa Ruzgar. Further, the view of John of the Cross I will be developing in this article has benefitted greatly from the scholarship of Cristobal Serran-Pagan, who counteracts the view of John of the Cross's thought as world-negating by locating his mysticism in the context of the *conversos*, those Jews and Muslims who converted to Christianity and who enriched the tradition of mysticism in the Abrahamic religions in Spain (see Serran-Pagan 2018).

I should note at the outset that, despite the conceptual similarities I am emphasizing among classical theists in all three Abrahamic religions, there are also obvious differences among them that will not be treated in the present article. For example, regarding mysticism, there is a tendency in Christian mystics like John of the Cross to describe infused contemplation in terms of "mystic union" with God, if not absorption into God. In this regard, John of the Cross's *deificacion o endiosamiento* is like the *theosis* of Eastern Orthodox mystics in emphasizing mysticism as a process whereby one becomes deified. By partial contrast, Jewish mystics tend to emphasize the concept of *devekuth*, a loving intimacy with God, but not union with God or absorption into God. Mystics in Islam are also intent on preserving the view of God as *totaliter aliter* (as totally other). However, the important conceptual similarities among classical theists in general, to be described momentarily, are undeniable. Further, when John of the Cross speaks of *igualdad de amor* with God he is obviously not trying to establish a literal equality between human beings and God, but is rather trying to remove any unnecessary distance between the two such that mystical experience can be facilitated; this consideration tends to lessen any huge gaps among mystics in the Abrahamic religions. That is, in this article I will be dealing with the *common* problem of reconciling the monopolar concept of God with the experience of God, a problem John of the Cross shared with other Iberian mystics like Moses de Leon (Moses de Leon 1988), Abraham Abulafia (see Cole 2007), Ignatius of Loyola (Ignatius of Loyola 1964), Teresa of Avila (Teresa of Avila 1976), Ibn Abbad of Ronda (Ibn Abbad of Ronda 1986), Ibn al-Arabi (see Bashier 2004), etc.

2. Monopolar vs. Dipolar Theism

It is widely assumed that "the God of the philosophers" is at odds with John of the Cross's God of mystical experience. However, it will serve us well to be skeptical of the claim that it is the philosophical concept of God per se that is the problem. We would be better served to see what is problematic about the classical concept of God from the perspective of John of the Cross's mystical experience, then explore the possibility of there being a concept of God more congenial to such experience. Granted, *as a theologian*, John of the Cross was heavily influenced by the classical concept of God, especially as articulated by Thomas Aquinas, but as an author of several classics in the history of *mystical experience*, John of the Cross gives us a quite different view of God. My hope in the present article is to try to close the gap between the concept of God and mystical experience. Although John of the Cross was brought to the Carmelite order by St. Teresa of Avila precisely because of his Thomistic expertise in philosophy and theology so as to give intellectual rigor to the order, it is ironic that the classical theistic view that he knew was at odds with the mystical theology for which he became famous. It must be admitted that there is a tension in his thought between systematic theology and mystical theology, and that he leans in the direction of the latter, but there is no good reason to assume that this tension cannot be relaxed by more carefully examining the monopolar systematic theology in question.

At the outset, I would like to make it clear that by "classical theism" I refer to a view of God in philosophy and theology, not to biblical theism. It is an interesting question, and an open one, whether classical, monopolar theism or neoclassical, dipolar theism (to be explained momentarily) does a better job of preserving the best insights regarding the concept of God in the Abrahamic scriptures. This classical theistic view involves at least the following five features:

1. Omnipotence (including the related claim that God created the world out of absolute nothingness).
2. Omniscience (in the sense of God knowing, with absolute assurance and in minute detail, everything that will happen in the future).
3. Omnibenevolence.
4. Eternity (in the sense not of God existing everlastingly throughout all of time, but rather of God existing outside of time altogether).
5. Monopolarity (again, to be defined momentarily).

It will be to our advantage to be as clear as we can on what we mean by the term "God." In this effort we will be able to see more clearly the strengths and weaknesses of the classical theistic view. I will use the term "God" to refer to the supremely excellent or all-worshipful being. A debt to Anselm is evident in this preliminary definition. It closely resembles Anselm's "that than which no greater can be conceived" (see chp. 2–3 of his *Proslogion*).

However, Anselm's ontological argument is not what is at stake here. Even if the argument fails, the preliminary definition of God as the supremely excellent being, the all-worshipful being, or the greatest conceivable being seems unobjectionable. To say that God can be defined in these ways still leaves open the possibility that God is even more excellent or worshipful than our ability to conceive. This allows us to avoid objections from those in mystical theology in the Abrahamic religions like John of the Cross who might fear that by defining God we might be limiting God to "merely" human language. I am simply suggesting that when we think of God, we must be thinking of a being who surpasses all others, or else we are not thinking of God. Even the atheist or agnostic would admit this much. When the atheist says, "There is no God", there is a denial that a supremely excellent, all-worshipful, greatest conceivable being exists (see Mahoney 2004).

The excellent–inferior contrast is the truly invidious contrast when applied to God. If to be invidious is to be injurious, then this contrast is the most invidious one of all when both terms are applied to God, because God is only excellent. God is inferior in no way. To suggest that God is in some small way inferior to some other being is no longer to speak about God, but about some being that is not supremely excellent, all-worshipful, or the greatest conceivable. The dipolar theist's major criticism of classical theism is that it assumes that all contrasts, or most of them, when applied to God are at least somewhat invidious.

Let us assume that God exists. What attributes does God possess? Consider the following two columns of attributes in polar contrast to each other (see Hartshorne 1953, pp. 1–25):

one	many
being	becoming
activity	passivity
permanence	change
necessity	contingency
self-sufficient	dependent
actual	potential
absolute	relative
abstract	concrete

Classical theism tends toward oversimplification. It is comparatively easy to say, "God is strong rather than weak, so in all relations God is active, not passive." In each case, the classical theist decides which member of the contrasting pair is good (on the left), then attributes it to God, while wholly denying the contrasting term (on the right). Hence God is one but not many, permanent but not changing, and so on. This leads to what can be called the monopolar prejudice. The classical theist believes that the categorical contrasts listed above are invidious.

The defects found in monopolar, classical theism are artificial. They are produced by the assumption that excellence is found by separating and purifying one pole (on the left) and denigrating the other (on the right). That this is not the case can be seen by analyzing some of the attributes in the right-hand column. Classical theists in the Abrahamic religions have been convinced that God's eternity means not that God endures through all time everlastingly, but that God is outside of time altogether, and is not, cannot be receptive to temporal change and creaturely feelings. Thomas Aquinas (following Aristotle, who was the greatest predecessor to classical theism) identified God as unmoved; and John of the Cross followed Thomas Aquinas in this regard in his *systematic theology*, but not in what he indicates regarding *mystical experience*. Yet both activity and passivity can be either good or bad. Good passivity is likely to be called sensitivity, responsiveness, adaptability, sympathy, and the like, as John of the Cross frequently indicates (e.g., *Spiritual Canticle* 32, 1–2). Insufficiently subtle or defective passivity is called wooden inflexibility, mulish stubbornness, inadaptability, unresponsiveness, and the like. Passivity per se refers to the way in which an individual's activity takes account of, and renders itself appropriate to, the activities of others. To deny God passivity altogether is to deny God those aspects of passivity that are excellences. Or, put another way, to altogether deny God the ability to change does avoid fickleness, but at the expense of the ability to lovingly react to the sufferings of others, a reaction that is central to the testimony of the great theistic mystics, including John of the Cross.

The terms on the left side also have both good and bad aspects. Oneness can mean wholeness, but also it can mean monotony or triviality. Actuality can mean definiteness, or it can mean non-relatedness to others. What happens to divine love when God is claimed by Thomas Aquinas to be *pure* actuality? God ends up loving the world, but is not intrinsically related to it, whatever sort of love that may be. Self-sufficiency can, at times, be selfishness.

The task when thinking of God is to attribute to God all excellences (left and right sides) and not to attribute to God any inferiorities (right and left sides). In short, excellent–inferior, knowledge–ignorance, or good–evil are invidious contrasts; but one–many, being–becoming, and the like are noninvidious contrasts (see Beng 2009).

Within each pole of a noninvidious contrast (e.g., permanence–change), there are invidious or injurious elements (inferior permanence or inferior change), but also noninvidious, good elements (excellent permanence or excellent change). The dipolar, process theist does not believe in two gods, one unified and the other plural. Rather, there is belief that what are often thought to be contradictories are really mutually interdependent correlatives. The good is unity-in-variety or variety-in-unity. Too much variety leads to chaos or discord; whereas too much unity leads to monotony or triviality (see Hartshorne 1948, 1970; Artson 2013; Ruzgar 2008).

Supreme excellence, to be truly so, must somehow be able to integrate all the complexity there is in the world into itself as one spiritual whole. The word "must" indicates divine necessity, along with God's essence, which is to necessarily exist. The word "complexity" indicates the contingency that affects God through creaturely decisions. In the classical theistic view, however, God is identified solely with the stony immobility of the absolute, implying non-relatedness to the world. God's abstract nature, God's being, may in a way escape from the temporal flux, but a living God is related to the world of becoming, which entails a divine becoming as well, if the world in some way is internally related to God. The classical theist's alternative to this view suggests that all relationships to God are external to divinity, once again threatening not only God's love, but also God's nobility.

A dog being behind a particular rock affects the dog in certain ways, and thus this relation is an internal relation to the dog, but it does not affect the rock, whose relationship with the dog is external to the rock's nature. Does this not show the superiority of canine consciousness, which is aware of the rock, to rocklike existence, which is unaware of the dog? Is it not therefore peculiar that God has been described by classical theists solely in rocklike terms: Pure actuality, permanence, having only external relations, unmoved, being and not becoming? These terms are very much at odds with the God described in John of the Cross's mystical experiences and in the experiences of other mystics in the Abrahamic religions. Granted, John of the Cross in some sense (in the divine existence and abstract essence) sees God as immutable (*immutable para siempre*), but he is also quick to point out that in another sense, God changes, indeed that God surrenders to us (see, e.g., *Ascent of Mount Carmel*, III, 21). That is, one must question exactly what it *means* to refer to divine immutability.

The dipolar nature of John of the Cross's theism is evidenced when we see him often describing God as just *and* merciful, powerful *and* loving, sublime *and* delicate, etc. This sort of pairing is a commonplace in his thought (see, e.g., *Living Flame of Love*, 3, 2 and 3, 6). In fact, he says that if we are seeking God, it is even more true to say that God is seeking us, is moving toward us (see, e.g., *Living Flame of Love*, 3, 28).

3. Some Criticisms

One may wonder at this point why classical theism has been so popular among Abrahamic theists when it has so many defects. One can imagine at least four reasons, none of which establish the case for classical, monopolar theism: (1) It is simpler to accept monopolarity than dipolarity. That is, it is simpler to accept one and reject the other of contrasting (or better, correlative, noninvidious) categories than to show how each, in its own appropriate fashion, applies to an aspect of the divine nature. Yet the simplicity of calling God "the absolute" can come back to haunt the classical theist if absoluteness precludes relativity in the sense of internal relatedness to the world, including those who enter into sanjuanistic mystical union with God.

(2) If the decision to accept monopolarity has been made, it is simpler to identify God as the absolute than to identify God as the most relative. Yet this does not deny divine relatedness, nor that God, who loves all, would therefore have to be related to all, or, to use a roughly synonymous term, be relative to all. God may well be the most relative of all as well as the most absolute of all, in the sense that, and to the extent that, both of these are excellences. Of course, God is absolute and is relative in different aspects of the divine nature.

(3) There are emotional considerations favoring divine permanence, as found in the longing to escape the risks and uncertainties of life. Yet even if these considerations obtain, they should not blind us to other emotional considerations, like those that give us the solace that comes from knowing that the outcome of our sufferings and volitions makes a difference in the divine life, which, if it is all-loving (as John of the Cross and other Abrahamic mystics attest), will certainly not be unmoved by the suffering of creatures.

(4) Monopolarity is seen as more easily compatible with Abrahamic monotheism. Yet the innocent monotheistic contrast between the one and the many deals with God as an individual, not with the dogmatic claim that the divine individual itself cannot have parts or aspects of relatedness with the world.

In short, the divine being becomes, or the divine becoming is. God's being and becoming form a single reality, and there is no reason to leave the two poles in a paradoxical state: God *always changes* and both words are crucial. There is no logical contradiction in attributing contrasting predicates to the same individual provided they apply to different aspects of this individual. Hence, the remedy for "ontolatry," the worship of being, is not the contrary pole, "gignolatry," the worship of becoming. God's *existence* is everlastingly permanent, but God's *actuality* (how God exists concretely from moment to moment) is constantly changing (see Hartshorne 1984).

The concept of God that I am defending is: (a) *Dipolar*, because excellences can be found on both sides of contrasting categories (i.e., they are correlative and noninvidious); (b) a *neoclassical* theism, because it relies on the belief that classical theists (especially Anselm and Gersonides—see Hartshorne 1953, pp. 75, 106, 112, 118, 189, 225) were on the correct track when they described God as the supremely excellent, all-worshipful, greatest conceivable being, but the classical theists did an insufficient job of thinking through the logic of perfection; (c) a *process* theism because it sees the need for God to *become* in order for God to be called perfect, but not at the expense of God's always (i.e., permanently) *being* greater than all others; and (d) a theism that can be called *panentheism*, which literally means "all in God." God is neither completely removed from the world—i.e., unmoved by it—as in classical theism, nor completely identified with the world, as in pantheism. Rather, God is: (i) World-inclusive, in the sense that God cares for all the world, and all feelings in the world—especially suffering feelings—are felt by God; and (ii) transcendent, in the sense that God is greater than any other being, especially because of God's love. Thus, we should reject the concept of God as an unmoved mover not knowing the moving world (see Aristotle's *Metaphysics*); as the unmoved mover inconsistently knowing the moving world (classical theism); and as the unmoved mover knowing an ultimately unmoving, or at least noncontingent, world (Stoics, Spinoza—see Spinoza 1992, pantheism). Indeed, Heschel refers to God as the *most* moved mover (see Heschel 1962).

Two objections may be raised by the classical theist that ought to be considered. To the objection that if God changed God would not be perfect, for if God were perfect there would be no need to change, there is this reply: In order to be supremely excellent, God must at any particular time be the greatest conceivable being, the all-worshipful being. At a later time, however, or in a situation where some creature who previous did not suffer now suffers, God has new opportunities to exhibit divine, supreme excellence. That is, God's perfection does not merely allow God to change, but requires God to change.

The other objection might be that God is neither one nor many, neither actual nor potential, and so forth, because no human concept whatsoever applies to God literally or univocally, but at most analogically. The classical theist would say, perhaps, that God is more unitary than unity, more actual than actuality, as these are humanly known. Yet one wonders how classical theists, once they have admitted the insufficiency of human concepts, can legitimately give a favored status to one side (the left side) of conceptual contrasts at the expense of the other. Why, if God is simpler than the one, is God not also more complex, in terms of relatedness to diverse actual occasions, than the many? Analogical predication and negative theology in Abrahamic thinkers can just as easily fall victim to the monopolar prejudice as univocal predication. To be agent and patient together is much better than being either alone. This is preeminently the case with God, and a human being is more of an agent and patient than is an ape, who is more of both than a stone. Stones can neither talk nor listen, nor can they decide or appreciate others' decisions. The problem is not with analogical discourse regarding God per se, but rather with analogical discourse when distorted in a monopolar way.

4. The Importance of Divine Love in the Abrahamic Traditions

It probably does not even occur to classical theists to seriously question the idea that God is wholly immutable and nontemporal in that it is simply assumed that mutability and temporality constitute the order of the created. Or again, Abrahamic classical theists do not see as problematic the seemingly obvious contradiction between a concept of God as not compassionate (because immutable) and the evidence of John of the Cross's mystical experience and of the experience of other Abrahamic mystics wherein God is eminently compassionate and loving. This evidence is found on almost every page of John of the Cross's writings in all four of his major works: *Ascent of Mount Carmel*, *Dark Night of the Soul*, *Living Flame of Love*, and *The Spiritual Canticle*. Somehow or other, the classical theist alleges by way of contrast, God helps those in misery without sympathizing with them. It is simply assumed in classical theism that not to suffer is better than to suffer, rather than to think through carefully the dipolar (rather than monopolar) logic of perfection.

In this regard, Jewish classical theists (Philo, Maimonides, etc.), Christian classical theists (Saints Augustine, Anselm, Thomas Aquinas, as well as Martin Luther, John Calvin, etc.), and Muslim classical theists (Al-Ghazzali, Averroes, Avicenna 2005 [Ibn Sina]—see Inati 1996) are saddled with a monopolar metaphysics that is at odds with the great works in mystical theology (especially John of the Cross's) in these various traditions. Classical theists, in general, have a tendency toward the naked worship of power on the analogy of the political form of coercive power found in the despot, rather than toward responsive love (see Bushlack 2020). Or at least they have a tendency to put the concept of responsive love in a position subservient to divine omnipotence (see Sanderline 1989, 1993).

The problems of Saints Augustine and Thomas Aquinas are those of *all* classical theists in creating a monopolar concept of God that is impossible to reconcile with John of the Cross's mystical experience or the mystical experience of other Abrahamic theists. The relations between human beings and God are, from the Thomistic point of view, real to the creatures, but not to God, despite the fact that in mystical experience the mystic senses real relations both ways. The classical theist is imprisoned in the half-truths of monopolarity.

Further evidence of John of the Cross's similarity to mystics in other Abrahamic faiths is not hard to find. For example, there is an "Oriental" layer to the biblical *Song of Songs* that is the basis for his work *The Spiritual Canticle*, as Gerald Brenan has argued (see Brenan 1973). There is also a Jewish sensibility in evidence in his mystical *knowing by acquaintance* (in contrast to the abstract *knowledge by description* found in systematic theology), which reminds one of "biblical knowing" in the Hebrew scriptures, as Leo Spitzer and Deirdre Green have argued (see Spitzer 1969; Green 1986). Camille Campbell has gone so far as to suggest that John of the Cross's panentheism was influenced by the Jewish Kabbalah (see Campbell 1982). In this regard, one is reminded of Teresa of Avila's own well-known Jewish roots and of the fact that Mt. Carmel itself, which was something of a spiritual "Mecca" for both John of the Cross and Teresa of Avila, was the site of both Jewish and Christian solitaries over the centuries (see Dombrowski 1992; also Dombrowski 1994, 2001, 2004, 2006, 2010). Additionally, because John of the Cross's ancestors on his father's side were, like Teresa of Avila's, Toledan silk merchants, there are reasons to suspect that John of the Cross may also have had Jewish roots. And John of the Cross's mother may have had Moorish roots, as Luce Lopez Baralt has emphasized (see Baralt 1985). Fernand Braudel also argues that John of the Cross's famous dark night may have had an Islamic predecessor in Ibn Abbad (see Braudel 1975; also see Nieto 1979). That is, despite surface differences among the Abrahamic faiths, philosophers and theologians in these traditions are, like an aquatic plant, rooted beneath the surface by their classical theistic, monopolar assumptions. Likewise, contemporary process theists in different traditions are united beneath the surface in their neoclassical, dipolar theism (see Bornstein 2019).

5. Metaphysical Considerations

The doctrine of divine monopolarity is integrally connected to substantialist thinking, with John of the Cross's mystical experience by contrast connected to the vision of God as *living* and processual. There is an inverse relationship between classical theism's inability to explain a concept of God compatible with mystical experience in the positive sense of "mystical" and its tendency to proliferate "mystery" in the pejorative sense of the term. Two examples of the latter are: (1) The nastiest version of the theodicy problem created by classical theism's version of divine omnipotence, wherein all of the evils in the world are either sent by, or at least permitted by, the classical theistic God; and (2) the "mystery" of how human beings could be free (and hence responsible for their actions) if God is omniscient even with respect to the outcome of future contingencies such that the classical theistic God knows with complete assurance and in minute detail *everything* that human beings will do even before they act. That is, classical theism inflates "mystery" in the pejorative sense of the term and deflates the positive aspects of mystical experience that are the lifeblood of religious belief.

None of the concrete religious experiences of God in the Abrahamic religions are explicated well by classical theistic abstractions. Luckily, the long reign of monopolarity and etiolatry (worship of divine causality and denigration of divine receptivity) need not last forever. The hope is that Christian and other varieties of mysticism would find a theoretical home. By focusing on divine eternity (in contrast to divine temporal everlastingness) and divine unchangeableness in a classical theistic manner, the very heart of religion in mystical experience is lost or at least denigrated.

The problems with classical theism are ultimately metaphysical in that they involve a defective understanding of the relationship between being and becoming, the abstract and the concrete. Although mystical experience itself is not to be identified with metaphysical thinking, it may very well be the case that the latter is needed to accurately illuminate the former. The proper indictment of classical theism need not imply a veto on mystical experience. Instead, such an indictment can be seen as part of the effort to purify mystical experience and belief and to insure that the rich history of mystical experience (especially evidenced in 16th c. Spain) will not only be remembered, but will inspire a flourishing of mystical experience into the long-run future.

A changeless being cannot love, at least if the love in question is even remotely analogous to what we human beings understand love to be. This is why the greatest conceivable being could not be changeless, despite the entrenched concept of God that is inherited by many religious believers in the Abrahamic religions. Granted, if we abstract from all of God's contingent qualities (as in God's particular responses to those who have had sanjuanistic mystical experiences), the rest of the divine reality is described somewhat accurately by classical theists in that some aspects of God (e.g., God's everlasting *existence*) are in fact unchanging. God is both contingent and necessary in different aspects of the divine life and classical theists are to be thanked for the intellectual progress made with respect to description of the latter aspects. The gradual collapse of classical theism from the time of the Enlightenment until the present is due in large measure, however, to classical theism's inaccurate description of God's contingent aspects. Indeed, these are denied.

God's superiority to us, on classical theistic grounds, is that God only acts and does not interact. This unfortunately eliminates *any* analogy between God and creatures, despite the insightful Thomistic emphasis placed on the doctrine of analogy. The lack of such interaction would, if it occurred, ring the death knell for sanjuanistic and other mystical experience. Luckily there is a dipolar alternative to monopolar classical theism.

To worship Being (or the absolutely independent) is to worship an aspect of God rather than God. When classical theists do precisely this, they should not be surprised that they do not provide a warrant for mystical experience. It will be remembered that Jewish, Christian, and Islamic philosophers alike were heavily influenced by Aristotle in this regard, with Islamic scholars and translators making such influence possible; further, this consideration is not at odds with the thesis that it was actually the Neoplatonists who influenced classical theists in these traditions when it is realized that, regarding the concept of God, the Neoplatonists could just as easily be referred to as Neoaristotelians (see Dombrowski 2005; also Plato 1999; Aristotle 1984). By confusing the divine fulness with an abstraction like being or absoluteness, classical theists no doubt think that they are doing the concept of God a favor, but what is more likely is that they will make classical theism increasingly unbelievable. Nonetheless, we should thank classical theists for offering us a first approximation, albeit a one-sided one, of the concept of God. At its worst, this concept leads to idolatry. Likewise, classical physics was a first approximation to truth in that discipline, with relativity and quantum theories supplementing the original theory in crucial ways.

If there can be nothing greater or more worshipful than preeminent love, as John of the Cross indicates, especially in *The Living Flame of Love*, then there is something crucially misleading in the classical theistic identification of divine love with strict independence and nonrelativity (i.e., absoluteness). This sort of being would not even be minimally loving, much less worshipfully so. The chief contribution of classical theism is its emphasis on permanence, but the permanence *of divine love* is not explicated well on a classical theistic basis. As before, God *always changes* and both words are crucial. If one prays to the classical theistic God, or if one enters into mystical union with such a God, the divine being is in no way influenced because the God of classical theism influences all, but is in no way influenced. Aristotle accurately saw the consequences of God as an unmoved mover and more consistently owned up to these consequences than classical theists. Aristotle's God knows only itself, does not care for others, and cannot be affected by them, whether positively or negatively. So much the less for the notion that God is an unmoved mover. How can a "God" who is in no way changeable, is not capable of growth in any sense, is in no way open to influence or enrichment by the creatures, and is wholly self-sufficient, nonsocial, and nonloving (or at least is not loving in any way we can understand or feel) nonetheless be the God of religion and mystical experience?

The basic axioms that underly divine monopolarity are seldom questioned by classical theists, which is why neoclassical theists provide such an important function in the effort to understand John of the Cross's mystical experience. In all cases of knowing that we are able to understand, the knower conforms to and partially depends on the known. However, in classical theism God is made an exception to this understandable view by making the known conform to and depend on the divine knower. To say that God knows the world in classical theism is to say quite ironically that the world is known by God.

The flourishing of various mystics in Judaism, Christianity, and Islam in *convivencia* and 16th c. Spain, despite the lack of theoretical support from classical theism for their religious experiences, indicates the vitality and resilience of mystical theology. The attempt to blend total self-sufficiency and nonreceptiveness, on the one hand, with a God of loving relatedness to creatures, on the other, is inherently futile. The fact that God is totally devoid of compassion (or even passion) is a skeleton in the classical theist's closet that is an embarrassment in the face of the long history of mystical experience in 16th c. Spain and elsewhere. Love and compassion need not be seen as signs of weakness: To the contrary! To say that God is an unmoved mover *and* a preeminently loving being is to reduce the analogy between human love and divine love to the vanishing point (see Payne 1990; Murphy 1996).

The classical theistic world is one devoid of chance due to its view of divine power. This mistaken view acts like a corrosive acid with respect to any effort to extricate the classical theistic God from responsibility for the intense suffering that pervades human history. Or more charitably, if the half-truths of classical theism were recognized as such, rather than confusing them with an approximation to the whole truth, then there could also have been a closer approximation of a concept of God compatible with the testimonies of the famous mystics like John of the Cross. It is quite amazing that the definition of God as immutable perfection without remainder has held sway for two millennia.

The view I am defending in this article is that classical theism is an incorrect translation of the central religious ideas (e.g., love) into philosophical categories. The goal is to preserve, and perhaps even enhance, mystical experience while avoiding contradictions and paradox. Theological paradoxes, it should be noted, tend to be contradictions applied to the divine case. To repeat an example from classical theism itself, if a dog is superior to a pillar because the former is capable of internal relations, whereas the latter is capable only of external relations, and if a human being is far superior to the dog in terms of capacity for internal relations, is it not a "paradox" that the greatest conceivable being was thought to be wholly devoid of internal relations? If God is purely absolute (i.e., purely unrelated), then the only being that God could enjoy is divinity itself and we cannot, even in mystical experience, be related to such a God.

The classics in mystical theology in the Abrahamic religions give testimony not to a God unrelated to creatures, but to a God *supremely* related to creatures. A rock has no internal relations to others, a dog has some, a human being is internally related to a far greater number than a dog and in a qualitatively superior way, and God is either unrelated to creatures in a reversion back to a rock-like existence (as in classical theism) or supremely related to creatures both quantitatively and qualitatively (as in neoclassical theism).

To be known or felt by all subjects is just as distinctive as to know or feel all subjects, on the dipolar view. Whereas classical theism legitimately emphasizes the status of God as the subject-for-all-objects, it is a distinctive feature of neoclassical theism to also see God as the object-for-all-subjects. Neoclassical theism contributes to mystical theology by emphasizing not only that God knows us, but also that we can know and feel God. Our contributions to divine awareness can even be enjoyed by us as such. Preeminent activity *and* passivity should be included in any adequate description of deity. The Jesuit motto derived from St. Ignatius of Loyola and with which John of the Cross was no doubt familiar due to his education under the Jesuits (*ad majorem Dei gloriam*—that we should contribute all for the greater glory of God), which might function well as a motto for Abrahamic theists in general, makes no sense if God cannot receive our contributions. On a classical theistic basis, it seems that we can only serve ourselves.

The seeking of simple explanations for complex issues can indeed prove to be useful as a first step on a long journey. However, there is no virtue in engaging in a method of tenacity in defense of traditional simplifications. Experience does not exhibit the essential inferiority of the contrasting terms to classical theism's preference for permanence and activity. That is, change and receptivity are often *experienced* as positive, as in listening very carefully to someone who thinks that her point of view has been trivialized or ignored in the past, or as in listening for the counterpoint in a difficult piece of music.

6. God as Personal

Whatever happens in the contingent world of becoming must be a matter of indifference to the entirely impassive God of classical theism. This is equivalent to denying that God can love. *Nothing* that we can do or suffer makes a difference to a God who is immune from all passivity and who is identified *simpliciter* with coercive power and sheer causation. Further, this classical theistic worship of naked power has had a negative effect on ecclesiastical polity in different religions, including John of the Cross's Christianity in Spain in the 16th c. Although classical theists in the Abrahamic religions have historically *said* that God is personal, and although they have traditionally derided pantheists because of their impersonal deity, such derision, although understandable, is misplaced given the classical theist's own inability to make room for personal traits in deity. That is, the personal qualities of deity *as articulated by* classical theists are pious fictions.

Another oddity in the classical theistic view is its attempt to appropriate for itself scriptural theism, with its obvious emphasis on divine passivity, love, mutability, and personality, as in the attribution of mercy to God in the *Psalms*, the *Koran*, and elsewhere. Although some scriptural passages can, in isolation and out of context, be used to support classical theistic monopolarity, the weight of the biblical tradition is better supported by the dipolar view. Anselm, although himself a classical theist, has offered a definition of God as that than which no greater can be conceived that facilitates the transition from classical theism to neoclassical theism: It is an open question whether divine perfection is necessarily to be equated with perfect stability or can include as well perfect changes. For Christians, Jesus himself was obviously changeable, hence the need in this tradition to figure out a way to relate his changeableness with the permanent aspect of deity. The process approach to this problem is to emphasize that deity is preeminent change as well as preeminent permanence, to the extent that, and in the ways that, both change and permanence are excellences. In addition to problems trying to render consistent classical theism with the best in scriptural theism, there are related problems trying to render it consistent with trinitarian theology in Christianity, where personhood is central. What sense can be made of the concept of a person who is in every sense unchanging? Wherein lies a believable analogy to human persons?

God must have a pattern of emotions if God is a person, a pattern that presumably would include both suffering and joy in light of the realization that the greatest conceivable being would share not some of, but all of, the sufferings and joys of the creatures. Whereas we are likely to be overwhelmed by the sufferings and joys of others, the greatest conceivable being would be able to sympathize with *all* of them in a qualitatively superior way. Whatever John of the Cross's theological views may have been, it is this sympathetic God that he experienced mystically. Even the least sorrows of creatures would be compassionately felt by the all-worshipful being, as in the biblical line regarding not even the fall of a sparrow escaping divine notice (*Matthew* 10: 28).

To extricate God from all suffering is to (perhaps unwittingly) attribute ignorance to God regarding the suffering of creatures. Process mysticism contains the idea that we are in fact accidents in the divine life, in contrast to the classical theistic view (inherited from Aristotle) that there are no divine accidents: God is pure actuality. There is quite a difference between claiming, as in neoclassical theism, that there is real process and becoming in God and claiming, as in classical theism, that there is only real process and becoming outside God. Although the language of the mystics like John of the Cross is sometimes cloudy and fanciful, they clearly indicate a view closer to neoclassical theism than to classical theism. Our decisions contribute to divine development. It is true that, lacking familiarity with the dipolar alternative to monopolar classical theism, many of the great figures in mystical theology, including John of the Cross, either give up on "the God of the philosophers" altogether or give half-hearted support to classical theism on the assumption that this was the only intellectual option (see DeWet 2008).

The task for the dipolar panentheist is to make sure that when contrasting predicates are applied to God they apply to different aspects of the divine life so that these contrasting predicates do not degenerate into contradictories, as in the aforementioned claim that on the neoclassical view divine *existence* may be permanent, but God's *actuality* (or *how* God exists from moment to moment) must be constantly changing. Likewise, by way of analogy, there is no contradiction involved in saying that a particular human being is always honest even if his/her individual utterances constantly change.

On the dipolar view, divine omniscience involves ideal memory such that God exhibits a meticulous care that nothing be forgotten, a point especially emphasized in Judaism. It must be admitted that the heavily sedimented concept of God is such that the very use of the word "God" immediately leads a great number of people to think in classical theistic terms, but an emphasis on these tender aspects of the divine process found in John of the Cross and other mystics can mitigate the damage done by classical theism. Failure to even consider a concept of God other than that found in classical theism is one of the greatest errors in the history of ideas in that the concept of God affects all others, either directly or indirectly, as in the long and dehumanizing history of determinism in the sciences, which is rooted in classical theistic omniscience. At base, all determinism is theological in that the determinist is claiming to have, or at least aspires to have, the sort of foreknowledge (with absolute assurance and in minute detail) of how future "contingencies" will be determined that is claimed for the God of classical theism.

It may be correct, as Joseph Van Ess argues, that the origins in Islam of doctrines like divine omnipotence and divine omniscience with respect to the details of the future (hence determinism) may have been the partial results of *political* disputes in early Islam (Van Ess 1975; also Murad 1987). The same could be said regarding influential envy of Caeser-like power in early Christianity, as Whitehead argues in *Adventures of Ideas*. However, in all of the Abrahamic faiths, there were also *philosophical* or *theological* disputes that led to these doctrines; classical theism is the unfortunate residue of these latter disputes.

Obviously, there have been many intellectually honest classical theists, but this does not mean that they have done an adequate job of fairly examining their basic assumptions. Thomas Aquinas, for example, vigorously played the devil's advocate for various theorems, but not for alternative fundamental axioms that underlay the theorems, as in the very abstract preference for permanence over change, hence the resulting idea that God is *completely* unchanging. Any sort of intimate mystical experience of God is at odds with such preference, hence my desire to articulate a concept of God that at least reaches rapprochement with sanjuanistic mystical experience.

When classical theists refer to "God," what they should say is "God in one aspect." With this simple move, the gap between classical theism and neoclassical theism could be narrowed significantly. Without this move, however, a large gap remains because in classical theism God is treated as an immobile, impersonal object, not as a subject. Unmitigated classical theism leads to contradiction at every turn: God is a power over the utterly powerless, a will that cannot change, a knower of the contingent yet wholly necessary in such knowledge, a lover who is completely unaffected by the subject loved, and so on. Only by distinguishing two different aspects of God—abstract existence and concrete actuality—can such contradictions be avoided.

7. Toward the Future

Another sort of limited thinking is found in the assumption that there are two and only two alternatives: Either belief in the classical theistic God or some sort of religious skepticism. This dichotomy does mystical experience a disservice because the conceptual difficulties found in classical theism from the Enlightenment until the present are assumed to discredit mystical experience as well. A conceptually cogent view of God would help to get mystical experience itself a fairer hearing than it has received in recent centuries. Contrary to popular belief, even in philosophical and theological circles, the problem of theism, including the problem of how to assess mystical experience, is a remarkably new one that has only come into focus as a result of the prominence in the past century of the higher synthesis offered by neoclassical theists in all three Abrahamic faiths, which preserves the best and eliminates the worst in both classical theism and religious skepticism, as well as preserving the best and eliminating the worst in classical theism and pantheism. In addition to the recent neoclassical theists from all three Abrahamic faiths mentioned at the beginning of the present article, we should also note the influential work of Henri Bergson, a Jew who converted late in life to Catholicism. Bergson views the transition from classical theism to neoclassical theism as the transition from closed to open religion, with only the latter really being conducive to mystical experience (see Bergson 1977). Bergson, it should be noted, was the inspiration for Mohammed Iqbal to bring the neoclassical concept of God to Islam (Iqbal 2013). Bergson also had an impact on several contemporary translators in Iran who are making the best works in neoclassical theism available in that part of the world.

My project in the present article is very much compatible with the work of the influential Japanese commentator on Islam, Toshihiko Izutsu. This scholar has had a very favorable reception in the Muslim world in part because he was unconnected to the Judeo-Christian world and hence could avoid the concern that he was an "orientalist." Izutsu was very much aware of the influence of Plato and especially Aristotle on classical philosophers in Islam and then, in turn, on how these classical philosophers in Islam were instrumental in the formation of scholasticism in the West. However, he was also very much attentive to the ways in which language is instrumental in the construction of the Sufi concept of being (*wujud*). It is not too much of a stretch to see Izutsu as a process philosopher in that, in addition to being influenced by Bergson, he was also interested in the ways in which philosophical thinking interacts with mysticism in *dynamic* ways. Further, the ontology that Izutsu finds in the Koran is living and dynamic in its concreteness, in contrast to the static and abstract ontology found in classical theism. The dynamism he and Bergson champion can also be found in the interplay between language and culture (see Izutsu 1965, 1970, 2008, pp. 2–3; Al-Akoub 2012, p. 127; Sayem 2020; Solihu 2010).

It is understandable why, upon first hearing, many classical theists are turned off by neoclassical theism. It is quite a cipher for them how God could be, say, maximally independent *and* maximally dependent. However, the apparent contradiction is removed when the neoclassical distinction between *existence* and *actuality* is kept in mind. There is no contradiction in God being maximally independent in existence, but *how* God exists (i.e., God's actuality) largely depends on the creatures, and this due to divine omniscience and omnibenevolence (as understood in neoclassical theism). God's existence, on this account, is less than the entirety of the divine life. The abstract fact *that* God exists is quite different from concretely *how* God exists, hence contrasting predicates applied to existence and actuality involve no contradiction.

The biblical command "Be ye perfect!" (*Matthew* 5: 48) obviously does not mean "Be ye immutable!" The reason for this is that there can be such things as *perfect changes*, say in response to others (see Enxing 2012). It is for this reason that we should hope not only to develop an adequate concept of God ("the God of the philosophers") and an adequate understanding of mystical experience ("the God of religion"), but *both together* as part of the overall effort to overcome the bifurcation of nature that has been the hallmark of modernity from the time of Rene Descartes and Galileo until the present. This effort also includes the neoclassical realization that the following features of classical theism tend to mutually reinforce each other, hence the need to rethink classical theism as a whole: The priority of being over becoming, the reduction of creaturely freedom to the mere reiteration of items decreed by divine fiat, the denial of chance or randomness in the world, and the complete immutability of deity.

In order to break up the block universe that is implied in classical theism (even if this implication is seldom noticed), it is necessary to affirm the theory of time as objective modality. Although the necessary features of reality that always are (as indicated in the proposition "Something exists") in a way escape the temporal flux (or better, are omni-temporal), the contingent features of reality are inherently temporal: The past is fixed and determinate, the future is at least partially open and indeterminate in that it is not here yet to be determined, and the present is that fleeting region wherein immediate future determinables are rendered determinate. An omniscient being, on this neoclassical view, knows everything that can be known, but no being, not even divinity, can know how future determinables will be rendered determinate before such determination occurs regarding the contingent features of reality. God is not to be conceived on the neoclassical view as a mere *eternal* spectator, but as an *everlasting* existence who enjoys and/or suffers all that occurs in temporal process. God is as great as possible at any particular time, but new events bring with them new possibilities to exhibit divine knowledge and benevolence. In this regard, it makes sense to see God as ideally perfectible. Concrete actuality, even in the divine case, is always contingent, even if divine existence is in itself abstract and immune to coming to be or passing away.

It is crucial to emphasize that God's dependence is just as unique and remarkable and admirable as God's independence. Therefore, also regarding the other dipolar contrasts. To cite another pair: God's passivity (say with respect to those who seek sanjuanistic mystic union with God) is just as unique and remarkable and admirable as God's activity. We are affected by only a relatively few fellow creatures and intermittently so and qualitatively in a manner mixed with our own egoistic concerns. God is affected by *all* creatures *all the time* and in a qualitatively superior way. As before, Heschel emphasizes the idea that God is the *most* moved mover, in contrast to the classical theistic view. Hence, it is one of the biggest mistakes in classical theism to think of independence-dependence and activity-passivity as logical contradictories, rather than as correlative pairs that mutually reinforce each other. This mistake is due to a failure to consider the possibility that these contrasting pairs apply to different aspects of the divine nature, thus avoiding dreaded logical contradiction. Further, there is nothing in dipolar theism that diminishes that aspect of God that classical theists did isolate and explicate rather well, as in God's envisagement of various conceptual objects, which makes our inferior envisagement of such conceptual objects possible.

Because an understanding of mystical experience has been hampered for many centuries by a defective metaphysics, it can be said that a central *religious* problem is metaphysical. An advance in metaphysical thinking that would aid in the understanding of mystical experience would involve a distinction within the Anselmian concept of that than which no greater can be conceived. There is a crucial difference between God being unsurpassable by another being (the neoclassical theistic view) and God being unsurpassable even by Godself (the classical theistic view). The latter involves the inability to change *simpliciter*, even if such change is by way of addition or enrichment. One model for deity ignored altogether by classical theism is that of dialogue. The classical theistic view has God speaking, but not hearing. This is the supposedly benevolent dictator (not accidentally male—see Christ 2003) view of deity that is surpassed by a view wherein God is both preeminent speaker *and* listener on the analogy of an ideal human dialogical partner. One is reminded here of John of the Cross's famous oxymorons regarding silent music (*musica callada*) and sounding solitude (*soledad Sonora*).

Monopolarity is actually a type of idolatry rather than a prominent feature of the worship of deity properly conceived. Even the suggestion that God is infinite in every respect and in no way finite is problematic. It makes sense to see God as infinite in the temporal reach of divine existence, but it also makes sense to see God as finite, yet preeminently so in the way that God relates to *particular* creatures here and now, as so often noticed by John of the Cross. The positive sense of finitude is related to the concept of determinateness and intimacy, in contrast to an indeterminate blob of relatedness to others. In this regard, the ancient Greeks are instructive in the way they see the infinite (*apeiron*) as *in a sense* inferior to the finite (*peras*). The greatest conceivable being, on the neoclassical theistic view, would be excellently infinite and excellently finite in the ways that both of these are excellences. Mere infinity is empty, formless; nor could the merely infinite (or necessary) know finite (or contingent) things. This is why "infinite" is not a synonym for "supremely good," nor is "finite" a synonym for "not supremely good."

A God who gives everything and receives nothing is a radically deficient being, as least if God is one whom we serve, as John of the Cross and other mystics believe. By concentrating too much on the question of the *existence* of God, rather than on the *concept* of God, philosophers and theologians have done a disservice to religious thought. In fact, without an adequate concept of God the very issue of the existence of God seems quite beside the point. The religious view of the world, I assume, is one where God not only loves, but also is loved by others, who contribute to the actuality of God in process. Although God is the necessarily existing individual who is better than those who do not exist necessarily, this does not mean that God is *exclusively* necessary because such a being would be incapable of love and being loved by others and hence would be at odds with the virtue of omnibenevolence. Although sanjuanistic mystical experience can give us insight regarding the quality of divine love and of God being loved, it cannot help us to clarify the thorny conceptual issues entwined within classical theism that give rise to the problem mystics have had historically with "the God of the philosophers" (see *Ascent of Mount Carmel* II, 8–10; *Spiritual Canticle* 38, 8). In a way, classical theists never really believed in what they said in that, from a pastoral point of view or when talking about the trinity, etc., they have always insisted on a personal (or tri-personal) God of love who reacts to us and to our suffering. However, such insistence cannot be based on their own metaphysical views.

Of course, just as the greatest being cannot be pure actuality, it is also true that a perfect being cannot be pure potentiality; likewise, regarding divine activity-passivity, etc. This highlights the great achievement of classical theism *within its own limited sphere*. However, the dual transcendence that characterizes neoclassical theism means that there is twice as much transcendence in this view than is found in monopolar theism. God is transcendently permanent as well as transcendently changing, transcendently active as well as transcendently passive, and so on. Contrary to the familiar charge made against process theism (counteracted in Loomer 2013), it is actually the classical theistic God who is too small by one-sidedly exhibiting abstractions rather than the fulness of deity. In the dually transcendent God, there is admittedly transcendent unity, but also a sort of divine inclusiveness of all of the diversity of creation; likewise, regarding permanence of existence plus endless novelty. To be universal cause *and* effect is far better than being either alone.

In sum, it is well known that religion, in general, and mystical theology, in particular, have had a rich history. The present article is an effort to ensure that they will have a rich future as well.

Funding: This research received no external funding.

Conflicts of Interest: The author declares no conflict of interest.

References

Al-Akoub, Eisa. 2012. Izutsu's Study of the Qur'an from an Arab Perspective. *Journal of Qur'anic Studies* 14: 107–30. [CrossRef]

Al-Ghazzali, I. 2000. *The Incoherence of the Philosophers*. Translated by Michael Marmura. Provo: Brigham Young University Press.

Anselm. 1962. *Basic Writings*. Translated by S. N. Deane. LaSalle: Open Court.

Aristotle. 1984. *The Complete Works of Aristotle*. Edited by Jonathan Barnes. 2 vols. Princeton: Princeton University Press.

Artson, Bradley. 2013. *God of Becoming and Relationship*. Woodstock: Jewish Lights Publishing.

Augustine. 1998. *Confessions*. Translated by Henry Chadwick. New York: Oxford University Press.

Averroes. 1954. *The Incoherence of the Incoherence*. Translated by S. van den Bergh. 2 vols. London: Luzac.

Avicenna. 2005. *The Metaphysics of the Healing*. Translated by Michael Marmura. Provo: Brigham Young University Press.

Baralt, Luce Lopez. 1985. *San Juan de la Cruz y el Islam*. Mexico City: Colegio de Mexico.

Bashier, Salman. 2004. *Ibn al-Arabi's Barzakh: The Concept of the Limit and the Relationship between God and the World*. Albany: State University of New York Press.

Beng, Peter Gan Chong. 2009. Union and Difference: A Dialectical Structuring of St. John of the Cross. *Sophia* 48: 43–57. [CrossRef]

Bergson, Henri. 1977. *The Two Sources of Morality and Religion*. Translated by R. Ashley Audra, and Cloudesley Brereton. Notre Dame: University of Notre Dame Press.

Bornstein, Pablo. 2019. An Orientalist Contribution to 'Catholic Science'. *Religions* 10: 568. [CrossRef]

Braudel, Fernand. 1975. *The Mediterranean and the Mediterranean World in the Age of Philip II*. New York: Harper and Row.

Brenan, Gerald. 1973. *St. John of the Cross: His Life and Poetry*. Cambridge: Cambridge University Press.

Bushlack, Thomas. 2020. Shadows of Divine Virtue: St. John of the Cross, Implicit Memory, and the Transformative Theory of Infused Cardinal Virtues. *Theological Studies* 81: 88–110. [CrossRef]

Campbell, Camille. 1982. Creation-Centered Carmelites: Teresa and John. *Spiritual Life* 28: 15–25.

Christ, Carol. 2003. *She Who Changes: Re-Imagining the Divine in the World*. New York: Palgrave Macmillan.

Cobb, John. 2007. *A Christian Natural Theology Based on the Thought of Alfred North Whitehead*, 2nd ed. Louisville: Westminster John Knox Press.

Cole, Peter, ed. 2007. *The Dream of the Poem: Hebrew Poetry from Muslim and Christian Spain, 950–1492*. Princeton: Princeton University Press.

DeWet, Chris. 2008. Mystical Expression and the Logos in the Writings of St. John of the Cross. *Neotestamentica* 42: 35–50.

Dombrowski, Daniel. 1992. *St. John of the Cross: An Appreciation*. Albany: State University of New York Press.

Dombrowski, Daniel A. 1994. Mysticism and Divine Mutability. *Process Studies* 23: 149–54. [CrossRef]

Dombrowski, Daniel A. 2001. Visions and Voices vs. Mystic Union. *Sophia* 40: 33–43. [CrossRef]

Dombrowski, Daniel A. 2004. Saint John of the Cross and Virtue Ethics. *Mystics Quarterly* 30: 7–14.

Dombrowski, Daniel A. 2005. *A Platonic Philosophy of Religion: A Process Perspective*. Albany: State University of New York Press.

Dombrowski, Daniel A. 2006. *Rethinking the Ontological Argument: A Neoclassical Theistic Response*. Cambridge: Cambridge University Press.

Dombrowski, Daniel A. 2010. Rival Concepts of God and Rival Versions of Mysticism. *International Journal for Philosophy of Religion* 68: 153–65. [CrossRef]

Enxing, Julia, ed. 2012. *Perfect Changes*. Regensburg: Verlag Friedrich Pustet.

Green, Deirdre. 1986. St. John of the Cross and Mystical 'Unknowing'. *Religious Studies* 22: 29–40. [CrossRef]

Griffin, David Ray. 2001. *Reenchantment without Supernaturalism: A Process Philosophy of Religion*. Ithaca: Cornell University Press.

Hartshorne, Charles. 1948. *The Divine Relativity*. New Haven: Yale University Press.

Hartshorne, Charles. 1953. *Philosophers Speak of God*. Chicago: University of Chicago Press.

Hartshorne, Charles. 1970. *Creative Synthesis and Philosophic Method*. LaSalle: Open Court.

Hartshorne, Charles. 1984. *Existence and Actuality*. Chicago: University of Chicago Press.

Heschel, Abraham. 1962. *The Prophets*. New York: Harper and Row.

Ibn Abbad of Ronda. 1986. *Letters on the Sufi Path*. Translated by John Renard. New York: Paulist Press.

Ignatius of Loyola. 1964. *The Spiritual Exercises*. New York: Doubleday.

Inati, Shams. 1996. *Ibn Sina and Mysticism*. London: Kegan Paul.

Iqbal, Mohammed. 2013. *The Reconstruction of Religious Thought in Islam*. Palo Alto: Stanford University Press.

Izutsu, Toshihiko. 1965. *The Concept of Belief in Islamic Theology*. Tokyo: Keio Institute.

Izutsu, Toshihiko. 1970. Review of Alfarabi's Philosophy of Plato and Aristotle. *Philosophy East and West* 20: 196–97. [CrossRef]

Izutsu, Toshihiko. 2008. *God and Man in the Qur'an: Semantics of the Qur'anic Weltanschauung*. Kuala Lumpur: Islamic Book Trust.

John of the Cross. 1948. *Biblioteca Autores Espanoles*. Madrid: Editiones Atlas, vol. 1.

John of the Cross. 1973. *The Collected Works of St. John of the Cross*. Translated by Kieran Kavanaugh, and Otilio Rodriguez. Washington, DC: Institute of Carmelite Studies.

Loomer, Bernard. 2013. The Size of the Everlasting God. *Process Studies Supplements* 18: 1–45.

Mahoney, Timothy. 2004. Understanding the Christian Apophaticism of St. John of the Cross. *Logos* 7: 80–91. [CrossRef]

Maimonides. 1885. *The Guide of the Perplexed*. Translated by M. Friedlander. London: Ludgate Hill.

Moses de Leon. 1988. *The Book of the Pomegranate: Moses de Leon's Sefer ha-Rimon*. Edited by Elliot Wolfson. Providence: Brown Judaic Studies.

Murad, Hasan Qasim. 1987. The Beginnings of Islamic Theology: A Critique of Joseph Van Ess' Views. *Islamic Studies* 26: 191–204.

Murphy, John. 1996. St. John of the Cross and the Philosophy of Religion. *Mystics Quarterly* 22: 163–86.

Nieto, Jose. 1979. *Mystic, Rebel, Saint: A Study of Saint John of the Cross*. Geneva: Droz.

Payne, Steven. 1990. *John of the Cross and the Cognitive Value of Mysticism*. Boston: Kluwer.

Philo. 1929–1962. *Philo*. Translated by F. H. Colson, and G. H. Whitaker. 10 vols. Cambridge: Harvard University Press.

Plato. 1999. *The Collected Dialogues of Plato*. Edited by Edith Hamilton and Huntington Cairns. Princeton: Princeton University Press.

Ruzgar, Mustafa. 2008. Islam and Process Theology. In *Handbook of Whiteheadian Process Thought*. Edited by Michel Weber and Will Desmond. Frankfurt: Ontos Verlag, vol. 1.

Sanderline, David. 1989. Faith and Ethical Reasoning in the Mystical Theology of St. John of the Cross. *Religious Studies* 25: 317–33. [CrossRef]

Sanderline, David. 1993. Charity According to St. John of the Cross. *Journal of Religious Ethics* 21: 87–115.

Sayem, Mohammed Abu. 2020. Religious Perspectives on Environmental Issues: A Dialogue with John Cobb. *Process Studies* 49: 254–74.

Serran-Pagan, Cristobal. 2018. *Saint John of the Cross: His Prophetic Mysticism in the Historical Context of Sixteenth-Century Spain*. Washington, DC: Pacem in Terris Press.

Solares, Carlos Conde. 2020. The Moral Dimensions of Sufism and the Iberian Mystical Canon. *Religions* 11: 15. [CrossRef]

Solihu, Abdul Kabir Hussain. 2010. The Linguistic Construction of Reality: Toshihiko Izutsu's Semantic Hermeneutics of the Qur'anic Weltanschauung. In *Japanese Contributions to Islamic Studies*. Kuala Lumpur: IIUM Press.

Spinoza, Baruch. 1992. *Ethics*. Translated by Samuel Shirley. Indianapolis: Hackett Publishing.

Spitzer, Leo. 1969. *Essays on English and American Literature*. Princeton: Princeton University Press.

Teilhard de Chardin, Pierre. 1971. *Christianity and Evolution*. Translated by Rene Hague. New York: THarcourt, Brace, and Jovanovich.

Teresa of Avila. 1976. *The Collected Works of St. Teresa of Avila*. Translated by Kieran Kavanaugh, and Otilio Rodriguez. Washington, DC: Institute for Carmelite Studies.

Thomas Aquinas. 1972. *Summa Theologiae*. Edited by Blackfriars. New York: McGraw-Hill.

Van Ess, Joseph. 1975. The Beginnings of Islamic Theology. In *The Cultural Context of Medieval Learning*. Edited by J. E. Murdoch and E. D. Sylla. Dordrecht: Springer.

Whitehead, Alfred North. 1967. *Adventures of Ideas*. New York: Free Press.

Whitehead, Alfred North. 1996. *Religion in the Making*. New York: Fordham University Press.

religions

MDPI

Article

The Active Life and the Contemplative Life in St. John of the Cross: The Mixed Life in the Teresian Carmelite Tradition

Cristóbal Serrán-Pagán y Fuentes

Department of Philosophy and Religious Studies, Valdosta State University, Valdosta, GA 31698, USA; cserranpagan@valdosta.edu

Received: 5 September 2020; Accepted: 1 October 2020; Published: 8 October 2020

Abstract: The prophetic and the mystical are two key theological concepts in St. John of the Cross. The aim of this article is precisely to shed light on the essential role that St. John of the Cross played in the history of Christianity by acknowledging the prophetic and the mystical dimensions of his life testimonies and writings. The notion of prophetic mysticism is not altogether foreign to the Carmelite tradition, especially following the prophetic example of Elijah. This article will then explore the intrinsic relationship that exists between the active life (Martha) and the contemplative life (Mary) in St. John of the Cross and in the Teresian Carmelite tradition.

Keywords: prophetic mysticism; mystical theology; St. John of the Cross; Christianity; Carmelite tradition; Elijah; contemplative life and active life; Teresian

1. Introduction

Generally speaking, scholars and pietistic people tend to classify St. John of the Cross (1540?–1591) under the category of "contemplatives", "poets", or "inspiring authors." Many Sanjuanist commentators, especially hagiographers, have been interested in giving the Carmelite saint a rigid, austere, and pious image. However, these categorizations into one group often overlook John's active involvement in the world. As Janet Ruffing notes:

> One of the most important—though often neglected—theoretical issues related to types, descriptions, or definitions of mysticism and the mystical is the relationship of mysticism to prophecy or the prophetic. The connection between these two phenomena is rarely explicitly discussed, although it may be evoked in some communities, in some authors, and in some situations. (Ruffing 2001, p. 7)

The purpose of this article is to show that many Sanjuanist commentators and biographers in the past err in portraying St. John of the Cross as a world-negating contemplative by calling him the doctor of the dark night of the soul, a phrase he never used (see Zimmerman 1928; Peers 1945; Doohan 1995; Perrin 1997; Serrán-Pagán 2017). Other Sanjuanist modern commentators have attributed the dark night to John's imprisonment in Toledo (Matthew 1995, p. 55; McGreal 1997, pp. 13, 25; Hardy 2004, p. 74) when, in fact, John clearly defined dark night as infused contemplation or mystical theology. This misconstrued view of dark night has led many of these modern biographers to imply that John heavily emphasized the contemplative life over the active life when, in reality, St. John of the Cross followed the mixed life of the Teresian reform (see Hardy 2004, p. 102; McGreal 1997, p. 8). Why do many commentators still portray St. John of the Cross only as a contemplative by overlooking John's active life and his preference for the mixed life which is one of the major reasons why he entered the Teresian Discalced Carmelite Order? "Probably in September of the same year [1567] Teresa met John at Medina del Campo and convinced him to join the reform because of its stress on contemplative prayer" (McGinn 2017, p. 232). Is this historical view accurate? Even the great contemporary historian Bernard McGinn implicitly participates in this Sanjuanist projection of John being portrayed merely as a contemplative but not as a contemplative in action, as we see in his treatment of St. Ignatius of Loyola in chapter 2 or St. Teresa of Avila in chapter 3 of his voluminous work *Mysticism in the Golden Age of Spain, 1500–1650*. None of these modern biographers subscribe to the view that John was a contemplative in action and when they do, they called it an apostolate of prayer. The "fugi mundi" of some Christian ascetics is attributed to John. For instance, one modern biographer states, "His [John] tragedy was that while his only ambition was to spend his days and nights in prayer and mortification, his fate was to be involved during most of his adult life in a fierce conflict with his brother friars, which ended in his disgrace and death" (Brenan 1973, p. 3). "He [John] did not hate the world or the senses, as so many religious persons have done, but sought rather to escape from them and to leave them behind him. His career, therefore, appears less as a struggle than as a flight—a vertiginous ascent away from everything and towards God" (Brenan 1973, p. 134).

For the Christian mystical tradition, in which the Carmelite saint was deeply immersed, the contemplative and the prophetic are two aspects of the same reality. Mary often symbolizes the contemplative mystic, while Martha best represents the active prophet. The Spanish mystic entered the Discalced Carmelite Order of St. Teresa of Avila not only to be a good contemplative but more importantly to become the great contemplative in action that she is well known for. This misinterpretation is informed by a failure to understand John's historical context (the problem of "conversos" in sixteenth-century Spain) and the rich mystical theology tradition which John inherited from Pseudo-Dionysius the Areopagite (5th–6th century AD) through medieval Christian mystics; then most surely through non-Christian sources from Sufi mystics such as Ibn 'Arabi of Murcia (1165–1240) and Ibn Abbad of Ronda (1333–1390); and from Jewish Kabbalistic mystics like Moses de León (1240–1305), who by then had left a great spiritual legacy in the Iberian Peninsula (see Cuevas 1972; Asín Palacios 1981, 1990; Swietlicki 1986; López-Baralt 1990, 1998; Satz 1991; Carabine 1995; Valdivia Válor 2002; De Tapia 2006; Girón-Negrón 2008; Conde Solares 2019). This study cannot cover all these topics at once (see Serrán-Pagán 2018). I set up my boundaries to prove once and for all that St. John of the Cross was a contemplative in action.

St. John of the Cross primarily is known today as a poet, mystic, and theologian, and these are but a few of his most important contributions. However, he was for a long time depicted as an abstruse thinker, lacking originality and intellectual vigor. Thus, John's intellectual capacities were almost forgotten or overlooked for several centuries. St. John of the Cross, the Doctor of "la Nada" (the No-thingness), is also the Doctor of "el Todo" (the All). The Carmelite saint has become an icon for the Catholic Church. Yet, John is often portrayed by past commentators only as a model of sanctity reached by practicing an austere, ascetic, contemplative life. Even worse, some commentators are inclined to believe that John did not play an active role in his time. For example, Bede Frost offers an odd argument for this view claiming that St. John of the Cross

> played no part, humanly speaking, in that immense and stirring drama which filled the stage of sixteenth-century Europe. Nothing in his writings or in the slight contemporary references to him reveals the faintest interest in that interplay of vast political, economic, social and religious forces, so inextricably mingled, which strove together in that world writhing in the processes of disintegration and reformation. His portrait occupies no place in that gallery upon whose walls there hang the great men and women of the second half of the sixteenth century. (Frost 1937, p. 1)

Frost claims that John's apparent lack of interest in the world was due to a long Christian practice of dying to the sins and to the things of the world. However, Frost may well be incorrectly interpreting the monastic idea of "contemptus mundi," even after studying John's life and thoughts. Thus, this Carmelite saint, widely known as a mystic, is not generally thought of as a reformer, even less a prophet. John the reformer is often characterized only as a pious monk whose.

> earnest desire was to strengthen the contemplative side of the constitutions. He feared lest the frequent coming and going, necessitated by sermons and conferences undertaken by the friars, should lead to relaxation of the solitude, prayer and penance essential to the Carmelite vocation.... He was also opposed to the undertaking of foreign missions. He considered that there were religious orders enough in the Church with that special end, whereas the Carmelite friars were intended, not to preach the gospel to pagan lands, but to promulgate its perfect practice in the home countries, by their example, their prayers, and their teaching. (The Sisters of Notre Dame 1927, pp. 97–98)

In the words of researcher Peter Slattery:

> St. Teresa worked hard with the support of the Prior General and other Church officials to spread the reform among the Carmelite order. St. John of the Cross took a less active role, but nevertheless earned the admiration of all as a faithful religious, renowned confessor and spiritual director.... Although St. John did not take much part in actively spreading the reform he was extremely influential as one of the founding friars, from the character of his writings and poetry and his unquestioned sanctity. Others, like Fray Nicholas (Doria) and Fray Jerome (Gracian), took a more active role in the struggles of the early years of the reform. St. John was several times elected prior, definitor and consultor. Apart from some letters mostly dating from the last years of the saint's life, primary sources contemporary to events are scarce. (Slattery 1991, pp. 70–71)

Another Sanjuanist commentator makes the following observation:

> Unlike Teresa, he [John] was singularly devoid of all those vivid and arresting features that one calls personality. We see an inward looking, silent man with downcast eyes, hurrying off to hide himself in his cell and so absent-minded that he often did not take in what was said to him. We note the immense tenacity of purpose that underlay his somewhat feminine sensibility, his strictness in matters of discipline and his entire and whole-hearted devotion to the contemplative life. (Brenan 1973, p. 83)

It is time to study St. John of the Cross as a whole person, bringing together his multifaceted dimensions and his historical context. Attention should be devoted to his active role in the Carmelite tradition as a religious reformer, an administrator, and a prophet; to his intellectual capacity as a mystical theologian and teacher; to his religious devotion as a poet, friar, hermit, spiritual director, confessor, and priest; and to his Moorish and possibly also Jewish roots (see Gómez-Menor Fuentes 1970; Swietlicki 1986; Satz 1991; Dombrowski 1992; Martínez González 2006; Poveda Piérola 2011; Serrán-Pagán 2018). It is my personal hope that future Sanjuanist scholars will find more clear evidence of the link that exists between St. John of the Cross and the shared prophetic mystical traditions of Judaism, Christianity, and Islam.

2. Historical Analysis and the Problem of *Conversos* in St. John of the Cross

Juan de Yepes y Álvarez was his secular name, son of Gonzalo de Yepes and Catalina Álvarez, and the youngest of three sons. John was born in Fontiveros (Ávila), at the heart of the Castilian region known as la Moraña. This region was heavily populated by Moorish converts, or, in Spanish, "moriscos". There are some disputes about the right date of his birth. Some scholars believe that John was born around 24 June 1540. However, the most accepted year among Sanjuanist scholars is 1542 due to the age recorded as 49 at the time of his death, on 14 December 1591. The story goes that Gonzalo's family, who were silk merchants in Toledo (a trade that usually belonged to Jewish converso families), disinherited John's father from their wealth for getting married to Catalina, a poor young woman who could also have inherited the social stigma of having been born into a Moorish family (McGreal 1997, pp. 2–3). This event in John's life is significant because if it is true, it could prove how close John was to the Muslim world (especially to the Sufi mystical tradition). Luce López-Baralt, the famous scholar from Puerto Rico, has done an intensive study on John's poetry and commentaries and has concluded that John's mystical poetry share many symbols with the Sufi language of the dark night and intoxicated love (López-Baralt 1998, pp. 17–18). It is not an exaggeration to affirm that the Carmelite saint felt at home using Muslim, Jewish, and Christian symbolic expressions. Three reasons come to mind: first, the Iberian Peninsula was historically influenced by the three monotheistic religions of the West, even after the expulsion of the Jews in the fifteenth century and the Muslims in the seventeenth century; second, John not only had access to the best university libraries in the kingdom of Castile (Salamanca, Alcalá de Henares, and Baeza), but he also knew people who were knowledgeable or in contact with non-Christian sources; and third, John himself could have been raised in a Jewish and Moorish household if it is accepted the theory that his parents were conversos (see Serrán-Pagán 2018, pp. 10–14, 61–77, 126–29). As Carlos Conde Solares points out in a recent article published in *Religions* 2020, 11, 15:

> St John of the Cross's interest in the language of Sufism can be traced in manifold credible ways. His biography provides plenty of opportunities for him to have engaged with the learned traditions of pre-Islamic, Islamic, Sephardic and Eastern spiritualities: we know of his culture, of his thirst for knowledge and of his studies in Salamanca. However, St John was also the ideal recipient of popular currents and their folkloric and oral expressions. The theory that St John would have been in direct contact with the popular expressions of not just Christianity but also other religions finds intuitive backing in his poetic works. (Conde Solares 2019, p. 11)

There are a good number of researchers (Cuevas 1972; Asín Palacios 1981; López-Baralt 1990; Satz 1991; Valdivia Válor 2002) who saw similarities in John's mystical poetry and his doctrine and Sufism, based primarily on their shared usage of mystical symbols and ascetic practices. So, a great number of connections between John and the world of Sufi mystics in the Iberian Peninsula can be made based on the theory of literary transmission by moriscos. Additionally, many of the places where John lived had a significant number of Moorish (moriscos) *conversos* inhabiting those cities from Avila to Granada (see De Tapia 2006). In sixteenth-century Spain, John and Teresa had to suffer the stigma of being new Christians, like all other conversos. Both Spanish Carmelites had to be vigilant at all times knowing that they were under the inquisitorial radar of old Christians who passed the purity of blood ("pureza de sangre") laws in the kingdom of Castile. According to Serafín de Tapia, the Calced Carmelites adopted these purity of blood laws in 1566 (De Tapia 2006, p. 206). According to this professor of history in the University of Salamanca, the Jewish converso population in Ávila was more than 20% and the Moorish converso population was between 10% and 15%. By 1570, many moriscos came from Granada seeking refuge after being dispersed in Las Alpujarras (De Tapia 2006, p. 210). This new data supports my suspicion that the problem of conversos was at the heart of the major persecutions that both Teresa and John had to suffer at the hands of old Christians, including some of his Carmelite brothers. This alone could very well explain why John and Teresa never mentioned their converso origins and why some people even within his own Carmelite Order treated them so poorly, even though both of them were known historically as being the co-founders of the Discalced Carmelite tradition.

What we can infer from all the new data available to us now is that most probably John's father came from a Jewish converso family (see Gómez-Menor Fuentes 1970; Márquez Villanueva 1998). In John's case, we can only speculate because we do not have any documentation to prove or disprove his Jewish roots following his father's lineage. However, we can deduce that there is enough evidence to link John's converso lineage not only through his father's lineage as a "marrano", which is easier to prove in this case knowing that John's paternal family came from the "oficio of mercader de sedas" (the business of trading and selling silk products), but in the case of John's mother we know she came from the occupation of being a weaver; this trade was typical of Muslim conversos, also known as moriscos (see Asín Palacios 1981; Garrido 1989; López-Baralt 1990; Poveda Piérola 2011).

Not only their families played those societal roles well, as did many other converso families, but more importantly, John and Teresa always lived in towns and cities where there was a large converso population that serves them well to protect themselves from the constant pressure exerted by old Christians and the holy inquisitors, whose major task was to make sure that the large Jewish and Muslim converso population in the kingdoms of Castile and Aragon will fully abide by the Christian laws and adopt the Roman Catholic faith without any hesitation or heretical deviation, as Teresa's family prove to have done in Toledo. As a matter of fact, we know Teresa's family moved from Toledo to Ávila to escape from any suspicion in the eyes of the "Santo Oficio" (Holy Office). In so doing, her family, coming from a wealthy middle-class family of merchants, was able to purchase an old Christian last name from an "hidalgo" family. In those days, it was a common practice among converso families to convert their old Jewish names into new Christian names; thereby, they were able to escape from any suspicion raised by old Christians living in their new adopted city of Ávila, where the Jewish and Muslim converso population was larger than other cities in Castile.

Could it be that the problem of conversos was one of the main reasons why Teresa's and John's writings were denounced to the Holy Inquisition? In John's case, he was kidnapped twice by his own Calced Carmelite brothers and had to suffer persecutions until the end of his life, including Doria's effort to send John in exile to Mexico. He ended up in Andalusia, where he received the mistreatment and abuse by his abbot in Úbeda. In other words, the traditional narrative does not take into consideration this important historical factor of the two co-founders of the Discalced Carmelite tradition sharing converso lineages, at least through direct Jewish bloodline. As Daniel A. Dombrowski points out, in one of the most notable Sanjuanist studies of the twentieth century:

> No treatment of the social world of John of the Cross and Teresa of Avila would be adequate without the mention of the latter's Jewish roots. After the nobles and the higher clergy, the Jews had been the wealthiest and most influential people in Spain, a status which, when combined with anti-Semitism, led to the pogroms of the late fourteenth century and to the "reconciliation" of many Jews to Christianity in the Inquisition. Teresa of Avila's ancestors were among these. In fact, because her ancestors were, like John of the Cross's, Toledan silk merchants, there are reasons to suspect that John of the Cross may also have had Jewish roots. (Dombrowski 1992, p. 38)

The lack of references in John's writings to the historical events and the social environment in which he lived may not be his fault but rather the result of circumstances apart from him. The problem of conversos could partly explain John's apparent silence. John, who was probably born in a new Christian family, of Jewish and of Moorish converso stock, wrote his prose commentaries, poems, counsels, and letters under great pressure. He would have feared being persecuted by the inquisitors and would have been subjected to constant censorship by his own Carmelite brothers. John was surely aware of the risks people took in his time, especially after seeing how Friar Luis de León (1528–1591), one of the most popular professors in the University of Salamanca, was arrested and imprisoned for several years. The historical reason for Luis's imprisonment was his translation from the Hebrew Scriptures of the *Song of Solomon* (or most commonly known in the Christian mystical tradition as the *Song of Songs*). Modern scholars have shed some new light on Friar Luis's Jewish converso background. As a matter of fact, we now know that Friar Luis was the one in charge of editing the whole collection of writings by Teresa of Ávila (1515–1582, and his nephew Fray Basilio Ponce de León (1570–1629) defended John of the Cross posthumously after the Inquisition was investigating John's works for being heretical. This proves there was a direct link and connection between all these great Spanish writers descended from converso families who were more inclined to form a network of friends to support each other under such harsh times in sixteenth-century Spain. For all of these converso families, it was a matter of survival to be associated with the right group of people in order to avoid persecution.

In fairness to the spirit of the Carmelite saint, it is the purpose of this article to carefully analyze John's life and writings in reference to the mixed life in the context of the Teresian Discalced Carmelite reform. I would suggest that poetry and theology must be brought together for a full understanding of

John's integral Christian humanism. John's poetry is intuitive, direct, and experiential, whereas his theology is analytical, reflective, and rational. One complements the other. As Willis Barnstone puts it:

> San Juan was a mystical poet because in a formal sense his poems were written, he himself states, as a result of mystical knowledge, and in his commentaries he endeavors to explain the poems, in great detail, as steps toward the mystical union. The question of whether mysticism is a valid religious experience or a form of hysteria, hallucination, psychedelic substitute, or sublimated sexual ecstasy, or even whether the poems themselves convey the mystical experience, is secondary and not the issue. The point I wish to make clear is that the appellation mystical poet-*Doctor Místico* as he was called-is correct in that San Juan was himself a mystic and the origin of his poems lies in the mystical experience. (Barnstone 1972, p. 25)

St. John of the Cross the mystical poet is often depicted in his ecstatic moments of union with God. The Carmelite mystic seems oblivious to the world. The poet's only concern is to be alone with God. Literary critics (see Brenan 1973; Thompson 2015) have praised John's poetry and his own creativity as a thinker. In mixing different literary styles and intellectual sources, John helped to enhance the Castilian language by bringing old and new elements together into the Spanish grammar, and also by incorporating a vast knowledge accumulated through his years as a student and friar. It is not entirely surprising at all to find out that John never had the popular appeal of other saints. As a mystic, John's popularity did not cross the Spanish borders until he was beatified, canonized, and proclaimed a Doctor of the Catholic Church. Although the Carmelite saint was not widely recognized outside Spain until recent times, most of his contemporaries knew him as a holy man, even before he was proclaimed a saint. John is recognized today as one of the most important mystics partly as a result of Pius XI (1857–1939) declaring him the Mystical Doctor of the Roman Catholic Church. Numerous studies on John's mysticism have been published since then. However, John is still characterized by some modern Sanjuanist commentators as a life-denying ascetic who loved the detachment of la Nada by calling him the doctor of the dark night of the soul. Some of these Sanjuanist commentators, especially following the writings of hagiographers, have been interested in giving the Carmelite saint a strict austere image. They have ignored altogether John's involvement in the world. F.C Happold states the following on John's ambivalent portraits:

> To some this gentle little Carmelite saint, as he has been called, the devoted companion of St Teresa, is utterly repellent. Of him one writer uses these words: 'terrible, sanglant et les yeux secs'; another labels him 'l'áscète terrible'. Yet others find him the most attractive of the contemplative saints. Why are there these apparently conflicting responses? The answer is not far to seek. St John of the Cross has two faces: he is both the apostle of absolute detachment and also the apostle of absolute love. He teaches a detachment so absolute that it appears, taken alone, to be a complete abandonment of everything earthly, a philosophy of world-negation of a most extreme type. But he can also write such sentences as: 'All the ability of my soul and body is moved through love, all that I do I do through love, and all that I suffer I suffer for love's sake,' and, in one of his letters: 'Where there is no love, put love in and you will draw love out.' And these two elements are intimately intertwined; the detachment is absolute because the love is absolute. (Happold 1988, p. 355)

The Sanjuanist doctrine of the dark night is often misrepresented and John is regarded as "a life-denying and world-hating ascetic when in reality his mysticism superabounds in love, vitality, and joy" (Merton 1968, p. 81). For some readers, the dark night means turning away from all created and sensible things so that you will come to know God, even at the expense of excluding fraternal union. For Thomas Merton, "this is bad theology and bad asceticism" (Merton 1990, p. 38). Actually, it is through this dark night that we can let God find us in total surrender by an act of grace. In other words, John's symbol of the dark night might be interpreted as absence of light in encountering the mystery of the Godhead. Yet the mystic receives infused knowledge from God's luminous revelations.

As a philosopher and theologian, John was underrated for many centuries. It was the works of twentieth-century French philosophers like Jean Baruzi (1881–1953), Henri Bergson (1859–1941), or Jacques Maritain (1882–1973) that resurrected John by placing him on top of the list of philosophers and mystical theologians of all times. Philosophers tend to be suspicious of mystical thinkers either because they subordinate reason to faith or simply because they want to transcend and move beyond the realm of reason altogether. Ironically, theologians are also very threatened by mystics because their dogmas and doctrines sometimes are not in tune with the language of the mystics.

The aim of this article is precisely to shed light on the essential role that St. John of the Cross played in the history of Christianity by acknowledging both the mystical and the prophetic dimensions of his life testimonies and writings. The notion of prophetic mysticism is not altogether foreign to the Carmelite tradition, especially following the prophetic example of Elijah. This study will then explore the intrinsic relationship between the prophetic and the mystical elements in St. John of the Cross. In the words of a great American religious philosopher, William E. Hocking (1873–1966):

> The prophet must know himself; and he must know his world, not in detail but in so far as it is relevant to his purpose: such knowledge as this must come to him through his relation to the absolute. The prophet is but the mystic in control of the forces of history, declaring their necessary outcome: the mystic in action is the prophet. In the prophet, the cognitive certainty of the mystic becomes historic and particular; and this is the necessary destiny of that certainty: mystic experience must complete itself in the prophetic consciousness. (Hocking 1955, p. 511)

By "prophetic mysticism" I do not simply mean to foresee the future as it is frequently understood in popular circles. Rather, the prophetic mystic is one who bears witness to truth, justice, and love. In addition, the mystic in action develops what Hocking called "the prophetic consciousness." He asserts:

> By the prophetic consciousness I do not mean a knowledge that something is to happen in the future, accomplished by forces beyond myself: I mean a knowledge that this act of mine which I now utter is to succeed and hold its place in history. It is an assurance of the future and of all time as determined by my own individual will, embodied in my present action. It is a power which knows itself to be such, and justly measures its own scope. (Hocking 1955, p. 503)

According to Wayne Teasdale, the prophetic voice demands witness and response to the most pressing moral and religious issues of our time:

> The prophetic voice vigorously acknowledges the unjust events and policies that cause enormous tension, misery, and dislocation in the lives of countless numbers of people. War; the plight of refugees (most of whom are women and children); unjust economic, social, and political conditions that enrich a small class of rulers while oppressing the masses; threats to the environment—all are matters that should evoke the moral voice and our willingness to respond. We no longer have the luxury of ignoring the many challenges to justice in all its forms. We have a universal responsibility to apply the moral or prophetic function wherever we see justice disregarded, threats to world peace, oppression by states against its people or a neighboring nation, or some other danger as yet unforeseen. (Teasdale 1999, pp. 157–58)

I will define "prophetic mysticism" in the historical context of the Judeo-Christian tradition, and especially following the lines established by John's Carmelite eremitical–prophetic tradition (see Reeves 1969; Kreisel 2001; Hvidt 2007; Fitzgerald 2017). Then, I will briefly sketch the history of the active life (Martha) and the contemplative life (Mary) following relevant biblical sources and Christian modern theological sources in an effort to better understand John and the Teresian preference for the mixed life. Finally, I will conclude with a selection of passages from John's writings to elucidate his theological commentaries on the active life and the contemplative life (also known as the mixed life in scholastic Thomism).

3. Prophetic Mysticism in the Historical Context of the Judeo-Christian Tradition

Christians, like Jews, share a rich prophetic and mystical tradition throughout time. These two religious traditions are familiar with the many prophetic figures in ancient Israel who gave testimony to their special covenant with the Godhead (Abraham, Moses, Isaiah, Jeremiah, Elijah, Ezekiel, Hosea, Daniel, and many others). Incidentally, John's Carmelite Order adopted the Hebrew prophet Elijah as their spiritual founder and patron, although there was no direct historical link between them. The Carmelites saw Elijah as their spiritual role model because he balances the eremitical life and the prophetic life. According to Peter Slattery,

> The spirit, the personality, and the work of Elijah dominate the sacred site of Mount Carmel. In his prayer and reflection the great prophet heard the call of God to bring his people back to him. With ardent zeal, prophetic courage, and a certain amount of passion he answered the call of God. The prophet is so present to God that God dominates his whole life. He is moved by the needs of the people who are being neglected and being misled... In this way Carmelite spirituality encourages people to lie continually in the presence of God, and like the prophet, to be attentive to the signs of the times, so that they may hear the cry of the poor. (Slattery 1991, pp. 136–37)

In addition to Elijah, Moses best represents the archetype of the prophetic mystic in Judaism and Christianity. Although Moses never saw God face-to-face because God is not an object among other objects, he became a mystic by virtue of his personal encounter with God. Moses came down from Mt. Sinai to free his people from their enslaved state. Moses, the mystic visionary, became the prophetic exemplar for having received the direct revelations from G-d. Yet it was Moses the prophet who raised his voice against the tyranny of Egypt against his people. He offered them a way out of their captivity by forcing them into a forty-year exodus in the desert (Beal 2013). Mystics often adopt paradoxical symbols to describe the indescribable. The divine is so utterly mysterious that even those who are called the friends of God prefer to speak of "Him" in terms of divine attributes (cataphatic mystics) or they try to explain what God is not (apophatic mystics). Both Jewish and Christian mystics refuse to define God even when they had felt the presence of God within. The Godhead of the mystics is beyond any thought or word. Yet mystics are full of words to describe their personal experiences of the divine. The mystics cannot fully express in words and in thoughts the true nature of the Godhead; nonetheless, they are the ones who have gained immediate loving knowledge of God and have described for us the divine attributes or earthly manifestations of God.

Christians borrow from the Jewish tradition the archetypical image of Moses as the prophetic mystic par excellence. As a result, Christians follow the example of Moses in his prophetic response to the divine calling by sharing the fruits of his contemplation with the rest of the world. Thus, Christians believe vision must follow action. Otherwise, the divine message never gets to the community and the mystic's response to God's plan is simply nullified by an act of cowardice, total passivity, or rebellion. Therefore, action must be rooted in contemplative wisdom and divine knowledge, not vice versa. In short, the Judeo-Christian God demands from each believer cooperation in an effort towards building the heavenly kingdom on earth.

To place St. John of the Cross within the Christian mystical tradition, one must understand the terms "mystical vision" and "prophetic voice". Christian mystics define mystical vision as the highest state of contemplation in this life. The mystic is one who has a direct experience of the divine. John defines mystical vision as the last step on the mystical ladder of divine love:

> The tenth and last step of this secret ladder of love assimilates the soul to God completely because of the clear vision of God that a person possesses at once on reaching it.... [And] this vision is the cause of the soul's complete likeness to God. St. John says: We know that we

> shall be like him [1 Jn. 3:2], not because the soul will have as much capacity as God—this is impossible—but because all it is will become like God. Thus it will be called, and shall be, God through participation. (John of the Cross 1991, p. 445; *Dark Night* 2.20.5)

John concludes that the mystic who attains a vision of God in this life becomes like God by participation. As St. Paul states in the Bible, God shall be "all in all" (1 Cor. 15:28). As a mystical theologian, John describes that contemplative or mystical state as an ascent to God. It is no wonder why the Carmelites chose Elijah as their archetype in his ascent to the mystical heaven through his chariot of fire. John writes: "Thus, by means of this mystical theology and secret love, the soul departs from itself and all things and ascends to God. For love is like a fire that always rises upward as though longing to be engulfed in its center" (John of the Cross 1991, p. 445; *Dark Night* 2.20.6).

By prophetic voice, the Christian mystics mean that prophets not only are entrusted with God's Word but they also announce the deepest troubles of society by denouncing those who commit injustices against the suffering people, even at the expense of dying as martyrs themselves. In fact, many prophetic mystics were killed because they were serving God, even when they knew that their lives were in danger. In this regard, John says:

> God truly grants the soul what it formally desired and what he promised it because the formal desire of the soul was not a manner of death but the service of God through martyrdom and the exercise of a martyr's love for him. Death through martyrdom in itself is of no value without this love, and God bestows martyrdom's love and reward perfectly by other means. Even though the soul does not die a martyr, it is profoundly satisfied since God has fulfilled its desire. (John of the Cross 1991, p. 219; *Dark Night Ascent* 2.19.13)

For John, the prophetic mystics are those who act in the world for the service and love of God. They are often called the friends or messengers of God. John understands that there is no greater love of God than the love of neighbor and the love of creation. Charity is the key element to those who are seeking eternal life. Without love, the seeker is condemned to live in darkness. Through love ("caritas"), the prophetic voice is linked to the path of apostolic action by serving God in all that he or she does, says, or thinks.

4. Biblical and Modern Theological Sources on the Active Life and the Contemplative Life

In the Christian mystical tradition, in which St. John of the Cross is deeply immersed, the active life and the contemplative life are well represented in the paradigmatic model of Martha and Mary. They are two aspects of the same reality. Mary of Bethany often symbolizes the contemplative life, while Martha best represents the active life. Mary and Martha are both sisters. The Teresian Discalced Carmelite tradition saw them as the Christian prototype of contemplatives in action, which may explain why St. John of the Cross joined the Teresian reform.

The biblical sources play an essential role in understanding the Christian message of St. John of the Cross because his contemplative desire to become one with God ultimately led him to embrace the world in his apostolic ministry. The biblical passage that creates the apparent dichotomy between contemplation and action in Christian circles comes from the Gospel of Luke. Jesus, answering to Martha's inquiry about why her sister Mary has left her to serve alone, says: "Martha, Martha, you are anxious and troubled about many things; but one thing is needed, and Mary has chosen that good part, which shall not be taken away from her" (Luke 10:41–42).

Some Christians have interpreted that passage as if Mary, who symbolically represents the life of contemplation, chose the best life. Such biblical interpreters believe the passage devalues the life of action, which is depicted in the story of Martha. Jesus never rejected the life of action. On the contrary, Jesus lived a very active life in his own time. The exegetical problem that Luke's biblical passage poses to the Christian tradition is whether or not Jesus intended to reject the active life altogether. It could be inferred from a careful reading of the Gospel of Luke that Jesus encouraged women to participate in the meetings if they wished. Otherwise, why did Jesus allow Mary to sit at his feet and hear his word when her duty as a Jewish woman was to take care of the household? (Luke 10:39).

The other historical problem that Christians faced in the past, and one that still is a debatable question in some Catholic and Protestant circles, is whether or not active apostolic work by and of itself is conducive to a more perfect union with God, and thereby may lead practitioners to salvation. However, the paradigmatic models of contemplatives in action today are Jesus of Nazareth, Bartolomé de las Casas (1484–1566), Saint Ignatius of Loyola (1491–1556), Saint Teresa of Ávila (1515–1582), Teilhard de Chardin (1881–1955), Dorothy Day (1897–1980), Martin Luther King, Jr. (1929–1968), Thomas Merton (1915–1968), and Oscar Romero (1917–1980), among others. These Christians found God in all things. They thought of apostolic service, which resulted from mystical graces from God, as a way of worshipping the Lord. Each one of them stood up as prophetic witnesses to justice and peace in their unique ways, asking for forgiveness and reconciliation in a time when they were fighting against all odds inside and outside the Church. Without doubt, the prophetic mystics did not turn their backs to the suffering inflicted on millions of people in different parts of the world. They did not withdraw completely from society in search for solitude alone. Instead, they protested against the individual and structural evils of their respective societies. Their spirituality was based on the ideal of building a compassionate world, including the love of the enemy. In the words of Wayne Teasdale:

> Socially engaged spirituality is the inner life awakened to responsibility and love. It expresses itself in endless acts of compassion that seek to heal others, contributing to the transformation of the world and the building of a nonviolent, peace-loving culture that includes everyone. (Teasdale 1999, p. 239)

Liberation theologians like Segundo Galilea (1928–2010) or Gustavo Gutiérrez (1928–present) have contributed a great deal in our time to recovering the prophetic element of Christianity. They have studied the life of the mystics and identified their prophetic dimension. For instance, Segundo Galilea was really interested in the study of the Spanish mystics, especially of St. Ignatius of Loyola, St. Teresa of Ávila, and St. John of the Cross. Galilea saw the Spanish mystics as great contemplatives in action, or as prophetic mystics. He writes:

> The service of the kingdom is the point at which the great mystics of the 16th century converge in their presentation of the ideal practice of effective love. All of them assumed responsibility in facing the history and needs of their time, and responded to them with a lucid and faithful Christian practice. In this respect they were prophets and their service of the kingdom was not ordinary, but prophetic. A prophet is a person who discerns the signs of the times in order to undertake the attitude and the response which the Spirit wills. Prophetism is an eminent form of the practice of effective charity. (Galilea 1985, pp. 64–65)

Galilea also recognized the social and religious reform taken by the Spanish mystics. The mystics played such an important role because they sanctified not only themselves but also the world. The Spanish mystics were not too worried about their own personal salvation. Their religious commitment was a testimony to their social awareness in a time of great troubles, especially in their mission of establishing religious monastic centers and universities. Both Jesuits and Carmelites were well known for their religious centers of prayer and for their missionary foundations all over the world. Even in the sixteenth century, Spanish missionaries included travels to Europe, Africa, Asia, and the Americas. Galilea points out that

> Ignatius, Teresa and John of the Cross have the same attitude and the same prophetic practice: to join extreme fidelity and adherence to the church with a practice featuring not words or criticisms, but rather, daring and significant deeds aimed at reforming the church from within. Their prophetism also manifests itself in their distrust in resorting to temporal means and powers and in their insistence above all on evangelical conversion and on the personal and collective holiness of the church. (Galilea 1985, p. 65)

Gustavo Gutiérrez, the founding father of liberation theology, was also interested in the Spanish mystics. He saw them as prophetic figures responding with their loving wisdom to the signs of the time. Having in mind the Spanish mystics, he says:

> A particular spirituality always represents a reorganizing of the fundamental foci of Christian life, on the basis of a central intuition or insight. The intuition is that of great men and women of the Spirit as they respond to the needs and demands of their age. Every spirituality is a way that is offered for the greater service of God and others: freedom to love. (Gutiérrez 1984, p. 89)

In other words, the mystic and the prophet are not two separate beings. Rather, the mystic in action integrates both the mystical and the prophetic elements in himself or herself, like St. John of the Cross or St. Teresa of Ávila who have greatly contributed to their world by founding new monasteries and convents, reforming the Carmelite tradition, and inspiring those who follow them through their teachings and writings. Clearly, their actions are the working effects of God's love in them. Similarly, Gustavo Gutiérrez describes the intrinsic relationship between the mystical and the prophetic elements in terms of language: "Mystical language expresses the gratuitousness of God's love, prophetic language expresses the demands this love makes" (Gutiérrez 1987, p. 95).

Thus, in the Judeo-Christian tradition, there is no clear gap between the mystical and the prophetic consciousness. The roles of the mystic and the prophet often intertwine. The mystic is the person who has a direct experience of God; the prophet is that person who, after having a vision of the divine, takes the initiative by partaking in the activities of the world without clinging to his or her own actions. The mystic has a vision of the divine; the prophet gives voice to that vision so that he or she becomes a witness and a vessel of truth, justice, and love in the world. The mystic is personally transformed by his or her encounter with the divine; the prophet attaches great importance to social commitment by becoming a "messenger" or "friend" of God. The mystic seeks out personal salvation; the prophet searches for the welfare of the community and the world at large. The mystic withdraws from the world to return more fully equipped to the world as a prophet, fully engaged and involved in his or her social environment. The mystic, therefore, is a spiritual master, a seeker of the really real; the prophet fulfills the task of a social and/or religious reformer, who is committed to building the kingdom of God on earth by fully conforming to God's vision for the world. The "healthy" mystic returns to the worldly activities fully equipped after having received special revelations from the Divine. At this meeting point, the mystic becomes the prophet. Interestingly enough, the genuine prophet is first of all a mystic, a messenger of God. He or she will not stop working until the message is delivered and implemented. According to José María Vigil (1946–present), "the prophet listens to the living God and then speaks in God's name" (Casaldáliga and Vigil 1994, p. 125).

Genuine prophetic mystics are those who commit themselves to an unrestricted desire to love God and persevere in their courageous effort to better serve the community and the world at large by sharing the fruits of their actions. By studying the life and thoughts of the prophetic mystics, one might realize that the mystics' actions spring from a deep contemplative life in which apostolic service is seen as an extension of their prophetic life. Christian mystical theologians traditionally understood the story of Martha and Mary as two complementary aspects of the divine life, the active and the contemplative. As Thomas Merton rightly observes:

> Contemplation and action necessarily have their part in every religious Rule. The two must always go together, because Christian perfection is nothing else but the perfection of charity, and that means perfect love of God and of men... But the active Orders would soon find that their activity was sterile and useless if it were not nourished by an interior spirit of prayer and contemplation, while the contemplative who tries to shut out the needs and sufferings of humanity and isolate himself in a selfish paradise of interior consolations will soon end up in a desert of sterile illusion. (Merton 1962, pp. 31–32)

St. Teresa and St. John of the Cross brought with them a renewed sense of what the Carmelite Order stands for by returning to the primitive spirit of Carmel which includes both the contemplative (or eremitical) life and the active (or apostolic) life. Thus, the Carmelite desert experience of silence and solitude led their mystics to become more aware of their special relationship with other creatures and the whole creation. Genuine mystics in action are those who are able to integrate a contemplative love for the glory and honor of God and an apostolic and social commitment for our neighbor and for all creation. As William Johnston declares in *The Inner Eye of Love*:

> I believe that the great prophets were mystics in action—their inner eye was awakened so that they saw not only the glory of God but also the suffering, the injustice, the inequality, the sin of the world. This drove them into action and often led to their death. And just as the great prophets were mystics, so the great mystics had a prophetic role. (Johnston 1982, p. 11)

Now we can turn to St. John of the Cross to explore more in depth what kind of mystic in action he was. The Spanish Carmelite was a great contemplative who served God and his community by holding many different administrative roles within his monastic Order. John's apostolic service proves his very busy active life and his prophetic mysticism attests to it.

5. St. John of the Cross on the Mixed Life

St. John of the Cross received his major training as a contemplative in action following the example set by his Discalced Carmelite tradition, and more specifically through the spiritual guidance of St. Teresa of Ávila, which strives to achieve a perfect balance between the eremitical (or contemplative) life and the apostolic service (or active) life. John, like Teresa, viewed the biblical story of Mary and Martha as two complementary sides in the Christian life. Genuine contemplation always leads to good deeds. The unselfish service that these two Spanish Carmelites offered to the world reminds us of the labor of love and the high price they paid embarking in their Discalced Carmelite reform.

The Carmelite Order, from its beginnings, was devoted to a contemplative and austere life. The holy prophet Elijah lived as a hermit on Mount Carmel. Elijah prayed in silence, listening to the still small voice within. Yet, Elijah was a prophetic mystic not only in the sense of witnessing God's Word but also one who announces the deepest troubles of his own society. He did not withdraw from the world for its own sake. Hermits traditionally received visits from friends and spiritual seekers in search of wisdom and practical advice. By the thirteenth century, the Carmelites observed the strict rules of the Order of St. Albert, striving to imitate the patron and founder of the Carmelite tradition, the prophet Elijah. The first Carmelites adopted the hermit life as a model of desert spirituality within their religious order in the pursuit of solitude and pure contemplation. However, this model was never intended to replace a life of prayer in community. The thirteenth-century Carmelites erected foundations "not only in desert places but in villages and towns, and thus abandoned the strictly eremitical life" (Zimmerman 1928, p. 3). The Teresian reform did not break with the thirteenth-century Carmelite tradition. On 24 August 1562, Teresa founded her first reformed convent, named after St. Joseph, in the inner domains of the city of Ávila where many Jewish conversos live, especially those who belong to the middle-class status. Scholars often argue that the Teresian reform was an attempt to return to the primitive rule of St. Albert given to the hermits on Mt. Carmel about 1206–1214. Ironically, Teresa founded the first Discalced Carmelite convent inside the famous medieval walls of Ávila, located at the heart of the inner city. Teresa moved away from her previous religious life as a nun in the Convent of the Incarnation, situated in the outer limits of Ávila, and built her first religious foundation in an urban setting.

Carmelite spirituality does not seek solitude and silence as ends in themselves. As a matter of fact, what the Carmelite prophetic mystical tradition sought from its beginning was not solitude but solidarity. By displaying acts of compassion and empathy towards others, the friar or the nun fulfills his or her religious vocation in the world. To be a Carmelite is to embody the Carmelite ideal. The etymological meaning of the Hebrew word "Carmel" (כַּרְמֶל) is often translated as paradise or garden. Carmelites are called to build the heavenly kingdom on earth. They make it possible for the ideal to become a reality. For them, the dichotomy presented to us between the life of contemplation and the life of action, between the mystical life and the prophetic life, is an artificial creation that impedes human beings from reaching their full potential, which has as its highest goal to love God, humanity, and creation. Thomas Merton puts it well when he says:

> There is no contradiction between action and contemplation when Christian apostolic activity is raised to the level of pure charity. On that level, action and contemplation are fused into one entity by the love of God and our brother in Christ. But the trouble is that if prayer is not itself deep, powerful and pure and filled at all times with the spirit of contemplation, Christian action can never really reach this high level... Without them our apostolate is more for our own glory than for the glory of God. (Merton 1990, p. 115)

The goal of the Christian mystic is to become God by participation so that the contemplative can share the fruits of his or her mystical vision with others by becoming a messenger of God on earth. St. John of the Cross clearly granted the possibility that some blessed souls become God by participation calling this transformative unitive experience of the human soul in God, "a lo divino." Mystical theologians called this process of divinization, "theosis" or pleromatization. Therefore, one might conclude that John seems to be closer to the Eastern (Orthodox) Church in his elaborated theology of partakers of the divine glory in God. Nonetheless, John's panentheistic mystical theology is perhaps rooted in the Pauline recapitulation of all things in Christ. As Manuela Dunn Mascetti notes:

> It is easy to forget that the Orthodox doctrine of deification was biblically based. In the famous saying of Peter, Christ saved us so that "we may become partakers of the divine nature" (2 Peter 1:4). Orthodox theologians leaned on this and other passages in creating this teaching. (See especially John 17:22–3 and 2 Corinthians 8:9). (Dunn Mascetti 1998, p. 99)

St. John of the Cross, as a cofounder member of the Discalced Carmelite tradition initiated by St. Teresa of Ávila, best exemplified the spirit of the mixed life. They both brought together the life of contemplation, which has as its goal a more perfect union with God through the daily practice of prayer following the eremitical model of the Carmelite tradition, and the life of apostolic action, which translates John's and Teresa's mystical vision into the realm of worldly activities. Yet John is often portrayed by many commentators only as a model of sanctity reached only by practicing an austere, ascetic, contemplative life. Furthermore, many commentators understand John's active life only in the context of an apostolate of prayer. Even worse, some commentators are inclined to believe that John did not play an active role in his time. The root cause of the problem lies in that some commentators saw the Carmelite tradition merely as a contemplative religious order. As E.W. Trueman Dicken (1919–2000) notes:

> Despite the Western origin of the Crusaders, the spiritual and monastic atmosphere of the Kingdom of Jerusalem had been largely that of the Eastern Church, and the life of the early Carmelites drew much from the traditions of Scete and Nitria. They lived as solitaries, bound together by their common eucharist and by the strict obedience enjoined by their rule. In the Eastern Church the notion of 'active' Religious orders is virtually unknown, and the weighty emphasis upon contemplation in the primitive Carmelite community is thus no more than one might expect to find. (Dicken 1963, p. 8)

Did the Teresian reform follow the primitive Carmelite spirit in the lines interpreted by Trueman Dicken? Most certainly not. Merton clearly understood the role of prophetic mysticism in the Spanish Carmelites when he declared: "Unless I am much mistaken, it was St Teresa and the Carmelite mystics of the sixteenth century who first brought into prominence the apostolic role and fruitfulness of the pure contemplative" (Merton 1980b, p. 61). The Carmelite reform is a perfect example of how central it is for Teresa and John to live a balanced life between interior prayer and apostolic action. As Merton observes in his essay, "What to Do - The Teaching of St. John of the Cross":

> The words of St. John of the Cross must be understood in the context of the saint's own life. He was not preaching an absolute repudiation of all duties and responsibilities and all works and labours for the Church of God or for other men. He and St. Theresa of Avila, the greatest contemplatives of their time, were also very active and laboured and suffered much for the reform of the Carmelite Order. (Merton 1981, p. 68)

It is important to understand John's writings in the historical context of sixteenth-century Spain. Needless to say, Sanjuanist scholars ought to reinterpret John's texts and his mystical thoughts in reference to the historical context in which he lived. Bringing out the multifaceted dimensions of St. John of the Cross illustrates more accurately the Carmelite saint's enormous contribution to the world. This study offers a new vision of St. John of the Cross as a contemplative mystic engaged in the socio-religious issues of his time. To illustrate this point, Merton rightly attributed the extremist austerity to John's age in general and to the new ideas implemented by Nicholas Doria (1539–1594) as the general definitor of the Discalced Carmelite order. He writes:

> [I]n the sixteenth century, within the Discalced Reform, there was also an extreme wing which sought solitude along with austerity and centralization: and this was the faction of Doria and the Friars of Pastrana, who eventually persecuted St. John of the Cross, and hounded him to his death. The curious thing is that St. John of the Cross, the defender of the pure Carmelite ideal of mystical contemplation, was himself not an extremist in favor of pure solitude, nor did he advocate extreme austerity, but took the middle way, favoring the combination of solitude and contemplation with preaching and the direction of souls.
>
> Hence it is evident that in the history of the Carmelites the pure and primitive spirit of the Order always remains incarnate in a kind of "prophetic" union of solitude and apostolate. When this balance is disturbed, when the shift is made too far in one direction or the other, then the primitive spirit is lost. That is to say that when too much emphasis is placed on apostolic action, the primitive spirit is of course weakened and eventually destroyed. But that does not mean that the return to the original ideal is a mere matter of abandoning the apostolate and embracing a solitary life that is primarily ascetical and austere. It seems likely that the apostolate in its own way encourages contemplation, just as contemplation is the source of a genuine apostolate. (Merton 1965, pp. 179–80)

Studying the Bible and the Christian tradition was a precondition for opening new Carmelite houses of study in the Teresian reform. As a result of achieving this final integration between the eremitic spirit of the earlier Carmelites and the apostolic fervor of the Teresian reform, John shows the sacramental link between the mysticism of action, wisdom, and devotion. St. John of the Cross wholly embraced a mystical theology of holiness in action by taking care of administrative businesses; serving as a spiritual director, priest, or confessor; and fulfilling his poetic talent and theological vocation. In other words, John consecrated his life to God by unfolding a special love for the world and for all creatures living in it.

On 28 November 1568, John and a senior friar named Antonio de Heredia (1510–1601) (later known in religious circles as Antonio de Jesús) founded at Duruelo the first male house following the Teresian ideal. From then on, John and Teresa would work together building new Carmelite monasteries and convents and traveling throughout Spain and Portugal. Among the administrative and religious tasks that John had to fulfill in the earliest stages of his monastic order were the establishment of new Carmelite houses for nuns and friars, and the spiritual direction of novices. The first Discalced Carmelite friary built in Duruelo was a tremendous victory for the Teresian reform. By extending the reform to male houses of the religious order, Teresa has fulfilled her dream by opening Carmelite houses for nuns and friars. No woman ever before Teresa was allowed to become the foundress of a religious order for friars. John was appointed subprior and novice master of the first Discalced Carmelite friary of Duruelo in 1570. John also held other important administrative positions within the Carmelite Order. In June of the same year, John became master of novices at the Carmelite foundations of Mancera de Abajo and Pastrana. He guided the Carmelite novices in their intellectual and spiritual development. The Teresian reform encouraged and almost required them to combine mental and spiritual exercises with apostolic service. They served in their monastic duties as confessors, preachers, and administrators, and were required to perform manual labor during certain hours of the day. This is further evidence that the Discalced Carmelites embraced the mixed life. In the spring of 1571, John was transferred to Alcalá de Henares. In April, John was appointed rector of the Carmelite College of Alcalá de Henares. His famous lectures and his spiritual guidance attracted people from all walks of life. He often met with lay university professors for discussions in public and in private. John integrated in his curriculum the study of theology and philosophy so that the students were ready to discern, with the help of a spiritual director or confessor, the mysteries of life. At Teresa's request, John left Alcalá to attend to the religious needs of the nuns in Ávila. In 1572, John was appointed the vicar, spiritual director, and confessor at the Convent of the Incarnation. John spent approximately five years working closely with Teresa. He held that office until 1577, when he was kidnapped in Ávila and later arrested in Toledo in that same year. He was held for nine months captive in a Calced Carmelite monastery.

Most commentators believe that the tug of war between Calced and Discalced Carmelites culminated in the incarceration of John in Toledo. He was accused of "rebellion and contumacy" against the Carmelite Order. He had been shut up in a cupboard, six feet by ten, through the bitter cold of the Toledan winter, and the scorching heat of the summer. "Imprisonment, flogging, fasting on bread and water were standards penalties in religious orders of the period," Peter Slattery notes (Slattery 1991, p. 71). In 1578, John escaped from prison after nine months of captivity. It is said that he finally managed to escape from the monastery at night, by tearing his bedding into strips to use as a rope to climb down the steep stone walls. He then made his way through the city to a convent of Reformed nuns, who sheltered him and nursed him back to a semblance of health. He then travelled to Madrid, and from there to El Calvario (where he became the vicar). He was now no longer in immediate danger, as the attempts to halt the Reform had, at least temporarily, slowed down. Although political disputes within the Carmelite Order were numerous, I believe there were other hidden motives behind John's persecutions. Why would he be persecuted by both Calced and Discalced Carmelite brothers? It cannot simply be said that John suffered persecution because he was on Teresa's side. I suspect that John's converso background played an important role, especially in the later persecutions at the hands of his Discalced Carmelite brothers.

I would argue that the Discalced Carmelite Order, cofounded by Teresa and John, affected not only the religious reforms of the sixteenth century but also the social fabric of the Castilian society, especially by providing a monastic refuge for women and conversos (both Jewish and Muslim converts). Needless to say, women and conversos were the social groups that suffered the most during Philip II's reign as the monarch, and some of his delegates persecuted those who departed from the newly established codes for the Kingdom of Castile. Women and conversos were seen as potential threats to the new social and religious identity adopted by old Christians in the so-called reconquest of the Iberian Peninsula. Some of these converso women had strong ties to so-called heretical sects ("iluminadas" or "alumbradas") and they were seen as a threat to the new social and religious policies of the Spanish empire. In the meantime, Teresa and John opened their convents and monasteries to those women and conversos who did not have a place in the new society by accepting them in large numbers and by allowing them to play a significant role inside the Carmelite tradition (see Márquez Villanueva 1998).

5.1. John the Mystic (or the Contemplative Life as Mary)

St. John of the Cross was certainly a mystic because he had a direct experience with the Divine. John's poetry is impregnated with many mystical symbols from the living flame of love to the dark night. John explicitly defined mysticism as contemplation when he said: "For contemplation is nothing else than a secret and loving inflow of God, which if not hampered, fires the soul in the spirit of love" (John of the Cross 1991, p. 382; *Dark Night* 1.10.6). He also defined contemplation as "an inflow of

God into the soul, which purges it of its habitual ignorances and imperfections, natural and spiritual, and which the contemplatives call infused contemplation or mystical theology" (John of the Cross 1991, p. 401; *Dark Night Dark Night* 2.5.1). For St. John of the Cross, infused contemplation meant experiential faith, or union with God. For John, the mystic is not simply someone who is seeking union with God but rather someone who has already experienced the loving wisdom of God at the deepest center of his or her soul. John lived his mystical experience of intimacy with God with great intensity. His devotional poems clearly illustrate his purpose of guiding the spiritual seeker to achieve union with God in this life. This spiritual journey in search of union with God is better expressed in his celebrated poem, the *Ascent of Mount Carmel*. However, it is in his poems, the *Spiritual Canticle* and the *Living Flame of Love*, where John fully expressed his deep intimate union with God using the nuptial imagery of the spiritual marriage between the lover and the beloved.

As a mystical theologian, John identified the highest degree of mystical union in this life with infused contemplation. Ultimately, it is only God who could grant the gift of grace to the human soul, although Christians prepare themselves to receive God within by fasting, meditating, or doing apostolic work for the service of God. John claimed that the mystical vision of God "is proper to the intellect" (John of the Cross 1991, p. 531; *Spiritual Canticle* 14.14). John, following St. Thomas Aquinas's (1225–1274) epistemology, explained his theory of divine knowledge and union with God through the concept of active and passive intellect. For him, the intellect was not associated with the rational faculty of the human soul but rather with intuitive understanding, a category that belongs to the realm of mystical theology. Moreover, this experiential loving wisdom was defined by John as the highest state of contemplation in this life. The blessed soul was able to hear the still small voice of God. "These are pure spiritual revelations or visions which are given only to the spirit without the service and help of the senses," as John observes (John of the Cross 1991, p. 531; *Spiritual Canticle* 14.15).

St. John of the Cross was a Christian mystic because he adopted the mystical language of Christianity as his own experience. He writes: "Insofar as infused contemplation is loving wisdom of God, it produces two principal effects in the soul: by both purging and illumining, this contemplation prepares the soul for union with God through love" (John of the Cross 1991, p. 531; *Spiritual Canticle* 14.15). Thus, the Carmelite saint, being rooted in the Christian mystical tradition, identified the purgative, illuminative, and unitive as the three stages of the mystical life. Here, below, are a few examples of how John followed the three stages in his writings.

The purgative stage corresponds to the beginners (or "principiantes"). John understood purgation as a spiritual discipline in which one prepares the soul to receive God by practicing asceticism. The ascetic person seeks God by fasting or praying. At the same time, the Christian monastic tradition requires from the practitioner or spiritual seeker great discipline and human effort (acquired contemplation). As a result of this effort, God will purge the human soul of all its bad habits (infused contemplation). John writes: "If you desire that devotion be born in your spirit and that the love of God and the desire for divine things increase, cleanse your soul of every desire, attachment, and ambition in such a way that you have no concern about anything" (John of the Cross 1991, p. 91; *Sayings of Light and Love*, p. 78). The illuminative stage corresponds to the proficient ones (or "aprovechados"). The religious believer desires to know God. With God's help, the seeker will experience union with God in this life by virtue of his or her own degree of receptivity of the divine light within. He or she who has been blessed by God will have the opportunity to know the hidden mystery of God. According to John, "a revelation is nothing else than the disclosure of some hidden truth, or the manifestation of some secret or mystery, as when God imparts understanding of some truth to the intellect, or discloses to the soul something that he did, is doing, or is thinking of doing" (John of the Cross 1991, p. 244; *Ascent* 2.25.1). God infuses wisdom and understanding to those who receive the divine life in their hearts. As John puts it, "God supernaturally illumines the soul with the ray of his divine light. This light is the principle of the perfect union that follows after the third night" (John of the Cross 1991, p. 156; *Ascent* 2.2.1). In another passage, John states:

> Jeremiah shows clearly that the soul is purged by the illumination of this fire of loving wisdom (for God never bestows mystical wisdom without love, since love itself infuses it) where he says: *He sent fire into my bones and instructed me* [Lam. 1:13]. And David says that God's wisdom is silver tried in the fire [Ps. 11:6], that is, in the purgative fire of love. This contemplation infuses both love and wisdom in each soul according to its capacity and necessity. It illumines the soul and purges of its ignorance, as the Wise Man declares it did to him [Eccles. 51:25–27]. (John of the Cross 1991, p. 422; *Dark Night* 2.12.2)

Finally, the unitive or transformative mystical stage belongs to the perfect or blessed souls (or "perfectos"). The goal of the Christian mystic is to become one with God. However, this mystical union is only possible by grace, not by nature. The human soul becomes God by participation. The one who is reborn in the Holy Trinity will be able to see the kingdom of God, which is the highest state of perfection in this life. As John explains in his theological commentaries:

> When God grants this supernatural favor to the soul, so great a union is caused that all the things of both God and the soul become one in participant transformation, and the soul appears to be God more than a soul. Indeed, it is God by participation. Yet truly, its being (even though transformed) is naturally as distinct from God's as it was before, just as the window, although illumined by the ray, has being distinct from the rays. (John of the Cross 1991, p. 165; *Ascent* 2.5.7)

John's mystical language of union echoes the experience of the Desert Fathers and the Greek Orthodox Christians, especially in their theology of participation (or theosis). As John points out, "[h]aving been made one with God, the soul is somehow God through participation. Although it is not God as perfectly as it will be in the next life, it is like the shadow of God" (John of the Cross 1991, p. 706; *Living Flame of Love* 3.78).

5.2. John the Prophet (or the Active Life as Martha)

St. John of the Cross uses the definition of prophecy in two different ways. On the one hand, the prophet is one who listens to the Word of God; and the blessed soul receives a message from God, that is, a divine revelation. John asks, "Whom will God instruct? And to whom will he explain his word and prophecy?" (John of the Cross 1991, p. 215; *Ascent* 2.19.6). On the other hand, the prophet is one who bears witness to truth, justice, and love (or charity). John quotes St. Paul to support his second definition of prophecy in the following passage:

> If I speak in human and angelic tongues and do not have charity, I am like a sounding metal or bell. And if I have prophecy and know all mysteries and all knowledge, and if I have all of faith so as to move mountains, and do not have charity, I am nothing, and so on [1 Cor. 13:1–2]. When those who esteem their works in this way seek glory from Christ saying: Lord, did we not prophesy in your name and work many miracles? He will answer: Depart from me, workers of iniquity [Mt. 7:22–23]. (John of the Cross 1991, p. 324; *Ascent* 3.30.4)

John's ascetico-mystical teachings are not designed to cause pain and anguish to the human soul as an end in itself. On the contrary, John's major task is precisely to free the soul from all personal and societal obstacles that impede union with God. Nevertheless, a human soul cannot feel totally free without experiencing some degree of pain and suffering in this life. Suffering is an integral part of life. Otherwise, human beings would not be able to grow. This could explain why John spoke so tenderly about his trials and afflictions. As he puts it in the *Sayings of Light and Love*, "Have great love for trials and think of them as but a small way of pleasing your Bridegroom, who did not hesitate to die for you" (John of the Cross 1991, p. 92; John's saying is catalogued as number 94). John understands that prophesy, in the sense of mystical revelations, must go hand in hand with the second definition of prophesy in the sense of serving God with true love, "for in charity lies the fruit of eternal life" (John of the Cross 1991, p. 324; *Ascent* 3.30.5). As Peter Henriot points out:

> "Faith without works is dead." This is the blunt answer given by the Apostle James to the perennial question about the relationship between belief and deeds. Today, we might phrase the question differently. We might ask about the relationship between faith and justice, prayer and action, spirituality and social commitment. But the answer is still the same. Faith without works is dead. (Dyckman and Carroll 1981, p. ix)

The question then becomes: Is John's faith fully alive without the works of peace, justice, and apostolic service? Clearly, John embraces the two definitions of prophecy in his writings. He says:

> The prophets, entrusted with the word of God, were well aware of this. Prophecy for them was a severe trial because, as we affirmed, the people observed that a good portion of the prophecy did not come about in accord with the letter of what was said to them. As a result the people laughed at the prophets and made much fun of them. It reached such a point that Jeremiah exclaimed: They mock me all day long, everyone scoffs at and despises me because for a long time now I have cried out against iniquity and promised them destruction, and the Lord's word has become a reproach to me and a mockery all the time. And I said: I do not have to remember him or speak any more in his name [Jer. 20:7–9]. (John of the Cross 1991, p. 222; *Ascent* 2.20.6)

Like the ancient prophets, John endured suffering. It is not shocking at all to find out that John was persecuted in his own time by inquisitors and by a few of his own Carmelite brothers. According to John's foes, he was holding unorthodox Christian ideas and practices. He was accused of being an illuminist, a quietist, and a rebel. In sixteenth-century Spain, mysticism was not welcomed in some religious circles. His foes were suspicious of new religious movements mainly composed of lay people, Jewish and Moorish converts, and Lutherans, because many of them rejected the position of the Catholic Church in matters of faith, salvation, sin, and sacraments. Some of these Lutheran figures believed that the human soul can find salvation without the mediation of the Catholic Church, meaning that humans do not need to confess before a priest, or receive certain sacraments, and so forth, because, eventually, God's grace alone will save the human soul.

John sided with those who suffered in his own time, especially the Carmelite nuns and Father Jerome Gratian or Gracián (1545–1614). John announced the deepest troubles of his Carmelite brothers and sisters. Because of that, he suffered persecution at the hands of some of his own Carmelite brothers. He was mocked and humiliated, put in prison for nine months, forced to leave his administrative offices, and sent to La Peñuela as a forced exile. Why did John get into so much trouble? Why was John forced to quit his administrative positions within the Carmelite Order? Why did Nicolas Doria want John to leave the Iberian Peninsula? Could it be viewed as an attempt to get rid of John by sending him to Mexico and, thus, remove him from the religious order? The plan never took place because John died in Úbeda before embarking with other Carmelites to the new world. However, it is well known today that some Carmelites working for Doria took the necessary steps to remove John of his Carmelite habit. As William Barnstone notes:

> he [John] was stripped of all office and exiled to la Peñuela, a desert house in Andalusia. Evidence was collected against him, some of it tragi-comic, such as a false accusation by a nun in Málaga that she had been kissed by Fray Juan through the grille of her window. At Beas de Segura, his favorite convent, the nuns destroyed all papers and letters from him, for fear of being implicated with the heretic monk. There was a move to expel him from the order, and only his sickness spared him this last step. (Barnstone 1972, p. 17)

John stood against Doria's efforts of centralizing the Discalced Carmelite Order, which left the nuns at his mercy, even at the expense of losing all his power within the Carmelite Order. John, the co-adjutor and co-founder of the Discalced Carmelite friars and a loyal friend to the Teresian reform, was released by Doria's decree of all his office duties before he died. John was sent to the hermitage of la Peñuela, away from the new Council formed by Doria. In 1591, John became very ill. He suffered from fevers and gangrenous sores on his foot. He moved from the convent of la Peñuela (today located in la Carolina) to Úbeda, where he received poor treatment from the prior, Friar Francisco Crisóstomo, who even denied him medical attention. John died on 14 December 1591. Because he acted on his prophetic vision and suffered for his actions, John was definitely a modern prophet, not the simple rebel as he was often accused of being. Unfortunately, John faced the martyrdom of rejection by his own Carmelite brothers. He paid the price of aligning himself with those who were oppressed. Like Jesus, he suffered persecution. Perhaps his death was linked to his enemies' constant oppression. Peter Slattery correctly understood John's prophetic dimension when he wrote:

> St. John, the poet, being a person of discernment, was sensitive to the injustices and exaggerations of his time, and in his innocence he made people aware of them. Poets are uncomfortable people to be with. Certainly, toward the end of his life those with power did not want him close to them. St. John, the poet, called on his fellow religious to examine stagnation in their lives and institutions—he did this by the force of the sanctity of his life and the power of his poetry. He was a silent contemplative who suffered, not only because of his own empathy, but because he threatened the powerful. Out of his silence he caressed and challenged all who read his poetry. (Slattery 1991, p. 74)

St. John of the Cross, like the ancient prophets, aligned himself with the oppressed or "anawim" (in that case, with the nuns, but John also stood behind the moderate Carmelite, Gracián, who was ultimately expelled from the Order). Thus, St. John of the Cross does not write his mystical writings in isolation from his religious and cultural environment. Rather, the Carmelite writer needs to be seen in the light of a continuous line of mystical thoughts that are primarily rooted in the Judeo-Christian biblical tradition, and yet not completely deprived of other possible influences coming from the non-Christian world (particularly the Greek, Latin, Jewish, and Muslim traditions). It is, therefore, not difficult to understand now why it took so many decades, even centuries, before John was beatified, canonized, and given the honorific title of Mystical Doctor of the Catholic Church. On 22 January 1675, Clement X beatified John of the Cross, a beatification that took longer than was expected, for St. Ignatius of Loyola and St. Teresa of Ávila (1614) were beatified right after their deaths. In 1726, Benedict XIII canonized him. On 24 August 1926, St. John of the Cross "was solemnly declared a Doctor of the Universal Church" by Pope Pius XI (Gabriel of St. Mary Magdalen 1954, p. xi). Pope Pius XI proclaimed St. John of the Cross "Doctor Ecclesiae" in his Apostolic Letter *Die Vicesima*. For many Sanjuanist specialists, the year 1926 marks the turning point of future critical studies once John has become the "Doctor Misticus" of the Catholic Church. In 1970, Teresa was the first woman to be granted this title by Pope Paul VI. In 1952, the Spanish Ministry of National Education named St. John of the Cross the patron of Spanish poets. Ironically, John's texts became the norm for testing the authenticity of what could be regarded as the model of sainthood and nowadays called by Pope Francis (1936–present) the model of holiness. Priests, monks, and theologians, especially in seminaries, still spend many hours studying John's writings. They see him as an authority in matters of faith. In an apostolic letter dated 14 December 1990, Pope John Paul II (1920–2005), who wrote his doctoral dissertation on St. John of the Cross, dedicates a whole document to the Carmelite saint calling him "a master in the faith" and a "witness of the living God." Today, people from all walks of life (particularly religious figures, poets, scientists, artists, philosophers, theologians, atheists, and so forth) have demonstrated a special interest in studying John's life and writings. The studies on St. John of the Cross have increased dramatically in the twentieth century. Researchers from an array of fields of knowledge have discovered new data on the historical background and the life events of the Carmelite saint, on the authenticity of his writings, and on the originality of his thoughts. This article has reconsidered the life and the thoughts of St. John of the Cross in light of these scholarly studies.

6. Conclusions

St. John of the Cross develops a holistic mystical theology by seeing the sacramental life as the body of religion in a twofold way which can be best expressed through the inner life of the liturgy during Mass, and through the outer life of carrying out the Christian message to the world by engaging with the social and apostolic issues of one's time; it also consists of seeing the theological life as the head of religion, and the mystical life as the heart and dynamo of religion. These three aspects of religion—the sacramental, the theological, and the mystical—are well integrated in the life events and thoughts of St. John of the Cross. The Sanjuanist mystical theology provides concrete and practical guidelines in addressing the physical, mental, and spiritual needs of the whole person. St. John of the Cross was a mystic of action who responded prophetically to the social and religious issues of his time. By being firmly rooted in the eremitical–prophetic Carmelite tradition, John was able to reach out to those in need by virtue of his apostolic ministry. In monitoring the progress made by friars and nuns in their respective ministries, John avoided the sorts of religious trapping of a contemplative Docetism which was prevalent in some religious circles of his time, especially in monastic communities. The Carmelite mystic recognized his social and religious responsibility to be morally engaged in his time. Because John had the courage to follow God's calling, he suffered persecution.

The prophetic mysticism of St. John of the Cross is sharply at odds with the old Carmelite picture of a contemplative who completely withdrew from society in search of God. John's mysticism does not reject the human condition in order to seek one's own individual salvation without manifesting any concern for the rest of humanity and all other creatures. Nor is his mysticism a matter of praying all day inside a monastic community for the sake of saving his soul. As Bernard McGinn notes:

> A large part of the secret of the Carmelite contribution seems to have been found in the ongoing tension between the desire for solitude—that is, withdrawal into the desert, especially the desert of the heart—and the need to be actively engaged in the work of spreading God's love in the world.... In an activist age and in a culture that tends to prize action above contemplation, this part of the Carmelite heritage is important both for the Carmelites themselves and for the witness they give to the rest of us. (McGinn 2017, p. 47)

From this study one thing becomes clear. St. John of the Cross was ahead of his time because he had the intellectual capacity to link the scientific, the literary, and the theological sources of his day. It has been suggested from previous sources that John belonged to the Renaissance age, although he was still living under the control of a medieval Church. In actuality, John played an important role as a pioneer thinker in the Christian humanist movement of the Iberian Peninsula. As a bridge builder, John understood quite well the relationship between primitive and medieval Christianity, especially in the context of monasticism. Yet John was a man of the Renaissance Age by having one foot at the gates of modernity and another foot anchored in the primitive Church. By incorporating new theories and ideas unheard of in his own time, John undisputedly contributed to the spiritual progress of sixteenth-century Europe. Proof of that is the fact that centuries later he was admired by people from all corners of the world, even in Asia where Eastern religious leaders have suggested that John was a sort of spiritual master. The contemplative John was also ahead of his time when he rightly saw the urgent necessity of courageously confronting the problems that were affecting the fragile thread of the Iberian family. John advocated for a non-violent way to shed light on the injustices committed in his own time, following the Christian principles of the Gospels. For John, the true contemplative was not only the blessed soul who achieves union with God in this life but also one who works for peace and unity in the world. "The finally integrated man is a peacemaker, and that is why there is such a desperate need for our leaders to become such men of insight," as Thomas Merton notes (Merton 1980a, p. 207). John was able to create in the midst of a harsh environment of hatred and resentment a non-violent, loving response to those who characterized themselves as his enemies. In his famous twenty-sixth letter addressed to Mother María de la Encarnación (1565–1618), John writes: "And where there is no love, put love, and you will draw love" (John of the Cross 1991, p. 760). John exposed in public the systematic expressions of sin that were part of the social and the religious establishment of his time. He chose writing, preaching, confession, and spiritual direction as the prophetic mediums for speaking out against such injustices as intolerance, hunger, illiteracy, or mistreatment of women. John was well rooted in the eremitical–prophetic tradition following the Carmelite tradition (particularly, the Teresian reform) which perfectly balances the contemplative life and the active life. In my humble opinion, Teresa's preference for the mixed life could easily explain why John left the Calced Carmelite Order, forgot his plans of joining the Carthusian Order (which is known for its austere life advocating long periods of silence and ascetic practices), and became the coadjutor and religious reformer, together with Teresa, of the Discalced Carmelite tradition. This article has demonstrated that St. John of the Cross was a contemplative in action within the parameters of his own monastic religious order.

Funding: This research received no external funding.

Conflicts of Interest: The author declares no conflict of interest.

References

Asín Palacios, Miguel. 1981. *Saint John of the Cross and Islam*. Translated by Howard W. Yoder, and Elmer H. Douglas. New York: Vantage Press.

Asín Palacios, Miguel. 1990. *El Islam Cristianizado: Estudio del Sufismo a Través de las Obras de Abenarabi de Murcia*. Madrid: Hiperión.

Barnstone, Willis. 1972. *The Poems of Saint John of the Cross*. New York: A New Directions Book.

Beal, Jane, ed. 2013. *Illuminating Moses: A History of Reception from Exodus to the Renaissance*. Leiden: Brill.

Brenan, Gerald. 1973. *St John of the Cross: His Life and Poetry*. Cambridge: Cambridge University Press.

Carabine, Deirdre. 1995. *The Unknown God. Negative Theology in the Platonic Tradition: Plato to Eriugena*. Louvain: Peeters Press.

Casaldáliga, Pedro, and José María Vigil. 1994. *Political Holiness*. Maryknoll: Orbis Books.

Cuevas, Cristóbal. 1972. *El Pensamiento del Islam: Contenido e Historia. Influencia en la Mística Española*. Madrid: Ediciones Istmo.

De Tapia, Serafín. 2006. Las huellas y el legado de las tres culturas religiosas en Ávila. In *Vivencia Mistica y Tejido Social*. Zamora: Ediciones Monte Casino, pp. 179–227.

Dicken, E. W. Trueman. 1963. *The Crucible of Love: A Study of the Mysticism of St. Teresa of Jesus and St. John of the Cross*. New York: Sheed and Ward.

Dombrowski, Daniel A. 1992. *St. John of the Cross: An Appreciation*. New York: State University of New York Press.

Doohan, Leonard. 1995. *The Contemporary Challenge of John of the Cross: An Introduction to His Life and Teaching*. Washington, DC: ICS Publications.

Dunn Mascetti, Manuela. 1998. *Christian Mysticism*. New York: Hyperion.

Dyckman, Katherine Marie, and L. Patrick Carroll. 1981. *Inviting the Mystic, Supporting the Prophet: An Introduction to Spiritual Direction*. New York: Paulist Press.

Fitzgerald, Brian. 2017. *Inspiration and Authority in the Middle Ages: Prophets and Their Critics from Scholasticism and Humanism*. Oxford: Oxford University Press.

Frost, Bede. 1937. *Saint John of the Cross. Doctor of Divine Love: An Introduction to His Philosophy, Theology and Spirituality*. London: Hodder & Stoughton.

Gabriel of St. Mary Magdalen. 1954. *St. John of the Cross: Doctor of Divine Love and Contemplation*. Westminster: The Newman Press.

Galilea, Segundo. 1985. *The Future of Our Past: The Spanish Mystics Speak to Contemporary Spirituality*. Notre Dame: Ave Maria Press.

Garrido, Pablo María. 1989. *San Juan de la Cruz y Francisco de Yepes: En torno a la Biografía de los dos Hermanos*. Salamanca: Ediciones Sígueme.

Girón-Negrón, Luis M. 2008. Dionysian Thought in Sixteenth-Century Spanish Mystical Theology. *Modern Theology* 24: 693–706. [CrossRef]

Gómez-Menor Fuentes, José. 1970. *El Linaje Familiar de Santa Teresa y de san Juan de la Cruz: Sus Parientes Toledanos*. Toledo: Gráficas Cervantes.

Gutiérrez, Gustavo. 1984. *We Drink from Our Own Wells: The Spiritual Journey of a People*. Translated by Matthew J. O'Connell. Maryknoll: Orbis Books.

Gutiérrez, Gustavo. 1987. *On Job*. Translated by Matthew J. O'Connell. Quezon City: Claretian Publications.

Happold, Frederick Crossfield. 1988. *Mysticism: A Study and an Anthology*. New York: Penguin Books.

Hardy, Richard P. 2004. *John of the Cross: Man and Mystic*. Boston: Pauline Books.

Hocking, William E. 1955. *The Meaning of God in Human Experience: A Philosophic Study of Religion*. New Haven: Yale University Press.

Hvidt, Niels C. 2007. *Christian Prophecy: The Post-Biblical Tradition*. Oxford: Oxford University Press.

John of the Cross. 1991. *The Collected Works of Saint John of the Cross*. Translated by Kieran Kavanaugh, and Otilio Rodríguez. Washington, DC: ICS Publications.

Johnston, William. 1982. *The Inner Eye of Love: Mysticism and Religion*. San Francisco: Harper & Row.

Kreisel, Howard. 2001. *Prophecy: The History of an Idea in Medieval Jewish Philosophy*. Norwell: Kluwer Academic Publishers.

López-Baralt, Luce. 1990. *San Juan de la Cruz y el Islam*. Madrid: Hiperión.

López-Baralt, Luce. 1998. *Asedios a lo Indecible: San Juan de la Cruz Canta al Éxtasis Transformante*. Madrid: Editorial Trotta.

Márquez Villanueva, Francisco. 1998. *El Problema Morisco (Desde otras Laderas)*. Madrid: Ediciones Libertarias.

Martínez González, Emilio J. 2006. *Tras las Huellas de Juan de la Cruz: Nueva Biografía*. Madrid: Espiritualidad.

Matthew, Iain. 1995. *The Impact of God: Soundings from St John of the Cross*. London: Hodder & Stoughton.

McGinn, Bernard. 2017. *Mysticism in the Golden Age of Spain (1500–1650)*. New York: A Herder & Herder Book.

McGreal, Wilfrid. 1997. *John of the Cross*. Liguori: Triumph.

Merton, Thomas. 1962. *The Waters of Siloe*. New York: Image Books.

Merton, Thomas. 1965. *Disputed Questions*. New York: A Mentor-Omega Book.

Merton, Thomas. 1968. *Zen and the Birds of Appetite*. New York: A New Directions Book.

Merton, Thomas. 1980a. *Contemplation in a World of Action*. Boston: Mandala Books.

Merton, Thomas. 1980b. *On Saint Bernard*. Kalamazoo: Cistercian Publication.

Merton, Thomas. 1981. *What Is Contemplation?* Springfield: Templegate Publishers.

Merton, Thomas. 1990. *Contemplative Prayer*. New York: Image Books.

Peers, E. Allison. 1945. *Spirit of Flame: A Study of St. John of the Cross*. New York: Morehouse-Gorham.

Perrin, David B. 1997. *For Love of the World: The Old and New Self of John of the Cross*. Bethesda: International Scholars Publications.

Poveda Piérola, Lola. 2011. *Conciencia Energía y Pensar Místico: El Hoy de Teresa de Jesus y Juan de la Cruz*. Bilbao: Desclee de Brouwer.

Reeves, Marjorie. 1969. *The Influence of Prophecy in the Later Middle Ages: A Study of Joachimism*. Oxford: Clarendon Press.

Ruffing, Janet K. 2001. Ignatian Mysticism of Service. In *Mysticism & Social Transformation*. Edited by Janet K. Ruffing. Syracuse: Syracuse University Press, pp. 104–28.

Satz, Mario. 1991. *Umbría Lumbre: San Juan de la Cruz y la Sabiduría Secreta en la Kábala y el Sufismo*. Madrid: Hiperión.

Serrán-Pagán, Cristóbal. 2017. Divine Mercy in Thomas Merton and St. John of the Cross: Encountering the Dark Nights in the Human Soul. *The Merton Annual* 30: 117–30.

Serrán-Pagán, Cristóbal. 2018. *St. John of the Cross: His Prophetic Mysticism in the Historical Context of Sixteenth-Century Spain*. Washington, DC: Pacem in Terris Press.

Slattery, Peter. 1991. *The Springs of Carmel: An Introduction to Carmelite Spirituality*. New York: Alba House.

Conde Solares, Carlos. 2019. The Moral Dimensions of Sufism and the Iberian Mystical Canon. *Religions* 11: 15. [CrossRef]

Swietlicki, Catherine. 1986. *Spanish Christian Cabala*. Columbia: Missouri Press.

Teasdale, Wayne. 1999. *The Mystic Heart: Discovering a Universal Spirituality in the World's Religions*. Novato: New World Library.

The Sisters of Notre Dame. 1927. *Life of Saint John of the Cross: Mystical Doctor*. New York: Benziger Brothers.

Thompson, Colin. 2015. Saint John of the Cross. In *Oxford Bibliographies*. Available online: https://www.oxfordbibliographies.com (accessed on 21 July 2020).

Valdivia Válor, José. 2002. *Don Miguel Asín Palacios. Mística Cristiana y Mística Musulmana*. Madrid: Hiperión.
Zimmerman, Benedict. 1928. The Development of Mysticism in the Carmelite Order. In *The Ascent of Mount Carmel*. Translated and Edited by David Lewis. London: Thomas Baker.

religions

Article

Don Quixote and Saint John of the Cross's Spiritual Chivalry †

Luce López-Baralt

Department of Spanish Studies, Río Piedras Campus, University of Puerto Rico, San Juan 00931, Puerto Rico; lucelopezbaralt@gmail.com

† Translated by Marcela Raggio, Gloria Martínez, Guadalupe Herce, Daniela Tornello, Agustina Fredes and Verónica Mastrodonato Pavetti, members of the Victoria Ocampo Literary Translation Porgram, Cuyo National University (Universidad Nacional de Cuyo). Translator's note (TN): In cases where there is no official English translation available, we have provided our own translation.

Abstract: Despite its ludic appearance, "The adventure Don Quixote had with a dead body" (part I, chapter XIX) is one of the most complex pieces of Cervantes' famous novel. In the midst of a dark night, the Manchegan knight errant confronts an otherwordly procession of robed men carrying torches who transport a dead "knight" on a bier. Don Quixote attacks them to "avenge" the mysterious dead man, discovering they were priests secretly taking the body from Baeza to Segovia. He wants to see face to face the relic of the dead body, but humbly turns his back, avoiding the "close encounter". Curiously enough, his easy victory renders him sad. Cervantes is alluding to the secret transfer of St. John of the Cross' body from Úbeda to Segovia, claimed by the devoted widow Doña Ana de Peñalosa. However, Cervantes is also establishing a surprising dialogue with St. John's symbolic "dark night", in which he fights as a brave mystical knight. Concurrently, he is quoting the books of chivalry's funeral processions and the curiosity of the occasional knight who wants to glance at the dead body. Furthermore, we see how extremely conversant the novelist is with the religious genre of spiritual chivalry, strongly opposed to the loose fantasy of the books of chivalry. Unable to look at St. John's relic, an authentic knight of the heavenly militia, Don Quixote seems to silently acknowledge that there are higher chivalries than his own that he will never reach. No wonder he ends the adventure with a sad countenance, gaining a new identity as the "Caballero de la Triste Figura".

Keywords: books of chivalry; books of spiritual chivalry; dark night of the soul; Caballero de la Triste Figura (Knight of the Sad Countenance); St. John of the Cross

Citation: López-Baralt, Luce. 2021. Don Quixote and Saint John of the Cross's Spiritual Chivalry. *Religions* 12: 616. https://doi.org/10.3390/rel12080616

Academic Editor: Cristobal Serran-Pagan Y Fuentes

Received: 22 June 2021
Accepted: 11 July 2021
Published: 9 August 2021

And so, Señor, it's better to be a humble friar, in any order at all, than a valiant knight errant (II, VIII: p. 508). [1]

1. A Bit of History: The Transfer of Saint John of the Cross's Remains from Úbeda to Segovia

We are in the middle of a *dark night* in the year 1593. It is literally the middle of a *dark midnight* since we have a record of the time of the events. The remains of Saint John of the Cross are furtively transferred across isolated and deserted lands from Úbeda, where he died, to Segovia. The Court Marshal, Don Juan de Medina Ceballos, is guarding the remains, now turned into relic, along with the guards and companions who are carrying it on a litter. They avoid the main path to Madrid so as not to be seen, and take different lanes and detours through Jaén, Martos, and Montilla. When they arrive in Martos, on a high hill, not too far away from the road, a man appears unexpectedly and shouts loudly: "Where are you carrying that corpse, you wicked mob? Leave the friar's remains you are taking away ... "[2] (Pasquau 1960, p. 2).[3] This startling appearance "made the Marshal and his companions feel so fearful and alarmed that their hairs stood on end"[4] (Fernández Navarrete 1819, pp. 78–79). Later, down the road, when they reach a deserted field, another man appears unexpectedly and, once more, the entourage is asked to give an account of what they are carrying: Medina and his companions answer they have superior orders to remain undercover, but the man keeps asking them questions. In the middle of these disturbing encounters, the person carrying the bier notices that the small box containing the Saint's remains is surrounded by shining lights.

The upsetting night scene sparkled with strange lights inevitably conjures up "the adventure Don Quixote had with a dead body" (I, XIX), which is transferred by men draped in mourning from Baeza to Segovia. In the middle of the dark night, the gentleman bursts onto the road where the entourage, muttering and holding burning torches, is carrying the mysterious "dead *knight*". I have quoted Don Quixote's literal words, and they should be noted since his way of referring to the enigmatic corpse as "knight" does not seem fortuitous. The gentleman couches his lance, positions himself in Rocinante's saddle, and, raising his voice, complains to the "shirted men" (making reference to the priests dressing surplices who escorted the dead body): "Halt, O knights, or whomsoever you may be, and give an account of yourselves: from whence you come, whither you are going, and whom you carry on that bier ... " (I, XIX, p. 136). Don Quixote's appearance and questioning closely concur with the testimony of the first witnesses who declared about the life of the venerable Friar John of the Cross in the deposition for his beatification process.

Did Cervantes know about the surreptitious transfer of the Reformer's remains that took place in mid-1593, two years after the Saint died?[5] That is what many Cervantists believe, based on the pioneer study of Martín Fernández Navarrete (1819), the first scholar who suggested the connection between the Reformer's historical transfer to Segovia and the episode in chapter XIX of the first part of *Quixote*. It is highly probable that Cervantes was aware of the events, since he was in Úbeda for the wheat harvest in 1592, the year after the friar's death and right before his remains were furtively carried to Segovia in the middle of the night (Sánchez 1990, p. 21). The events around the death of the future Saint John were so notorious that they would inevitably reach his ears: on the one hand, he was Saint Teresa's Reform companion and confessor, who died in the odor of sanctity, and, on the other hand, he was the author of a profound mystical work with no precedents in the Peninsula. The clandestine transfer of his body aroused a heated dispute between Úbeda and Segovia, where he was secretly taken after dying unexpectedly from a "pestilential fever". Let us remember the cause of his death, for I will refer to it later.

Doña Ana de Peñalosa, who received spiritual direction from the Saint and was the addressee of *The Living Flame of Love*, is the person who plans the removal of the body from the convent in Úbeda where he was buried. The devoted widow, to whom Saint John addresses his last letter, had arranged with Friar Doria, Prior General of the Carmelite Order, that wherever Saint John died, his body would be transferred to Segovia. She wanted him to rest in the monastery he had founded in his hometown along with his brother, the Royal Council judge, Don Luis de Mercado. Naturally, the task would not be easy as Úbeda was logically reluctant to resign the Saint's corporeal relic. However, after he died, Doña Ana made the appropriate diligences with Friar Nicolás de Jesús María, Vicar General of the Reform, so that the body could be transferred without any suspicion to Segovia, his hometown (Rodríguez Marín 1949, chp. IX, pp. 226–30).[6]

The secret transfer was planned to be performed nine months after the Reformer's death; yet, the people involved found out that the body was "so incorrupt, fresh and intact, and with such a wonderful fragrance and aroma, that the transfer was postponed, and the body was covered with lime and soil so it could be verified later without any issues"[7] (Fernández Navarrete 1819, p. 78). Already in mid-1593, Court Marshal Medina Ceballos, who was sent from Madrid "with a high sense of justice", found out that the body was leaner and drier, but with the same "fragrance and aroma, so he placed the remains in a suitcase to hide them better"[8] when they removed the body from the convent (Ibid.). The "suitcase" was actually a wooden box, but as it "was [...] small, they folded the legs for the body to fit in, and thus he was carried".[9] However, the story continues, because Úbeda did not surrender the body and established a dispute with Segovia. The serious misunderstanding over the Reformer's remains reached Rome: in 1596, Clement

VIII issued a Papal Brief *Expositium nobis fuit* that ordered for the body to be brought back to Úbeda. Even though the request was quite explicit, Úbeda could not manage to get the body back: Bishop Don Bernardo de Sandoval y Roxas promised he would comply with the Brief, but he considered it convenient to treat the complex issue in a friendly way. The diplomatic negotiation brought so many delays that the final decision about the transfer was prolonged. Úbeda finally settled for a hand and a tibia of the Reformer's body (Pasquau 1960, p. 2).[10]

2. Cervantes and the *mystici majores* of the Golden Age: St. Teresa of Jesus, Fr. Luis de Leon, and St. John of the Cross

Such a bitter litigation, which became known even in Rome, was probably noticed by Cervantes, especially if we consider that he was an admirer of the Spanish mystics, who were his contemporaries. Through Calliope (Cervantes Saavedra 1903, *Galatea*, Book VI), Cervantes sings a panegyric that leaves no doubt about his literary devotion to Friar Luis de Leon:

FRAY LUIS DE LEÓN it is I sing,

Whom I love and adore, to whom I cling. (*Galatea*, p. 240)

The same can be said about Mother Teresa of Teresa of Jesus (1946), whose works had been edited by the famous Augustinian: on the occasion of her beatification in 1614, while he was writing the second *Quixote*, Cervantes composes in his old age a song in which he celebrates, quite knowingly, the Reformer's mystic ecstasies:

One could say that you were born in Alba;

since the just are born where they die.

From Alba, o Mother! You left for Heaven:

Pure, beautiful dawn,[11] *followed by the clear day of immense joy;*

that you enjoy Him is just fair,

in ecstasies divine,

on all roads

where God knows how to guide a soul,

to give her as much of Him as she can hold,

and even broadens and dilates and makes her bog

and with soft love

to Him and of Him holds her and enriches [12]

Yet, and despite the fact that he wrote saintly poetry during his Algerian captivity, "the precise contours of Cervantes' religiosity have yet to be identified" (Iffland 1995, p. 2630). Few aspects are more elusive than the intimate spirituality of Spain's main novelist. In any case, the mystic revelations of St Teresa that Cervantes celebrates in verse form let us know that he was not a total stranger to the high mysteries of the soul. This familiarity with the spiritual world is also evident in Don Quixote's descent into the Cave of Montesinos (*Quixote*, II, XXII–XXIV). In spite of its being a parody, critics have read the adventure as an initiatory scene. Cervantes subsumes the hidalgo from La Mancha in an altered state with oneiric overtones and has him discover that inside the cave—his own inner self—there is a transparent crystal castle.

Of course, there are plenty of shiny castles in chivalry novels (*Amadís de Gaula* and *Florisel de Niquea* (Silva 1584), among others, come to mind); yet knights errant do not find them in states of deep introspection, but in the midst of their adventures instead. Don Quixote's fortress is, on the contrary, an "interior castle", and the precise lexicon and stylistic turn in the description resemble too closely that of St. Teresa's opening lines of her

Interior Castle. Where she says, "I began to think of the soul as if it were a castle made of a single diamond or of very clear crystal, in which there are many rooms" (*Interior Castle*, p. 201), Cervantes says, "Then there appeared before my eyes a royal and sumptuous palace or castle whose walls and ramparts seemed to be made of clear and transparent crystal." (*Quixote*, II, XXIII, p. 605) Such is the very place where Montesinos takes Don Quixote; thus, the strange psychic adventure takes place, just as St. Teresa's, inside a crystal castle.

The apparent parody of the Teresian symbol is a strange homage that Cervantes pays to the Saint. I insist on the intertextual dialogue with St. Teresa because a while later, behind the crystal walls of the castle, Don Quixote glimpses a strange procession. Belerma heads the pageant of mourning maidens. The charmed lady who leads the procession wears a long white turban and carries a relic in her hands—Durandarte's mummified heart—while she sings lamenting dirges along with her maids. It could be a convent procession, only that the strange nuns' headdresses have become sinister Turkish turbans within the enchanted cave.[13] What is more, Belerma's face is that of a middle-aged woman, beetle-browed, brittle colored, and with large eye bags. Such a description could recall Fr. John Misery's 1576 description of St. Teresa, when she was already 61 years old.[14] Legend holds that after seeing the painting, the saintly woman exclaimed, "God forgive you, Fr. John, since you had to paint me, you could have made me less ugly and eye-crusted". What cannot be denied is that the painting shows a mature woman's face, brittle colored, particularly thick brows, red lips, and deep eye bags. Such irreverence should not shock us: as José Gaos has pointed, Cervantes' transformation of mystic writing into picaresque may seem irreverent, but it allows us to see that the novelist did not limit his parody to chivalry novels (Gaos 1979, p. 74). James Iffland observes this, and analyzes the carnivalization of the dead body episode: "This is exactly the point. And since Cervantes probably harbored affection, at a certain level, for the very genres he parodied, why cannot the same hold true when he turns his attention to mystical texts?" (Iffland 1995, p. 264). Even if we assume, with Américo Castro ([1925] 1972), that Cervantes must have been in love with the chivalry genre he parodied—no one reads passionately and with such attention to detail a genre that bores him—it is not difficult either to understand that behind these religious parodies there is an unconfessed love, an equivocal admiration. Maybe, also, an unconcealed spiritual nostalgia. I shall return to this point later.

I suspect that St. John's texts, which circulated in manuscript, just as Fr. Luis', would call Cervantes' attention. Even if the Reformer's works saw light in 1618 (except for the "Canticle"), it was usual practice to distribute copies of St. John's writings among the Carmelite convents (Brenan 1973, p. 166). It is also possible that the manuscripts—or news about their content—may have reached Cervantes through her older sister, Luisa of Belén, who was a nun in Alcalá de Henares, where St. John was a Rector (Cannavagio 1987, p. 37).

The truth is Cervantes' obsessive allusions to the cryptic lexicon of the dark night suggest he had a somewhat precise knowledge of St John's work. Vicente Gaos (1971) and Arturo Marasso (1954) find echoes of St. John's nocturnal vocabulary in Don Quixote's night excursion in search of adventure, in Chapter II of the first part. Is this a spiritual knight who begins a *peregrinatio animae* in search of God? Don Quixote, as we will prove, seems to know about such nightly divine chivalry; Gaos even considers that the Reformer's nocturnal lexicon is also present, though in a parodic fashion, in Maritornes' meeting with the mule driver, which happens in the middle of a dark night.[15]

In his 1995 essay, Iffland carries out a detailed textual study that connects the dead body scene with the technical vocabulary in St John of the Cross's *Dark Night*. His meticulous analysis excuses me from going further into this, though I must say that the obsession with St. John's technical vocabulary was evident to me, too, each time I read it. It is difficult for a reader well-acquainted with the saint's verses to not recognize that Cervantes must be alluding to the works of the Carmel Reformer. Of course, he did so between the lines, since it was a dangerous matter to quote from St. John of the Cross, whom St. Teresa referred to as "*mi senequita*",[16] in the same carnivalesque fashion as that employed in Don

Quixote's and Sancho's *funny book* (Russell 1969). Cervantes bordered on impropriety and even sacrilege.

He was risking the possibility of being found suspect of heterodoxy. This is because in those times, it was not easy to allude to St. John of the Cross, who was impeached *post mortem* due to the parallels found between his writings and those of the *alumbrados*.[17] Several followers of St. John kept his name in strict silence when they quoted his poems.[18] Friar Agustín Antolínez (1554–1626) comments on St. John's *Canticle*, his *Flame*, and the *Night* without ever mentioning the author; and up to this day, St John's scholars—Dom Philippe Chevalier, Ángel Custodio Vega, Jean Krynen, among others—debate such a strange omission. The omission is repeated by Sister Cecilia del Nacimiento (1570–1646), a Carmelite nun from Valladolid whose *Liras de la transformación del alma en Dios* (*Lyre on the Transfomation of the Soul in God* [our translation]), commented on in two volumes, clearly show an imitation of St. John of the Cross. Like Friar Agustín, the nun never mentions her poetic mentor. They all ran the risk of self-incrimination, as the "heresiarch" Miguel de Molinos knew well when he in turn does not mention the name of the "doctor de las nadas" ("doctor of Nothings") to whom he owed much of his contemplative thought.[19] It is thus highly probable that Cervantes knew he was stepping on dangerous grounds by approaching St. John's writing, at once sacred and dangerous.

In spite of his literary discreetness, just like Molinos, Friar Agustín Antolínez, and Sister Cecilia, Cervantes did not hesitate to establish a hidden dialogue with the Carmel Reformer in the "dead body" episode. The obsessive and apparently unnecessary repetitions of the word "night" betray a conscious literary reference, even if it may have parodic overtones. It is hard to think that Cervantes' nocturnal *leit motiv* would indicate only the hour marking the beginning of the adventure of the shirted men carrying the mourning litter. The novelist seems to be imitating the special textual pattern of the *Dark Night* in which St. John's repetition of the word "noche" again and again leads us to understand that he is alluding to a technical mystic symbol and not just to the end of the day. What is more, the word "adventure", repeated throughout the passage, resembles phonetically the "venture" with which St. John's Bride begins her nocturnal journey, searching for that Beloved she knew so well.

Of course, the joy of the poet from Fontiveros—"dichosa ventura" or "happy chance" (*Dark Night*, vol. 1, p. 29), "happy night" (ibid., vol. 2, p. 30), or "night more lovely than the dawn" (ibid., vol. 5, p. 30)—contrasts with the "horror and fear" felt by Cervantes' characters when they plunge into the dark night. Sancho trembles as if he were under the effects of mercury and as "if he had quatrain fever" (*Quixote*, I, XIX, p. 136). Both characters believe the strange scenery to be supernatural: the mourners who carry torches and pray softly recall the apparitions (*"estantiguas"*),[20] that is, a nightly procession of souls wandering in the wood. Don Quixote's hairs stand on end, believing those are "demons from hell" (I, XIX, p. 138). I will say more about the gentleman's fear, which is as unusual as it is eloquent.

Iffland (1995, pp. 243–44), on the other hand, observes that Cervantes' narrator underlines the night's involving darkness: "and so night fell, bringing some darkness with it" (I, XIX), "the dark of the night did not allow them to see anything at all" (I, XX, p. 141), "the night, as we have said, was dark" (I, XX, p. 141), and "the darkness of this night" (I, XX, p. 142). He rightly concludes that "this is by far the darkest night of the entire work, including part II" (Iffland 1995, p. 243).[21] As may well be expected, St. John also insists on the darkness of his Night—the wandering Bride advances blindly, unseen and not seeing anything, only, paradoxically, she rejoices in the thick darkness.

Cervantes' nocturnal black hole is studded with otherworldly, threatening lights. "A great multitude of lights that looked like nothing so much as moving stars" (I, XIX, p. 135) pluck the thick darkness and paralyze master and squire with fear. The "lumbres" (lights) seem "phantoms" to Sancho; though we should not forget that the term is also associated to the sect of the "alumbrados", condemned by the Holy Office. It is obvious that the strange floating starts recalling St. John of the Cross' "Dark Night" poem: "Without light or guide save that which burned in my heart" (*Dark Night* vol. 3, p. 30). St. John alludes to a strange supernatural light, suspended inside the nocturne Bride's soul, which guides her towards herself in an impossible circular and mystic road. However, Cervantes seems to be making reference not only to the mysterious light which turns St. John's "dark night" into a prodigious *chiaroscuro* that reminds us of Rembrandt and El Greco) but also to the experience of the historical transfer of the Saint's body. The "burning torches" of Cervantes' shirted men who carry the "dead knight" across the fields would be the same as those brought by the guards of St. John's body lighting the path to Segovia. Yet, we should not forget the most significant parallel of all: the "very bright lights" suspended around the small, improvised box. They remind the reader of the lights the driver of Fontiveros' priest testified he had witnessed. That historical midnight was also crowned by supernatural stars, and contemporaries of the event must have commented on the portents of this strange sight.

The specific details of Don Quixote's night adventure—one of the few in which he is victorious—points to the fact that the whole episode is a reflection upon religious and ecclesiastical themes and even upon mystic themes. We know that the gentleman, defying fear, his lance at the ready, confronts the mourners with questions. He believes someone has killed the "knight" they carry, and that the misdeed calls for chivalrous vengeance, or perchance the shirted men themselves have done a wrong deed that calls for punishment.

The main mourning "knight" spurs his horse to advance and evade the impertinent stranger, but the mule is "skittish" and throws him to the ground. Furious at not getting specific answers, Don Quixote attacks the marchers with his spear, but the shirted men, fearful and disarmed, ran away with their lighted torches. The scene reminds the reader of a carnivalesque act where chaos reigns and the identity of the revelers remains blurry.

Yet the scene, as said earlier, soon acquires ecclesiastical overtones that render it still more unsettling: Don Quixote is battling priests, and now "sacrilegiously" charges against Bachelor Alonso López—a counterpart of the historical court bailiff Juan de Medina—who has a broken leg and lies on the floor (a curious ailment indeed: St. John of the Cross died precisely of an infection in his left leg). The Bachelor, who holds sacred orders, lets Don Quixote know that they are carrying the knight, who died in Baeza, to be buried in Segovia. Cervantes did not want to mention Úbeda explicitly, so he sets the death in nearby Baeza. Apart from this, attention should be paid to the identity adjective Alonso López uses to refer to the dead man: he is a *knight*. It could well be that he is another priest, since he was guarded by a whole custody of them. Yet, this is not the case: being a "knight", he suddenly becomes a soul brother of the Manchegan gentleman, a veritable knight errant. I will return to this significant point.

Don Quixote insists on his questions and has a burning desire to avenge his alter ego's death; but López informs that no one has killed the knight, since he died of a "pestilential fever" (I, XIX, p. 138). These are the same "fevers" that caused St. John of the Cross' death; probably a septicemia that spread from the leg to the rest of his body. Don Quixote is paralyzed by the answer: "since he was killed by the One who killed him, there is no other recourse but to be silent." (Ibid.) Not even a chivalrous battle is possible against God.

The Bachelor warns Don Quixote that he is now excommunicated "for having laid violent hands on something sacred" (I, XIX, p. 140).[22] He quotes Trento's dispositions in Latin, and it is strange that Don Quixote says he does not know Latin, for on other occasions it seems that he does. This is one of the proofs of the growing "Sanchification" of the literary character. Using a parodic ecclesiastical casuistic in a strict sense, Don Quixote argues he has not put his hand but just his spear on the corpse. Probably, the hidalgo suspects he has gone too far in his casuistic sarcasm, and immediately, he protests of his orthodox Catholicism.[23] However, his exaggerated declarations do not seem convincing enough. Don Quixote has not been in a church since the outset of his adventures, and even if he commends himself to God and to Dulcinea before starting his most daring adventures, he is not really pious: he neither prays fervently nor searches for spiritual guidance like his hero Amadís. As we shall see, Don Quixote's spirituality in this scene is not orthodox—Cervantes submits orthodoxy to parody—but it is rather associated with the sacred. As we shall see, in brief, the Manchegan knight errant will confront sanctity itself.

Meanwhile, Sancho, with his well-known appetite, has taken advantage of the situation to unload the generous provisions of a pack mule the priests were bringing. Iffland (1995, p. 254) considers that such a carnivalesque detail may imply a satire against some of first followers of the Carmelite reform, who betrayed St. John of the Cross's ascetism. Might Don Quixote be avenging the Saint and defending the Carmelite Reformation *al suo modo*? Is the Manchegan gentleman liberating St. John, prisoner of a Church entrenched in tradition and given to excess? Does the knight errant's Erasmian soul feel offended by the traffic of a corpse that has turned into a relic? Everything is possible: *todo puede ser.*

3. The Chivalrous Overtones of the "Dead Body" Episode. Cervantes and Spiritual Chivalry Errant

3.1. Traditional Chivalry Novels

So far, we have analyzed the intertextual dialogue in which Cervantes engages with Saint John of the Cross's nocturnal verses and with the historical transfer of the Saint's remains to Segovia. Critics have also been examining the close similarities that exist between the adventure depicted in chapter XIX and chivalry novels for a long time. Since it is a widely studied subject, I will refer to it briefly. Diego Clemencín (1947) follows Fernández Navarrete (1819), who was the first one to suggest the possibility that the secret transfer of Saint John's corpse inspired the adventure that Cervantes portrays in chapter XIX, and highlights a possible intertextual dialogue that Cervantes has with chapters LXXIII and LXXIV of *Palmerin of England*, with chapter CXXVII of *Amadis of Gaul* (Rodríguez de Montalvo 1803), and with chapter XLIII of the third part of the *chronicles of Don Florisel de Niquea*. If we read chapter LXXIV of *Palmerin of England*, "Of what befel [sic] Florian of the Desert in the adventure of the dead body in the litter" (Morais 1807, *Palmerin of England*, vol. II, p. 5), the first similarities are revealed:[24] the title of the adventure, alike Cervantes', hints at a "dead body". Florian of the Desert, Palmerin's brother, is wandering around a deserted place when he observes that three sorrowful squires are approaching; they are carrying a litter with a corpse draped in mourning. When he removes the pall, he uncovers the stiff body of a gravely wounded knight and wonders (as curiously as but less violently than Don Quixote) who it was. He soon learns that it was Sortibran the Strong, killed by four knights in an act of treachery, whose death was still unavenged. It is a good opportunity for Florian of the Desert to avenge his death, the same heroic deed that Don Quixote would have wanted to perform in favor of his "dead knight". Critics have noticed the names of Palmerin's knights: "the Strong" and, in particular, "Florian of the Dessert", which Cervantes associated with the ascetism of the Discalced friars, who meditated with "strong" spiritual courage in "deserted" places (Iffland 1995, p. 247). As we know, Cervantes does not give the "dead knight" a specific name: he seemed to be as discreet as the contemporary writers were regarding Saint John of the Cross.

Yet, there are possible similarities, which are less apparent, between the above-mentioned Quixotic scene and the novel *Florisel de Niquea*. In chapter XLIII (p. 43), Third Part, "four horses were carrying a litter mounted by four dwarfs. The litters were covered with a rug [. . .] and there were two heavily armed, strong and robust men leading the way, and twelve resembling knights followed behind."[25] Clearly, this procession of physically disproportionate beings may not seem to be closely related to the adventure of Don Quixote; however, the physical abnormality of the escort of the dead body being transferred would ring a bell in attentive readers of chivalry errant literature. Is this another intertextual joke made by Cervantes about the mourning priests that were on their way to Segovia?

It is relevant to stress the fact that both this scene of *Florisel* and the previously mentioned scene of the *Palmerín* take place in broad daylight, while the "dead body" scene described by Cervantes is characterized just by the frightening dark night. In this respect, there is another interesting literary source that Arturo Marasso (1954) wields regarding the nocturnal Quixotic scene: the *Aeneid*, which was translated into Spanish by Hernández de Velasco, an author widely read by Cervantes. The text describes the mourning procession of the recently deceased Turno, organized by Aeneas, which winds its way with burning torches in the midst of a "silent night" (Sánchez 1990, p. 16).

Alberto Sánchez (1990, p. 21) concludes that *Don Quixote* unequivocally depicts not only traits of chivalry myths, but also narrative details that coincide with the historical transfer of Saint John of the Cross's remains to Segovia. I agree with my former professor Sánchez: Cervantes seems to establish a simultaneous dialogue with Saint John's life and literature and, concurrently, with chivalry novels. Both intertextual dialogues coexist in a harmonious yet quite complex literary fusion.

Having said that, scholars (except for Iffland 1995, pp. 247, 257, 265) have not focused on another possible literary source of the nocturnal scene of Cervantes: spiritual chivalry. As we will see, Cervantes also invokes this genre, which is of the essence of understanding his mysterious nocturnal scene.

3.2. Spiritual Chivalry: An Essential Literary Source for Chapter XIX of Don Quixote

When chivalry novels, a fantastic literary genre, captured the Spanish editorial market early in the 16th century, the defenders of the religious genre of spiritual chivalry reacted adversely (Herrán Alonso 2005). An illustrative example of these first pious renaissance books is the *Libro de la Cavallería Cristiana* (*Book of Cristian Chivalry* [our translation], 1515), written by Franciscan friar Jaime de Alcalá. He had an edifying goal in mind, since his hero is a Christian knight of exemplary virtues. Clearly, the idea of a paradigmatic knight, who is foreign to the sexual superfluity of Tirant Lo Blanc and to the loose morals of Amadís, was already known in the Peninsula. It is important to remember the *itinerarium sacri amoris* represented by Ramon Lull's *Blanquerna*, which includes the delicate *Book of the Lover and the Beloved*, written by the Majorcan Blessed in imitation of the Sufi marabouts. Through this work, Llull portrays his own eremitic experiences and many of his mystical intuitions. Furthermore, in his *Book of the Order of Chivalry*, he claims that the "ofici de cavaller és de mantenir e defendre la santa fe catòlica" (Llull 1936, p. 100).[26] It is important to also consider the *Book of the Knight Zifar*,[27] which is thought to be the first Spanish book of chivalry. It was written during the first part of the 16th century by a still unknown author[28] who, like Llull, proves to be acquainted with the Islamic tradition of spiritual chivalry.[29] The prologue of the work reveals a pious pattern: "This knight was baptized with the name of Zifar and afterwards was called the Knight of God because he was ever close to God and God was always with him in all his deeds" (Nelson 1983, *The Book of the Knight Zifar*, p. 6). Even though some scholars classify the Zifar as a "divine book of chivalry", Felicidad Buendía disagrees with this classification since it includes the pious renaissance works that are contrary to the depraved morality and absurd imagination of traditional books of chivalry (Buendía 1960, p. 43). Buendía presumes that Cervantes must have read *Zifar* during his youth, since Ribaldo's practicality and expressiveness seem to anticipate Sancho Panza's simplicity.

During the 16th century, the vicious opposition to books of chivalry by pious moralists grows. They consider these imaginative novels to be inordinate, useless, vain, and morally doubtful, and thus its counterpart, edifying chivalry, is continuously strengthened. In spite of the plentiful output of this contestatory new genre, spiritual chivalry is far from being deeply studied. So much so that Enric Mallorquí Ruscalleda (2016, p. 380) believes that it is "one of the most forgotten and darkest chapters of Spanish literature."[30] Jorge Checa (1988, p. 50), however, offers "the starting signal for the study of the genre"[31] in 1988 (Mallorquí Ruscalleda 2016, p. 374) with his study on *Caballero del Sol* (*Knight of the Sun*), by Pedro Hernandez de Villaumbrales. From that moment onwards, the studies that deal with spiritual chivalry proliferate due to the contributions of scholars such as Estrella Ruiz-Gálvez Priego, Pierre Civil, Pedro Cátedra, and Emma Herrán Alonso, among other critics.

By now we have a representative *corpus* of these Renaissance "divine" chivalry novels, which will definitely grow in the future with new editions and studies. Some examples of the most outstanding works, either written or translated into Spanish, are the previously cited book *Libro intitulado Peregrinación de la vida del hombre, puesta en batalla debajo de los trabajos que sufrió el Caballero del Sol (Book called "Pilgrimage of the life of man, sent into battle through the lifeworks that the Knight of the Sun had to endure)*[32] (1552), by Pedro Hernández de Villalumbrales; *Le Chevalier Deliberé* (*The Resolute Knight*) by Olivier de La Marche, translated into English by Lois Hawley Wilson & Carleton W. Carroll (1999); *Libro de caballería celestial del pie de la Rosa Fragante (Celestial Chivalry from the Foot of the Fragant Rose)* (Amberes 1554); *Libro del caballero cristiano* (*Book of the Christian Knight*) by Juan Hurtado de Mendoza (1570–1577?); *Batalla y triunfo del hombre contra los vicios (Battle and Triumph of Men Against Vices),* written by Andrés de la Losa (1580); *The Pilgrimage of Human Life* (1992) and *Historia y milicia del caballero Peregrino, conquistador del Cielo* (*History and Militia of the Pilgrim Knight, Conqueror of Heaven*) (1601), by Friar Alonso de Soria. Some of these narratives are weaved as spiritual epic poems whose protagonists are abstract forces in order for illiterate Christians to understand them. An example of this is *The Pilgrimage of Human Life*, which tells of the initial journey of a pilgrim errant, accompanied by the beautiful lady Grace of God, who knights him as the "Caballero de las virtudes" or Knight of Virtue.[33]

This fictionalized spiritual militia is related to the literary–doctrinal tradition of the *homo viator* and the *peregrinatio animae* (Herrán Alonso 2007). It is also connected to the ancient Arthurian tales and to the search for the Holy Grail as a representation of a lost world. Connections can also be found with the ascetic simile of the ascending mountain. These themes send us back to previous renaissance divine chivalry novels, such as the book *In Praise of the New Knighthood*, written by Bernard of Clairvaux, *The Divine Comedy*, by Dante, and the *Handbook of a Christian Knight*, by Erasmus. Certainly, the peninsular writers excelled in writing manuals on spiritual warfare: let us remember Friar Luis de Granada, Friar Alonso de Madrid, Francisco de Osuna, and even Saint John of the Cross, whose work *Ascent of Mount Carmel* was influenced by *Ascent of Mount Sion*, written by Bernardino de Laredo. Perhaps the most representative example of this mystical militia is Saint Ignatius of Loyola, the spiritual soldier that, together with St. Teresa, was an avid reader of books of chivalry. As well as St. Teresa, he knew that it was feasible to make them "divine" and apply them to the heroic pilgrimage of their own souls.

In Spain, we can still find echoes of the literary tradition of these dauntless pilgrims of spiritual paths in Moorish works such as *Las coplas del alhichante* (*The Couplets of the Alhichante)* (meaning "pilgrim"), written by Puey Monzon, which narrates a pilgrimage to Mecca and its transcendent significance. The quintessential example of this genre is *Tratado de los dos caminos (The Treaty of the Two Paths)*, written early in the 17th century by an anonymous Moorish author who was a refugee in Tunis. The protagonist, or *homo viator*, of the long allegorical novel has to choose between two symbolic and forked paths: the path of virtue, full of sufferings and hardships, and the delightful path of perdition (Galmés de Fuentes et al. [1975] 2005).[34]

Even though the connection between the theme of the soul's pilgrimage and the divine books of chivalry still needs to be studied (Mallorquí Ruscalleda 2016), undoubtedly, these literary genres have a lot in common, so much so that Dámaso Alonso ([1951] 2008, p. 224) stated with certainty that "there exists a divine chivalrous climate in 16th-century Spain."[35]

4. Cervantes and Saint John of the Cross's Nocturnal Spiritual Chivalry

It is hard to believe that this vast literature of spiritual chivalry did not attract Cervantes's attention. For a start, he knew well the *peregrinatio animae* leit motiv because both Cervantes and Lope de Vega, author of the *Pilgrim of Castile*, experimented with the pilgrimage of religious and amatory overtones that characterized the Byzantine novel. His posthumous *Persiles and Segismunda*, Counter-Reformist in nature, describes the pilgrimage of the two protagonist sweethearts that travel from the northern isles (read Protestant), where they are originally from, to Papal Rome, where they get their "legitimate" canonical marriage.

However, the author of *Quixote* shows us that he is even more deeply familiar with the spiritual chivalry genre. In Chapter VIII, part II, the noble knight of La Mancha, together with Sancho, ponders his role as knight errant, and he states that true knights must seek eternal glory rather than worldly fame: "Christians, Catholics, and knights errant must care more for future glory, eternal in the ethereal and celestial spheres, than for the vanity of the fame achieved in this present and transitory world" (*Quixote*, II, VIII, p. 506). Suddenly, Don Quixote, who has not been very devout throughout his adventures, closes ranks with a spiritual devotion nearer to that of Knight Zifar's or Knight of the Sun's, or the dead "knight" Saint John of the Cross's.

Let us not forget that the episode of the dead body deals precisely with the transfer of a valuable corporeal relic and that critics have associated Don Quixote's libertarian and lay conduct with Erasmus's religious attitudes (Bataillon 1966). When the gentleman of La Mancha ponders with Sancho on religious issues like the veneration of the saints' relics, he treads, as was to be expected, on thorny ground. When he talks with his squire about the issue, it is not surprising when Sancho, a veritable "cristiano viejo" or "old Christian", defends the traditional ecclesiastical stance: "[...] these prerogatives [... are] what the bodies and relics of the saints have, and with the approval and permission of our Holy Mother Church, they also have lamps, candles, shrouds, crutches, paintings, wigs, eyes, and legs, and with these they deepen devotion and increase their Christian fame; the bodies of saints or their relics are carried on their shoulders by kings, and they kiss the fragments of their bones, and use them to decorate and adorn their private chapels and their favorite altars" (*Quixote*, II, VIII, pp. 507–8).

Sancho illustrates his words by referring to the beatification or canonization of "two discalced friars, and the iron chains they used to bind and torture their bodies are now thought to bring great good luck if you kiss and touch them, [...]" (*Quixote*, II, VIII, p. 508). He concludes: "And so, Señor, it's better to be a humble friar, in any order at all, than a valiant knight errant" (Ibid.) There is no other interpretation for this ethical truth: Sancho invites Don Quixote to exchange his chivalrous trade for the religious one, advising him "that we should begin to be saints, and then we'll win the fame we want in a much shorter time" (Ibid.). The peasant apparently becomes the spokesperson of the Counter-Reformation Church: he is a friend of relics, which curiously belong to the "discalced friars" like Saint Teresa's Senequita,[36] whose veneration ensured blessings of all kinds. However, the most significant issue around this episode is that the squire unexpectedly defies Don Quixote to become a saint, thus bringing up his chivalry errant to the highest position in the spiritual order.

This is such a far-reaching challenge that Don Quixote, no matter how valiant he is, knows well he cannot accomplish. And he gives up beforehand, admitting to Sancho his most intimate truth: "[...] we cannot all be friars, and God brings His children to heaven by many paths: chivalry is a religion, and there are sainted knights in Glory" (*Quixote*, II, VIII, p. 508). Sancho does not give up and replies that there are more friars than knights errant in heaven. Let us notice that he continues emphasizing the religious term "friar" instead of "priest": the squire, who often shows hints of wisdom, surely knew well why he posed his subtle distinction. Don Quixote, who usually had the last word in the verbal exchanges with his servant, concludes: "[There are] many [knights errant] *but few who deserve to be called knights*" (Ibid.). It is impossible to forget that the "dead body" of chapter XIX is referred to as "knight". Not only does Don Quixote name him this way, as was to be expected, but also does Bachelor Alonso López, who led his funeral procession. He thus distinguishes his minor orders from the those of the deceased he escorted, of "lay" condition though socially higher. Curiously enough, the body of a *knight* was transferred by clergymen: the text never clarifies this incongruity.

However, those who called the deceased "a knight" were right. It was precisely this way that the friar of Fontiveros referred to himself. In his prose commentary to his poem

"Dark Night of the soul", he admits having fought as a symbolic knight during "that war of the dark night" (St. John of the Cross 1990, *Dark Night of the Soul*, book II, chp. XXIV, vol. 2, p. 203).[37] Both Saint John and Don Quixote fight in the middle of the night. It is appropriate to remember that the discussion the knight of La Mancha and his squire had about the issue of sainthood and chivalry takes place again in the context of a night. Don Quixote was searching for the impossible in the midst of darkness: he was trying to find his beloved Dulcinea of Toboso. Ultimately, he wanted to reach Transcendence in the midst of the dark corporeal world. Whether at random or not, this adventure, like the one of the mysterious *knight's* "dead body", occurs in the middle of a *dark night*, and just like the historical transfer of the remains to Segovia, the Toboso adventure takes place at *midnight*. "It was on the stroke of midnight" (*Quixote*, II, IX, p. 533), the narrator solemnly announces at the beginning of the next chapter, resorting to the first verse of an old ballad, "Count Claros of Montalbán's romance".

Curiously, both Quixotic night paths end up in the Church: "We have come to the church, Sancho" (*Quixote*, II, IX, p. 509). Additionally, although we know that he has come across the church building in Toboso, we know he has crashed against a lot more: against the dogmatic structure of the hardened ecclesiastic institution of his time, represented in Part I, Chapter IX, by the priests draped in mourning, well-versed in casuistry and bearers of generous saddlebags.

Saint John of the Cross was, however, some other kind of ecclesiastic, one that Cervantes could have liked instinctively. He was an alienated dissident that, as his mentor Saint Teresa, fought like a hard-working knight to reorganize the monastic structures of his time and to bring them to an ascetic life and, above all, to purest contemplation. Cervantes must have known about this when he "avenged" the saintly little friar on his guardians, who were highly greedy, cowardly, and orthodox: Saint John of the Cross's fictionalized followers themselves would have betrayed the Reformation and all the heroic purity that it implied. Does Cervantes also avenge, after Erasmus's position, those who would reach glory by possessing the saintly dead body relic? As is well known, the "relic" of St John's body was pulled apart so that both cities, Úbeda and Segovia, could boast having a part.

I have already pointed out that Saint John of the Cross, in line with a spiritual chivalry, had declared himself a "striving knight". However, differently from Don Quixote, his is an *ad intra* mystical combat that occurs deep in his inner soul. During the *dark night* of his pilgrimage, he confronts "a painful disturbance, involving many misgivings, imaginings and strivings which the soul has within itself [...] (*Night*, book II, chp. IX, vol. 7, p. 133). To defend himself, the knight errant soul entrenches itself in the symbolic interior castle of his impregnable spirit, protected by fences and walls, and from there he holds an allegorical strife against the devil, enemy of the soul, that in the "Spiritual Canticle" is called "Aminadab".[38] This evil spirit is defeated in the apotheosis-like final verse of the poem, where, once again, Saint John depicts this inner fight as a chivalrous combat much like Saint Gregory's and, especially, much in the style of the Maghrebi Sufis that inspired Ramon Llull:[39] "For none saw it/Neither did Aminadab appear/And there was a rest from the siege/And the cavalry/came down at the sight of the waters" (St. John of the Cross 1961, *Spiritual Canticle*, book II, chp. XL, p. 484). When the poet tells us that the castle fence or fortified walls "rested", he means that the passions and appetites of the soul have been defeated, and it is no longer fought by "opposing parties" (*Canticle*, book II, chp. XL, vol. 4, pp. 485–86). The cavalry, on the other hand, "comes down", that is, it "rests", it obliterates itself, and it vanishes before the purest waters of the soul when in total union with God (*Canticle*, book II, chp. XL, vol. 5, p. 486). Nobody dares to trespass the sacred sphere of the transforming ecstasy: Saint Teresa knew well about this, and she declared the innermost mansions of her symbolic *Interior Castle* to be impregnable. When, in his poem "Dark Night", the poet tells us about "the turret breeze", insisting on another term associated with fortified castles, he suggests again that the soul is safe within the symbolic fortified castle, breathing God's high "breeze". As we all know, the "breeze" or "breath" is a common *leit motiv* that different spiritualities use for representing the mystical experience of Wholeness: the *Logos*, the breath, the Spirit, the *pneuma*, the *prana*, the *ruah* of the contemplative Jews, or the *ruh* of the Sufis. At the end of the "Night", St John indirectly states that nobody can combat the soul or defeat it, because it is in union with God. The poet's transcended chivalric knight errant is always victorious.

Yet, the combat is always rigorous. So much so that the Reformer of the order of Mount Carmel compares it to the pitched battle of a knight against a metaphorical dragon or a seven-head "beast":

> which makes war therewith against each one, and strives therewith against the soul in each of these mansions, wherein the soul is being exercised and is mounting step by step in the love of God. And undoubtedly if it strives faithfully against each of these heads, and gain the victory, it will deserve to pass from one step to another, and from one mansion to another, even unto the last, leaving the beast vanquished after destroying its seven heads, wherewith it made so furious a war upon it. (*Ascent of Mount Carmel*, book II, chp. XI, vol. 10, p. 211)

The Reformer outlines this beast allegory in the *Ascent of Mount Carmel*, one of the treatises he uses to comment on his poem of the "Dark Night of the Soul". The ascetic fight with the seven-head monster precisely takes place in darkness when "the soul sings of the happy chance which it experienced in stripping the spirit of all spiritual imperfections and desires for the possession of spiritual things. This was a much greater happiness to, by reason of the greater difficulty that there is in putting to rest this house of the spiritual part, and of being able to enter this interior darkness, which is spiritual detachment from all things, whether sensual or spiritual, and leaning on pure faith alone and an ascent thereby to God." (*Ascent*, book II, vol. 1, p. 163).[40] In the *dark night*, the body is obliterated and becomes, metaphorically, "dead": let us remember that the Cervantine "dead body" adventure also occurs in a dark night.

Had Cervantes heard of the symbolic *motif* of this night strife that the mystical knight had fought in his inner self against a seven-head "beast"? It is quite curious that Don Quixote takes for himself the same ascetic combat simile as Saint John does when he assures Sancho that the knight errant must kill the seven deadly sins symbolized by wicked giants. What the Manchegan gentleman explains to his squire reveals a detailed knowledge of chivalrous asceticism:

> We must slay pride by slaying giants; slay envy with generosity and a good heart; anger with serene bearing and tranquility of spirit; gluttony and sleep by eating little and watching always; lust and lasciviousness by maintaining our fealty toward those whom we have made mistresses of our thoughts; sloth by wandering everywhere in the world, seeking those occasions when we may become famous knights as well as Christians. (*Quixote*, II, VIII, p. 506)

It seems Don Quixote wants to model his combatant chivalry towards "the divine", hoisting his sword against the seven deadly sins. Likewise, let us not forget that, just as Saint John, Don Quixote fights ascetically, or at least he dreams of doing so, in a dark midnight.

It is not an easy contest either for Don Quixote or for Saint John. The Spanish Reformer, a knight errant at night, confesses that "deep is this warfare and this striving, for the peace which the soul hopes for will be very deep" (*Night*, book II, chp. IX, vol. 9, p. 134). Saint John was not beyond the terrifying fears that tormented Don Quixote and Sancho when they came across the "dead body" in an utterly dark night. Saint John interprets the "awakening night fears" of his "Spiritual Canticle" as the emotions aroused by the demons against which the soul strives: "because by means of [the terrors of the night] the devil tries to diffuse darkness in the soul, [so that the devil] may obscure the Divine light wherein it is rejoicing" (*Canticle*, book II, chp. XX, vol. 9, p. 363). However, the soul is already in its sweet inward sleep, and the "night terrors" cannot awaken it, "since it is deeply recollected and closely united with God" (ibid.). That is why it moves around confidently, "in darkness and secure", in the "guiding night", "a night more lovely than the dawn". This jubilant serenity in the darkness of the night is something that will never be granted to Don Quixote, who moves through the night with shivers down his spine. As we will see later on, he will end up acquiring a metaphorical *Sorrowful Face.*

5. A Little More about Saint John of the Cross's Initiatory Dark Night Strife

I should refer at least briefly to the quite complex symbol that Saint John of the Cross and the mystic Muslims[41] called "the dark night of the soul" since certain shades of this notorious nocturnal simile are of great interest to Don Quixote's adventure, which I have been following closely. Generally speaking, the reader is more acquainted with the spiritual night in its purgative and purifying sense, and in fact, it is this similar mystic dimension that has been precisely studied by experts such as Evelyn Underhill (1961), William James ([1925] 1986), and Juan Martin Velasco (Martín Velasco 1999).[42] In this arduous stage of mystic life, some contemplatives—though not all of them—undergo an almost unbearable inner dryness and spiritual grief, though an immensely enriching one, because it implies a process of growth. Both the sensual desires and an appetite for the flesh are purged here, and the soul is strengthened on its highest level. In modern times, this stage is associated—*toutes proportiones gardées*—with a depression or a state of spiritual exhaustion: the soul has undergone its ecstatic experiences in such an intense manner that it is struck by an alternate state of desolation. Both Saint Teresa of Jesus and Saint John of the Cross repeatedly refer to the torments and doubts that occur during this hard stage of the mystic journey. However, Saint John knows well how useful this purifying experience in darkness is: "the soul becomes enlightened in the midst of all this darkness" (*Night*, book II, chp. XIII, vol. 1, p. 149), assures the *Dark Night of the Soul*, and he adds in *the Ascent of Mount Carmel* that "here we call them (the purgations or purifications of the soul) nights, for in both of them the soul journeys, as it were, by night, in darkness" (*Ascent*, book I, chp. I, vol. 4, p. 56). It is within the purifying parameters of "this spiritual night" that we should understand the chivalrous ascetic strife that Saint John as well as Don Quixote undertake at night against the seven-deadly-sin beast. Both of them intended to purge the soul of its vices to gain the celestial Jerusalem.

However, St. John's symbolic *night* does not end here. It has a much deeper symbolic dimension that somehow can bring more light upon the scene of the "dead knight". This divine night of the unknowable leads to a converging ecstasy, or better, it is part of the experience of ecstasy. It is by turning off—by "darkening"—the senses and the reason that we may have the direct experience of God. Even language has to be annihilated. God is experienced only when we are blind to the spatial–temporal coordinates of this corporeal world. Let us remember that the mystic experience is an altered state of consciousness in which the contemplative experiences a Wholeness beyond time, space, and language. The transcendental knowledge of the living God is not gained by means of discursive reason. Reason is in a state of confusion, for it has been "darkened".

It is not by chance that when Saint John of the Cross begins celebrating his ecstasy in the middle of the "Canticle", it suddenly becomes dark. The "crystalline font" that springs up in the Bride's path, a symbol of mystical initiation, twinkles in the darkness because it has a "silvered surface". There, the traveler stops his pilgrimage to see reflected in the water the "eyes desired" that she "bears outlined in her inmost parts." Saint John knows very well that "the soul becomes enlightened in the midst of all this darkness, and the light shines in the darkness" (*Night*, book II, chp. XIII, vol. 1, p. 149). It is worth noting that the poet no longer says that the soul "is purified" but instead "enlightened". In fact, when "the soul has remained in darkness as to all light of sense and understanding, going forth beyond all limits of nature and reason" (*Ascent*, book II, chp. I, vol. 1, p. 88), it "will see supernaturally" (*Ascent*, book II, chp. IV, vol. 7, p. 95) and will enter "this narrow path of obscure contemplation" (*Ascent*, book II, chp. VII, vol. 13, p. 105). This darkness is celebrated, not endured. "The higher and more divine is the Divine light, the darker is it to our understanding" (*Ascent*, book II, chp. XIV, vol. 13, p. 126): this is precisely the reason why all mystics become aphasic in the midst of the ecstatic trance, because their language is insufficient. In his *Coplas a lo Divino or* the *Ballads "a lo divino"*, Saint John refers to this phenomenon: "Just when this flight of mine had reached its highest mark/my eyes were dazzled so/I conquered in the dark".[43] However, this nocturnal darkness is not frightening, and that is why the soul can really rejoice and say "in darkness and secure". I think that the verses in both the "Dark Night of the Soul" and the "Canticles" are the happiest ones in Spanish literature, and this is confirmed in the *poet's* lexis of overflowing joy: "Oh, night that guided me, Oh, night more lovely than the dawn". Certainly, this is not Cervantes's terrifying "dead body" adventure. Strangely enough, St. John's symbolic soul is disguised—"by the secret ladder, disguised"—to celebrate her secret wedding night. However, the masks behind which the mysterious protagonist of the poem hides strongly point to the immense abysses of her true identity. The entire poem revolves around this joyful encounter with herself, with the culmination of her transcended identity in God. The soul discovers that she is infinite, like the limitless night that masks her and blinds her mercifully in a corporeal manner. As Nietzsche said: "Whatever is profound loves masks. Every profound spirit needs a mask." The mask, indeed, "bares more than what it covers"[44] (Vélez Estrada 1989, vol. 190, p. 304). The Bride, as we know from the comments on the poem in the treatise of the Dark Night, had "changed [her] garments" and "disguised [herself] with three liveries and colours" (book II, vol. 15, p. 159). If we uphold the definitions by the Royal Spanish Academy and María Moliner (1994, vol. 2, p. 252), the livery was not only "a suit handed over to servants",[45] but also "the uniform worn by squads of knights in public celebrations."[46] Therefore, the Bride or nocturnal soul has acquired a new "chivalrous" identity that announces her new or, rather, newly obtained self. The "livery and disguise" that she wears are, according to Saint John, of three allegorical colors, white, green, and red, that represent the chromatic emblems of the three theological virtues of faith, hope, and charity. Covered with the chivalrous, allegorical, and distinctive livery worn by those who have ended their journey on Earth, the soul rests securely in utter darkness, safe from any vicissitude and extremely joyous. Let us not forget her condition: blessed. Her Beloved will recognize her, invested in the emblematic colors acquired on her nocturnal journey.

I have lingered on the disguise, symbolic of the new profound identity that the St. John's Beloved acquires in her pilgrimage because, strangely, Don Quixote will also take on a new identity among the shadows. Instinctively, we all know ourselves better in the dark: to reflect and pray undisturbed from any distraction, we usually close our eyes. That is to say, we usually darken the outer senses. It is Sancho who "reveals" to his master his new chivalrous identity in the middle of the dark night. The gentleman emerges as The Knight of the Sorrowful Face because, when, from among the shadows, the squire looks at him thoroughly, under the light of a burning torch, he warns him that "[his] grace has the sorriest-looking face [he has] seen recently, and it must be on account of [his] weariness after this battle, or the molars and teeth [he has] lost" (book I, vol. 19, p. 139). It is quite curious that Don Quixote, in one of the very few chivalrous battles in which he has easily triumphed without even a scratch, displays such a sad face before Sancho. He does not brag about his triumph nor does he celebrate it. The "weariness after this battle" seems to have undermined his soul, and this sad despondency reflects metaphorically on his face. We already know that it was not a simple strife: the Manchegan gentleman dared to walk on the delicate path of spiritual chivalry, and in doing so, he addresses unspoken dimensions of his own inner soul. He is measuring himself against the mystery of Transcendence.

Don Quixote meekly takes on his new identity, and he thinks the author in charge of writing down his "famous" deeds has inspired Sancho. As it is known, ancient knights took their names from their military victories—in that way, they received nicknames such as "The Knight of the Burning Sword", "The Knight of the Unicorn", or "The Knight of Death". The Manchegan gentleman identifies so deeply with his state of sorrow that he promises to depict on his shield "a very sorrowful face" as his emblem (book I, vol. 19, p. 139). Sancho, who now has the final say, dissuades him, assuring him that all he must do is to uncover his face: Don Quixote represents himself; he is the living symbol of his own ontological sorrow.

However, far before Don Quixote, Deocliano called himself "The Knight of the Sorrowful Face" in the third book La historia del muy esforzado y animoso caballero don Clarián de Landanís (The Story of the Very Strong and Spirited Knight Don Clarian of Landanis). Nevertheless, Deocliano was not ungraceful or inelegant, since the painting on his shield showed a damsel of strange beauty, whose expression was very sad: "and as a sign of this, one hand was on the heart and the other cleaned the crystalline tears shed by those beautiful eyes".[47] It is indeed curious that Don Quixote takes on a female chivalrous identity. However, so did Saint John, as the typical rhetoric of spiritual literature always uses the female gender to refer to the soul. Therefore, the literary alter ego in the "Night" and the "Canticle" is always a damsel. Nonetheless, the symbolic "damsels" that work as distinctive mask to Don Quixote and Saint John present very different characteristics: one is "sorrowful" in her "face" while the other one could not be happier. "Kindled in love" and "happy" like the night that covers her, St. John's damsel culminates her night wandering "transformed in the Beloved." Thus, she shares God's own "face", whose vast beauty is instilled in all living creatures: [He] "Left them, by his glance alone, clothed with beauty". The "Sorrowful Face" has no place in this mystical and joyful space. Cervantes must have surely known about this.

6. Saint John of the Cross: Inverted Mirror of Don Quixote?

Let us dig still further. Once the night adventure has finished, Don Quixote comes up with an odd idea: to confront the "dead body" of the knight in the litter. The "spiritual knight" Saint John of the Cross, turned into a coveted relic, would preserve, according to the devout tradition, some of the sacred halo that is typical of a saint. It is not the first time that the Manchegan knight measures himself up in the mirror of his interlocutors: it is impossible to forget when he fixes his gaze on Cardenio, his true ontological brother. Crazed like Don Quixote, ambushed like him by love, and with a truly mistreated appearance, Cardenio holds such an ontic relationship with *The Knight of the Sorrowful Face* that the encounter turns out to be very revealing. With a gallant air, the aged knight errant heads for the ambushed Cardenio and embraces him "as if he had known him for some time. [...] The other man, whom we can call *The Ragged One of the Gloomy Face*—as Don Quixote is *He of the Sorrowful One*—allowed himself to be embraced, then stepped back, placed his hands on Don Quixote's shoulders, and stood looking at him as if wanting to see if he knew him, not less astonished, perhaps, at the face, form, and arms of Don Quixote than Don Quixote was at the sight of him" (I, XXIII, p. 182). Both knights seem to examine themselves until they discover they are geminated in an "ontological" mirror. They recognize each other, and they identify with each other, for both their vital misfortune and their ruined physique are the same.

Don Quixote now wants to fix the same inquiring gaze of ontic overtones on the body of the dead "knight". Cervantes dialogues closely with the *Palmerin of England*, since, in Chapter 74 of the first part, the Florian knight did likewise: he raised the pall that covered the mortuary remains that the three squires carried in the litter, and he discovered a body lying in green armor, soaking with blood and horribly mutilated by the severe blows he had received in battle. This moves him to great compassion, and he asks who the dead knight is. He turns out to be, as we know, Sortibran the Strong. "Strong", as we know, also was the symbolic striving knight Saint John of the Cross.

It seems that Don Quixote, mimicking Florian, wants to know more about the identity of the "dead knight." However, there is no one he can ask, since the "shirted" priests have fled. The gentleman gets ready to look at the knight's body, as if he wanted to measure himself up against its owner and recognize himself in his chivalrous symbolic mirror. To accept this adventure in the order of the being requires extreme bravery, but Don Quixote, as it is known, has never been daunted by danger. The moment appears propitious, as the litter with the remains was abandoned by the runaway priests. Nothing stands between the *Knight of the Sorrowful Face* and the spiritual knight's corpse: everything is ready for the unparalleled encounter. The narrator states succinctly: "Don Quixote wanted to see if the body on the litter was actually bones or not" (I, XIX, p. 140). What is this? Don Quixote of La Mancha rummaging through relics? Did he know, perchance, that St. John's body had not been transferred to Segovia before due to its freshness, and that its smell was so good that the bailiff of the court, Medina Ceballos, had to throw lime at it and wait for another year so that its transfer to Segovia was viable? Did the anachronistic knight errant want to compare the saintliness of the "knight" by observing the state of the "incorrupt" corporeal relic? Clearly, we have moved away from the scene of the *Palmerín*, gruesome but yet typical of the rhetoric of the harsh combats of the cavalries. I suspect that there is more to it, since it seems incongruous that Don Quixote becomes a simple examiner of abandoned bones. By laying his eyes on the "dead knight", what the Manchegan knight will actually do is examine the inscrutable mysteries of death,[48] of the afterworld, and of saintliness. Furthermore, by symbolically confronting Saint John of the Cross, he is measuring himself up against the saintliness of the "divine" mystical "knight". Don Quixote confronts himself with the limits of his own chivalry errant. When we look in depth at the *Other*, we discover our own true identity. It is when we make use of someone else's mirror that we manage to understand ourselves better.

However, the challenge is excessive, and Don Quixote does not accept it. The magnitude of this adventure seems to intimidate him. Sancho, momentarily turned into the "master" of his will, prevents the ontological encounter due to practical reasons: he reminds the gentleman that they are already safe; the people draped in mourning have been defeated, and hunger is pressing: "as they say, let the dead go to the grave and the living to the loaf of bread" (I, XIX, p. 140). Sancho puts his faith in this world, not in the other one. Of course, not wanting to look at a dead person at night and in a deserted area falls in line with his profile of a fearful farmer: let us remember the mortal terror that tormented him when he believed he was seeing spook lights and otherworldly apparitions. Don Quixote, who had defeated his ecclesiastic "adversaries" so rapidly, is now defeated by the "Sanchification" to which his own squire subdues him: "since it seemed to Don Quixote that Sancho was right, he followed him without another word". Astonishing, but true: Don Quixote, silenced, seems to measure himself better with Sancho, hungry for thick corporeal life, than with the dead knight in the litter. "Don Quijote is literally 'anchored' by Sancho Panza".[49] It is easy to imagine his sense of personal defeat when he moves away from the litter of the unknowable corpse to resume his uncertain path. It is at this moment that he truly earns his nickname "The Knight of the *Sorrowful* Face". He is a sorrowful defeated man in the spiritual order.

I do not suppose much. It is Don Quixote himself who, in the second part of the novel and after observing some saints carved in relief, humbly accepts his subordinated place before the mystery of saintliness. The character displays a broad culture regarding spiritual chivalry, since he comments knowingly on the images that some farmers designed for the altarpiece they were erecting in their village.[50] When reaching Saint George's carving, Don Quixote informs that he was one of the best knights errant the "divine militia" ever had; he thinks that Saint Martin, who divided his cape with the poor man, was another "Christian seeker of adventures".[51] For him, Saint James was "one of the most valiant saints and knights the world has ever had, and that heaven has now" (II, LVIII, p. 834). As for Saint Paul, he was "a knight errant in life." In the Don Quixote's imaginary, saints are knights, just like the anonymous knight surrounded by priests whose sacred bones—those that Saint Teresa claimed "would perform miracles"—he dared not look at. It seems that now that the gentleman has observed saintliness at length—even tempered by its representation in wooden reliefs—he feels ready to admit the inferiority of the chivalry errant he professes in the face of heavenly chivalry: "these saints and knights professed what I profess, which is the practice of arms; the difference, however, between me and them is that they were saints and fought in the divine manner, and I am a sinner and fight in the human manner" (II, LVIII, p. 834).[52] Don Quixote is then a knight of Earth, not of Heaven; a sinner, not a saint; a warrior of the day, not a spiritual fighter in the dark night. His itinerancy, no matter how earnest it is, constitutes a chivalrous pilgrimage *ad extra*, not *ad intra*, like Saint John of the Cross's.

By opening before us the deep soul of the Manchegan gentleman, always frugal when sharing his authentic spiritual concerns, the episode of the "dead body" has given us unexpected clues. In this bewildering adventure, Cervantes merges with extraordinary skill the historical event of the transfer of Saint John's body to Segovia, arranged by Doña Ana de Peñalosa, with the literary *leit motiv* of the transfer of a dead knight, present, as we have seen, in more than one knighthood novel.

The novelist pays tribute to the spiritual chivalry errant, in which he seemed to be as conversant as his anachronistic gentleman was. The "dead knight" of the litter—i.e., Saint John of the Cross—closed ranks with the "divine" literary warfare tradition that was so in fashion back then, the very one that opposed the books of chivalry, full of literary fantasy, that Don Quixote intended to resurrect in the seventeenth century. The night adventure turns out to be a literary piece of the highest complexity: between lines, Cervantes is pressing the Manchegan knight's fantastic dreams to the limits. The *hidalgo* silently acknowledges that there are higher chivalries than his own that he will never reach. Don Quixote, who "fights in the human manner", could not measure himself up against the mystical chivalry of the humble friar who was his contemporary. With the passing of time, this mystic would be reach sainthood as an authentic knight of the heavenly militia. No wonder then that Don Quixote, unable to approach face to face the mystery of the Sacred which St. John represented, ends the adventure with a sad countenance. From now on, *et pour cause*, he will be known as the "Caballero de la triste figura".

Funding: This research received no external funding.

Conflicts of Interest: The author declares no conflict of interest.

Notes

1 All quotes from *Don Quixote* have been extracted from the English edition of *Don Quixote* (Cervantes Saavedra 2003) translated by Edith Grossman. The first Roman number indicates Part I or Part II of *Quixote;* the second Roman number indicates Chapter (TN).

2 Our translation.

3 The quoted words were said by Francis of St. Hilarion during the deposition of the Saint's Beatification. Fernández Navarrete (1819, pp. 78–89) quotes with a different version the words said by the first man who interrupted the entourage: "where do you take the Saint's remains? Leave them where they were ... " [Our translation]. The scholar also explains that when the guards tried to silence the second shouting person with money, "they found that he had disappeared" (Ibid.) [Our translation].

4 Our translation.

5 For extended bibliography about chapter XIX of the first part of *Quixote*, see (Jaime [1995] 2008).

6 Rodríguez Marín (1949) based his thesis on chapter XVI of *Historia del Venerable Friar Ivan de la Cruz, primer descalzo carmelita* (History of the Venerable Friar Ivan of the Cross, first Discalced Carmelite) by Friar Jerónimo de San Joseph (1641).

7 Our translation.

8 Our translation.

9 Apud Pasquau (1960, p. 2); our translation. The author quotes a manuscript kept in the Carmelite convent of Úbeda.

10 Gerald Brenan (1973, pp. 101–2) specifies that Úbeda received an "arm, a foot and a few fingers".

11 Alba is the name of St. Teresa's birthplace, and it also means dawn.

12 Sánchez (1990); our translation. The original poem is "A los éxtasis de Teresa de Jesús" ("To St. Teresa's Ecstasies") (Cervantes Saavedra 1981).

13 The descent into the cave of Montesinos has clear Islamic overtones, as several critics have agreed. One of the most recent studies is Wilnomy Zulayka Pérez (2015)'s "*La representación del viaje iniciático sufí: una simbología cifrada en la Cueva de Montesinos del Quijote*" (18–19 May 2015), presented at the Congreso Cervantes, el Islam y los moriscos, organized at Murcia University by Arabist scholar Pablo Beneito: Pérez's paper is sill unpublished, but is available at https://tv.um.es/video?id=72711&serie=15661&cod=a1b1c1d3&idioma=es. (accessed on 18 April 2017).

14 This curious coincidence was suggested by Prof. Dennis Madrigal during a course on *Don Quixote* I taught at the University of Puerto Rico some years ago.

15 Alberto Sánchez (1990), following Martín Fernández Navarrete's pioneering work (1819), proposes that what matters more than these stylistic and formal parallels, is the memory of the historical event when the Carmelite priest was transported from Úbeda to Segovia in 1593, as stated above.

16 Senequita or "little Seneca", after the saint's wide range of knowledge and small height (TN).

17 Friar Basilio Ponce de León defended the Reformer *post-mortem* from these and other accusations.

18 About the theme of St. John of the Cross' name being silenced among close followers, see (López-Baralt 1985, 1998, p. 87ff).

19 About Molinos, Eulogio Paco's *Defensa de la contemplación* (Molinos 1988) and José Ángel Valente's *Guía espiritual* (Molinos 1989) may be consulted.

20 See Redondo (1983, 1998) and Alvar (2009) for the tradition of "estantiguas".

21 Iffland (1995, p. 255) expands this observation on the importance of the night in Chapter XX, where Don Quixote and Sancho, still in the midst of darkness, hear the sound of water. This brings us to the "fountain well which flows and runs" (St. John of the Cross' "Song of the Soul Rejoicing in the Knowledge of God by Faith", *Complete Works* (Cruz 2015), p. 415) and even to St. John's "fount of crystal", which having "silvered surface", necessarily appears in the middle of the night (St. John of the Cross *Canticle*, p. 44). It is difficult to think that there would not be in these chapters a conscious intertextual dialogue with St. John of the Cross, on Cervantes' part.

22 Curiously, such excommunication links Don Quixote to St. John of the Cross, who was excommunicated several times by his own Carmelite order.

23 About Don Quixote's excommunication, see (Lumbreras 1952).

24 Some of these similarities have been highlighted by Martín de Riquer (1962, 1967), Luis Andrés Murillo (Murillo 1978) and Avalle Avalle Arce (1979), among others.

25 Our translation.

26 Regarding this book by Llull (Johnston 1990).

27 Critics vacillate between spelling Cifar or Zifar.

28 Mallorquí Ruscalleda (2016) suggests that the possible author is the cannon from Toledo Ferran Martínez. For authorship purposes, see the edition of the *Book of the Knight Zifar* by Charles L. Nelson (1983). The literary genre of Islamic mystical chivalry is highly significant, and its influence on the European genre is far from being studied. Mallorquí Ruscalleda (2016, p. 383) acknowledges that "the concept of spiritual chivalry poses an additional problem since it could be mistaken for a concept that arises from Sufism, *futuwwa* [...]; the extent and manner in which such Islamic influence is revealed still need to be defined." [Our translation].

29 There is a traditional epic-chivalry narrative "developed by oral narrators during the early stages of Islam, which has been collected by the genre known as *al-sîra*" (Galmés de Fuentes et al. [1975] 2005, part I, p. 16). This tradition reached Spain in the 16th century, as evidenced in *El libro de las batallas* (*The Book of Battles*), which was edited by Galmés in two volumes (Galmés de Fuentes et al. [1975] 2005). Such book is stylistically and thematically connected with the European books of chivalry, such as *Amadis* and the *Palmerin*. Predictably, this chivalry genre soon undergoes a mystical turn, widely studied by Henry Corbin (1995), among other authors. I have particularly explored the remarkable coincidences between the technical vocabulary employed by this mystical chivalry and the one used by the Spanish mystic, in particular that of Saint John of the Cross's (López-Baralt 1985, 1998, 2000). Iffland considers these studies in his quoted essay of 1995. This ancient Islamic divine chivalry literature is so renowned that even the Peruvian writer Luis Enrique Tord makes use of it in his fable *Cide Hamete Benengeli, coautor del Quijote* (*Cide Hamete Benengeli, Coauthor of Don Quixote*) (Tord 1987). In this fable, he pictures the mysterious Cide Hamete Ben Gelie meeting Cervantes in the prison of Seville, to whom he narrates many divine chivalrous tales from his Islamic tradition and therefore helps to inspire *Don Quixote*.

30 Our translation.

31 Our translation.

32 This book by Pedro Hernández de Villalumbrales should not be confused with the book *Pilgrimage of the Life of Man* by John Lydgate.

33 I have used the quoted studies of Herrán Alonso and Mallorquí Ruscalleda to assemble the short representative list of this divine chivalrous narrative.

34 Ancient Islamic tradition, both in Arab and Persian, is incredibly rich; it describes the returning journey of the soul of the meditative person, during which it experiences dangerous adventures through foreign imaginable spaces. It is well-known that European tradition owes a lot to the Islamic tradition (this is the case of Dante, Llull, Knight Zifar, Saint John of the Cross) but it has not been deeply studied yet, as Mallorquí Ruscalleda (2016) stated.

35 Our translation.

36 The identity of these discalced friars is still unknown, although some scholars think one of them could have been the Franciscan Saint Diego de Alcalá. Saint John of the Cross's beatification process had not begun yet, but it was not difficult to predict it would start soon, given his general saintly fame.

37 All the quotations from the *Dark Night of the Soul*, *Spiritual Canticle*, and *Ascent of Mount Carmel* throughout Sections 4 and 5 have been extracted from Allison Peers's editions and translations of these works [TN].

38 About the origin of this name, see (López-Baralt 1985, vol. 89, p. 273).

39 About these thematic parallels, see (López-Baralt 1985, p. 89, 2000), among others.

40 For the parallels between this warlike imagery and the Sufi knights errant, see (López-Baralt 1985, p. 90, 2000).

41 Miguel Asín Palacios (Asín Palacios 1933) was the first one to associate the *dark night* simile with Sufi spirituality, especially with Ibn Abbad de Ronda's work. See (López-Baralt 1985, p. 89, 2000) and especially (López-Baralt 1998, p. 147ff).

42 For the lexical variations of the Sanjuanist *dark night* symbol, see (Mancho Duque 1982) and, for the Cervantine adventure darkness, see Casalduero (1966).

43 Translation by Frederick Nims (1959).

44 Our translation.

45 Our translation.

46 Our translation.

47 [Our translation] I quote from the notes made by Luis Murillo (Murillo 1978) in note 18 of the edition of *Don Quixote* that I have been working with.

48 For more information on the subject, see H. R. Patch (1956), and, for additional bibliography, consult Carlos Alvar (2009).

49 Iffland (1995, p. 257) refers to the fact that Sancho ties Rocinante's forelegs together with his donkey's halter (part I, chp. XX, p. 141), but his comment applies closely to this failed encounter with Saint John of the Cross's body.

50 It is interesting that "the saints", both in these carvings and in the body of the divine knight, are, first, covered, since the carvings are protected by cloths that prevent them from being seen, and the zealous guards that carry Saint John's bones protect them from the "men who were recklessly curious" in the road. The carvings and the "dead body" were indeed moved from one place to another so that the encounter between them and Don Quixote is necessarily short: a brief encounter with the sacred.

51 Cervantes treats the scene with sarcasm, as Don Quixote says that Saint Martin gave the poor man just half of his cape for it was winter.

52 Iffland (1995, p. 265) quotes this passage in note 9.

References

Alonso, Dámaso. 2008. *Poesía Española. Ensayo de Métodos y Límites Estilísticos*. Madrid: Gredos. First published 1951.

Alvar, Carlos. 2009. Don Quijote y el más allá. In *'El Quijote': Letras, Armas, Vida*. Madrid: Sial/Trivium, pp. 123–41.

Asín Palacios, Miguel. 1933. Un precursor hispano-musulmán de san Juan de la Cruz. *Al-Andalus* 1: 1–79.

Avalle Arce, Juan Bautista, ed. 1979. *Don Quijote de la Mancha*. Madrid: Alhambra.

Bataillon, Marcel. 1966. El erasmismo en Cervantes. In *Erasmo en España. Estudios Sobre la Historia Espiritual del Siglo XVI*. México: Fondo de Cultura Económica, pp. 777–803.

Brenan, Gerald. 1973. *St. John of the Cross*. Cambridge: Cambridge University Press.

Buendía, Felicidad. 1960. *Nota Preliminar al Caballero Cifar*. Madrid: Aguilar, pp. 43–47.

Cannavagio, Jean. 1987. *Cervantes*. Translated by Mauro Armiño. Madrid: Espasa Calpe.

Casalduero, Joaquín. 1966. *Sentido y Forma del 'Quijote'*. Madrid: Ínsula.

Castro, Américo. 1972. *El Pensamiento de Cervantes*. Madrid: Casa Editorial Hernando, Barcelona: Noguer. First published 1925.

Cervantes Saavedra, Miguel de. 1903. *Galatea*. Edited by James Fitzmaurice Kelly. Translated by Hoelsner Welford. Glasgow: Gowans & Gray.

Cervantes Saavedra, Miguel de. 1981. *Poesías Completas*. Edited by Vicente Gaos. Madrid: Castalia, vol. 2, p. 385.

Cervantes Saavedra, Miguel de. 2003. *Don Quixote*. Translated by Edith Grossman. New York: Ecco.

Checa, Jorge. 1988. El Caballero del Sol de Hernández de Villaumbrales y el género de las novelas de caballería a lo divino. *Crónica Hispánica* 10: 49–66.

Clemencín, Diego, ed. 1947. *Don Quijote de la Mancha, Edición V Centenario, Enteramente Comentada por Clemencín*. Madrid: Ediciones Castilla.

Corbin, Henry. 1995. *El Hombre y su Ángel. Iniciación y Caballería Espiritual*. Barcelona: Ediciones Destino.

Cruz, san Juan de la. 2015. *Obra Completa*, 3rd ed. Edited by Luce López-Baralt y Eulogio Pacho. Madrid: Alianza Editorial, 2 vols.

Fernández Navarrete, Martin. 1819. *Vida de Miguel de Cervantes Saavedra*. Madrid: Real Academia Española.

Galmés de Fuentes, Álvaro, Juan Carlos Villaverde Amieva, and Luce López-Baralt, eds. 2005. *Tratado de los dos Caminos por un Morisco Refugiado e Túnez*. Madrid: Instituto Universitario Seminario Menéndez Pidal, Oviedo: Seminario de Estudios Árabo Románicos, Universidad de Oviedo. First published 1975.

Gaos, Vicente. 1971. *Claves de la Literatura Española*. Edad Media—Siglo XIX. Madrid: Guadarrama, vol. 1.

Gaos, Vicente. 1979. La "Noche oscura" y la salida de don Quijote (San Juan de la Cruz, fuente de Cervantes). In *Cervantes: Novelista, Dramaturgo, Poeta*. Madrid: Planeta.

Herrán Alonso, Emma. 2005. "La Caballería Celestial" y "Los Divinos". La Narrativa Caballeresca Espiritual del Siglo XVI. Unpublished Ph.D. dissertation, University of Oviedo, Oviedo, Spain.

Herrán Alonso, Emma. 2007. Entre el homo viator y el miles Christi. Itinerarios narrativos de la alegoría espiritual en la imprenta áurea. *Cahiers D'études Hispaniques Médiévales* 30: 107–21. [CrossRef]
Iffland, James. 1995. Mysticism and Carnival in Don Quijote I, 19–20. *Modern Language Notes* 110: 240–70. [CrossRef]
Jaime, Fernández S. J. 2008. *Bibliografía del 'Quijote' por Unidades Narrativas y Materiales de la Novela*. Alcalá de Henares: Centro de Estudios Cervantinos. First published 1995.
James, William. 1986. *The Varieties of Religious Experience. A Study in Human Nature*. New York: The Modern Library. First published 1925.
Johnston, Mark. 1990. Literacy, Spiritual Allegory and Power: Lull's Libre de ordre de cavalleria. *Catalan Review International* 4: 357–76. [CrossRef]
López-Baralt, Luce. 1985. *San Juan de la Cruz y el Islam*. México: Colegio de México y, Madrid: Hiperión.
López-Baralt, Luce. 1998. *Asedios a lo Indecible. San Juan de la Cruz Canta al Éxtasis Transformante*. Madrid: Hiperión.
López-Baralt, Luce. 2000. *The Sufi Trobar Clus and Spanish Mysticism: A Shared Symbolism*. Lahore: Iqbal Academy Pakistan.
Llull, Ramon. 1936. *Obres Originals del Illuminat Doctor Mestre Ramon Lull*. Palma de Mallorca: Edité par Estampa d'Amengual i Muntaner, vol. 1.
Lumbreras, Pedro O. P. 1952. ¿Quedó excomulgado don Quijote? In *Casos y Lecciones del "Quijote'*. Madrid: Ediciones Stvdium de Cultura.
Mallorquí Ruscalleda, Enric. 2016. El conocimiento de los libros de caballerías españoles a lo divino (1552–1601). Estado de la cuestión y perspectivas futuras de estudio. *eHumanista* 32: 374–412.
Mancho Duque, María Jesús. 1982. *El Símbolo de la Noche Oscura en san Juan de la Cruz*. Salamanca: Ediciones Universidad de Salamanca.
Marasso, Arturo. 1954. *La Invención del Quijote*. Buenos Aires: Hachette.
Martín Velasco, Juan. 1999. *El Fenómeno Místico. Estudio Comparado*. Madrid: Trotta.
Moliner, María. 1994. *Diccionario del Uso del Español*. Madrid: Gredos.
Molinos, Miguel de. 1988. *Defensa de la Contemplación*. Edited by Eulogio Pacho. Madrid: FUE/Universidad de Salmanca.
Molinos, Miguel de. 1989. *Guía Espiritual*. Edited by José Ángel Valente. Madrid: Alianza Editorial.
Morais, Francisco de. 1807. *Palmerin of England*. London: Longman, Hurst, Rees and Orme.
Murillo, Luis Andrés. 1978. *El ingenioso Hidalgo Don Quijote de la Mancha*. Madrid: Castalia, vol. 2.
Nelson, Charles L. 1983. *The Book of the Knight Zifar. A Translation of El Libro del Caballero Zifar*. Kentucky: University Press.
Nims, Frederick. 1959. *The Poems of St. John of the Cross*, 3rd ed. Chicago: The University of Chicago Press.
Pasquau, Guerrero Juan. 1960. *San Juan de la Cruz y la 'Aventura del Cuerpo Muerto'*. Madrid: ABC, Available online: http://www.vbeda.com/articulos/indexoa.php?num=63&48titulo=SAN_JUAN_DE_LA_CRUZ_YLA_%ABAVENTURA_DEL_CUERPO_MUERTO%BB_ (accessed on 18 April 2016).
Patch, Howard Rollin. 1956. *El Otro Mundo en la Literatura Medieval. Seguido de un Apéndice: La Visión de Trasmundo en las Literaturas Hispánicas*. Translated by María Risa Lida de Malkiel. México: Fondo de Cultura Económica.
Pérez, Wilnomy Zuleyka. 2015. La representación del viaje iniciático sufí: Una simbología cifrada en la Cueva de Montesinos del Quijote. Paper presented at Cervantes, el Islam y los moriscos Congress, Murcia, Spain, May 18–19.
Redondo, Agustín. 1983. La Mesnie Hellequin et la Estantigua: Les traditions de la 'Chasse sauvage' et leur résurgence dans le Don Quichotte. In *Traditions Poulaires et Difffusion de la Culture en Espagne (XVIe ey CVIIE Siècles)*. Bordeaux: Presses Universitaires de France, pp. 1–27.
Redondo, Agustín. 1998. Las tradiciones hispánicas de la estantigua. In *Otra Manera de Leer el 'Quijote'*. Madrid: Castalia, pp. 101–19.
Riquer, Martín de, ed. 1962. *Don Quijote de la Mancha*. Barcelona: Planeta.
Riquer, Martín de. 1967. La aventura del cuerpo muerto o de los encamisados. In *Nuevas Aproximaciones al "Quijote"*. Barcelona: Teide, pp. 97–99.
Rodríguez de Montalvo, Garci. 1803. *Amadis of Gaul*.
Rodríguez Marín, Francisco, ed. 1949. *Don Quijote de la Mancha*. Nueva Edición Crítica, con el Comento Refundido Dispuesta por F. Rodríguez Marín. Madrid: Ed. Atlas, pp. 226–30.
Russell, Peter E. 1969. Don Quixote as a Funny Book. *The Modern Language Review* 64: 312–26. [CrossRef]
San Joseph, Friar Jerónimo de. 1641. *Historia del Venerable Friar Ivan de la Cruz, Primer Descalzo Carmelita*. Madrid: Diego de la Carrera.
Sánchez, Alberto. 1990. Posibles ecos de san Juan de la Cruz en el Quijote de 1605. *Anales Cervantinos* 28: 9–21. [CrossRef]
Silva, Feliciano de. 1584. *Florisel de Niquea*, Zaragoza.
St. John of the Cross. 1961. *Spiritual Canticle*. New York: Image Books.
St. John of the Cross. 1990. *Dark Night of the Soul*. Edited by E. A. Peers. New York: Doubleday.
Teresa of Jesus, St. 1946. *The Complete Works of St. Teresa of Jesus*. Translated by E. A. Peers. London: Sheed and Ward.
Tord, Luis Enrique. 1987. *Cide Hamete Benengeli, Coautor del Quijote*. Lima: Premio Copé, pp. 249–64.
Underhill, Evelyn. 1961. *Mysticism*. New York: Dutton & Co.
Vélez Estrada, Jaime. 1989. Origen y sentido existencial del carnaval. In *Revista de Estudios Generales*. San Juan: Universidad de Puerto Rico.

MDPI
St. Alban-Anlage 66
4052 Basel
Switzerland
Tel. +41 61 683 77 34
Fax +41 61 302 89 18
www.mdpi.com

Religions Editorial Office
E-mail: religions@mdpi.com
www.mdpi.com/journal/religions

www.ingramcontent.com/pod-product-compliance
Lightning Source LLC
LaVergne TN
LVHW070649100526
838202LV00013B/919
* 9 7 8 3 0 3 6 5 2 1 2 7 5 *